Electronic Nose: Algorithmic Challenges

Lei Zhang · Fengchun Tian
David Zhang

Electronic Nose: Algorithmic Challenges

Lei Zhang
College of Microelectronics and Communication Engineering
Chongqing University
Chongqing, China

David Zhang
School of Science and Engineering
Chinese University of Hong Kong (Shenzhen)
Shenzhen, Guangdong, China

Fengchun Tian
College of Microelectronics and Communication Engineering
Chongqing University
Chongqing, China

ISBN 978-981-13-4741-2 ISBN 978-981-13-2167-2 (eBook)
https://doi.org/10.1007/978-981-13-2167-2

Softcover re-print of the Hardcover 1st edition 2018

This Springer imprint is published by the registered company Springer Nature Singapore Pte Ltd.
The registered company address is: 152 Beach Road, #21-01/04 Gateway East, Singapore 189721, Singapore

Preface

Electronic nose is an electronic olfactory system constructed to mimic the biological olfactory mechanism, which is also an important scientific field of artificial intelligence. Recently, electronic nose has attracted worldwide attention and a number of research findings have been developed in related areas including sensor arrays, bionic systems, and pattern recognition algorithms. These findings proved the feasibility and effectiveness of electronic nose in odor recognition, environmental monitoring, medical diagnosis, food quality monitoring, etc. However, there are several fundamental challenges in electronic nose community, such as sensor drift issue, disturbance issue, and discreteness issue, which seriously prohibit the scientific progress and industrial development of olfactory intelligence. Unfortunately, there are very few researches contributing to the above-mentioned challenges. Therefore, this book aims to systematically describe and define these challenges in electronic nose community and present specific methods, models, and techniques to effectively and efficiently address these important issues. To our best knowledge, this will be the first book to review and address the newest algorithmic challenges.

Specifically, this book focuses on four algorithmic challenges. First, in Part II, for general electronic nose, the algorithmic challenge in pattern recognition and machine learning (challenge I) will be described, including the odor recognition algorithms and concentration estimation algorithms. Second, in Part III, the sensor drift issue (challenge II) that would decrease the aging of E-nose will be introduced and described, with specific solving schemes in models and algorithms. Third, in Part IV, the disturbance issue (challenge III) that would break the general use of E-nose will be introduced and described, with specific approaches and techniques. Finally, in Part V, the discreteness issue (challenge IV) that would prohibit the large-scale industrial application of E-nose instruments will be introduced and described, with the effective models and algorithms proposed to address the issue.

All of the technologies, algorithms, and applications described in this book were applied in our research work and have proven to be effective in electronic nose. In this book, there are totally 20 chapters. Chapters 1 and 2 will describe the background of electronic nose with the unsolved issues and the literature review. From Chaps. 3 to 8, the odor recognition algorithms and gases concentration estimation algorithms will be introduced, which address the first odor recognition challenge in general E-nose. From Chaps. 9 to 14, different models, algorithms, and techniques for addressing the second long-term sensor drift compensation challenge will be gradually described. From Chaps. 15 to 17, specific approaches and strategies for addressing the third background disturbance elimination challenge will be presented. From Chaps. 18 and 19, the detailed algorithms and schemes for the fourth large-scale discreteness correction challenge will be presented and discussed. Finally, Chap. 20 gives the book review and future work. This book will benefit the researchers, professionals, graduate and postgraduate students working in the field of signal processing, pattern recognition, artificial intelligence, electronics and instrumentation science.

Our team has been working on the electronic nose research over 10 years. Under the grant support (Grant No. 61401048, No. 61771079) from National Natural Science Foundation of China (NSFC) and the grant support (Grant No. cstc2017zdcy-zdzx0077) from Chongqing Science and Technology Commission, we had started our studies on this topic. We would like to express our special gratitude to Dr. Zhaotian Zhang, Dr. Xiaoyun Xiong, and Dr. Ke Liu in National Natural Science Foundation of China (NSFC) for their help and support. The authors would also like to thank Dr. Zhifang Liang and Lijun Dang for their useful contribution to some chapters in this book.

Chongqing, China Lei Zhang
Chongqing, China Fengchun Tian
Shenzhen, China David Zhang
December 2018

Contents

Part IV E-Nose Disturbance Elimination: Challenge III

Part I
Overview

Chapter 1
Introduction

Abstract This chapter provides an overview of E-nose research and technology. We first review the progress of E-noses in applications, systems, and algorithms during the past two decades. Then, we propose to address these key challenges in E-nose, which are sensor induced and sensor specific. This chapter is closed by a statement of the objective of the research, a brief summary of the work, and a general outline of the overall structure of this book.

Keywords Electronic nose · Odor recognition · Drift compensation
Discreteness alignment · Disturbance elimination

1.1 Background of Electronic Nose

Electronic olfactory system constructed with a model nose was proposed for the first time to mimic the biological olfactory mechanism as early as in 1982 [1], which presented two key assumptions of mammalian olfactory system: (1) There is no requirement for odor-specific transducers; (2) odor signals from the transducers can be learnt. One key characteristic of model nose is that the odorant detectors (i.e., the primary neurons) respond to a wide range of chemicals. In 1994, Gardner et al. [2] showed a new definition for artificial olfactory system: "An electronic nose is an instrument, which comprises an array of chemical sensors with partial specificity and an appropriate pattern recognition system, capable of recognizing simple or complex odours." In other words, electronic nose (abbreviated as E-nose) can be recognized to be an intelligent sensor array system for mimicking biological olfactory functions. An excellent review of E-noses can be referred to as [3].

L. Zhang et al., *Electronic Nose: Algorithmic Challenges*,
https://doi.org/10.1007/978-981-13-2167-2_1

1.2 Development in Application Level

E-nose has been applied in many areas, such as food analysis [4–13], medical diagnosis [13–17], environmental monitoring [18–23], and quality identification [24, 25]. For *food analysis*, Brudzewski et al. [4] propose to recognize four different types of milk, Lorenzen et al. [5] tend to differentiate four types of cream butter, Bhattacharyya et al. [6] proposed to classify different types of black tea, Chen et al. [7], Dutta et al. [8], Hui et al. [9], and Varnamkhasti et al. [10] propose to predict tea quality, apple storage time, and the aging of beer. The reviews of the existing work in food control and analysis by using E-noses are referred to as [11–13]. For *medical diagnosis*, inspired by [14] that human breath contains some biomarkers that contribute to disease diagnosis, Yan and Zhang [15] proposed a breath analysis system to differentiate healthy people and diabetics, Di Natale et al. [16] proposed a lung cancer identification system with a quartz microbalance (QMB) sensor array, and Pavlou et al. [17] designed a 14-conducting polymer sensor array for diagnosis of urinary tract infections. For *environmental monitoring*, Getino et al. [18] and Wolfrum et al. [19] proposed to detect the volatile organic compounds (VOCs) in air, Zhang et al. [20, 21] proposed a portable E-nose system for concentration estimation using neural networks, targeting at real-time indoor air quality monitoring, and Dentoni et al. [22], Baby et al. [23], and Fort et al. [24] proposed tin oxide sensor-based systems for monitoring air contaminants, including single odor and mixtures of different odors. For *quality identification*, Chen et al. [7] and Dutta et al. [8] proposed to discriminate and predict tea quality, Gardner et al. [25] proposed to monitor the quality of potable water, and Cano et al. [26] proposed to discriminate counterfeits of perfumes.

Other applications are referred to as tobacco recognition [27], coffee recognition [28], beverage recognition [29], and explosives detection [30], etc. To this end, we refer to interested readers as [31] for extensive applications of E-noses.

1.3 Development in System Level

After observing a number of E-nose applications, we then present different types of E-nose systems, like conventional E-nose [32–35], differential E-nose [27, 28, 30], temperature-modulated E-nose [36–41], active E-nose [42–44], and LabVIEW E-nose [45, 46]. *First*, the conventional E-nose are constructed with a sensor array worked at constant temperature voltage. Zhang et al. [32, 33] proposed a 6-metal oxide semiconductor gas sensor system, Hong et al. [34] proposed a 6-thick film oxide sensor system, and Rodriguez-Lujan et al. [35] proposed a 16-screen printed MOX gas sensing system. *Second*, the differential E-nose proposed by Brudzewski et al. [27, 28, 30] is with two sensor arrays: One is for gas sensing, and the other one is for baseline measurement. *Third*, the temperature-modulated E-nose proposed by Lee and Reedy [36], Llobet et al. [37], Martinelli et al. [38],

Hossein-Babaei and Amini [39, 40], and Yin et al. [41] are with an idea that the heating voltage of each sensor is dynamic rather than constant. The adaptive change (ramp, sine wave, rectangular wave, etc.) of heating voltage is termed as temperature modulation. The rationality behind is that one sensor with multiple heating voltages would produce multiple patterns, such that temperature modulation can effectively lower the cost of sensor array [41]. *Fourth*, the active E-nose proposed by Gosangi et al. [42, 43] and Herrero-Carrón et al. [44] are evolution of temperature modulation systems. The "active" concept shows an adaptive optimization of operating temperatures, because not all heating voltages contribute positively to classification. *Lastly*, Imahashi and Hayashi [45] developed a computer-controlled odor separating system based on LabVIEW, which consists of adsorbing and separating cells. It separates the detected odorants in terms of the properties of installed adsorbents. Jha and Hayashi [46] also developed a LabVIEW-based odor filtering system. The advantages of LabVIEW-based E-noses are low cost, high efficiency, and heuristic for laboratory use.

1.4 Related Technologies

Based on the aforementioned statement, as shown in Fig. 1.1, the present book focuses on four algorithmic challenges of E-nose, including odor detection, drift compensation, disturbance elimination, and discreteness correction. Specifically, the following objectives will be achieved:

- Odor detection
  - Propose discriminative classification algorithms for qualitative odor recognition.

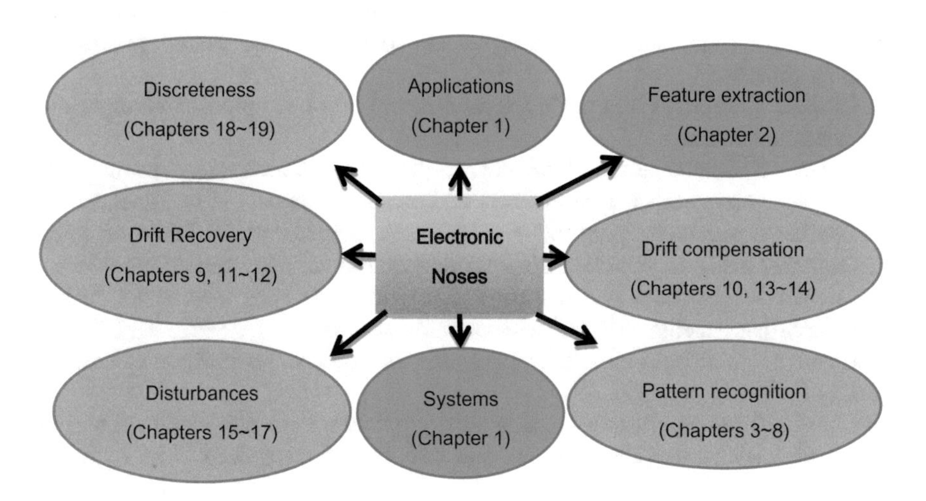

Fig. 1.1 Developments (*blue*) and challenges (*green*) of E-noses

  - Propose robust, bio-inspired, and heuristic regression models for quantitative odor concentration prediction.
- Drift compensation
  - Propose chaotic time series-based neural network method for drift signal prediction.
  - Propose transfer learning and domain adaptation-guided classifiers for long-term drift compensation.
- Disturbance elimination
  - Propose pattern recognition-based interference odor detection and counter-action method.
  - Propose self-expression error mismatch-guided abnormal odor detection models.
- Discreteness correction
  - Propose a global affine transformation-based linear alignment method for signal uniqueness.
  - Propose an effective batch correction scheme based on affine transformation for large-scale E-nose instruments calibration.

1.5 Outline of the Book

The organization of this book is presented as follows.

Chapter 2 reviews the existing work in electronic nose and the algorithms to address the new challenges.

Chapter 3 describes a heuristic and bio-inspired neural network model for gases concentration estimation and guides readers to optimize the neural network in training process.

Chapter 4 describes a new chaos-based neural network optimization algorithm for gases prediction.

Chapter 5 systemically presents how to predict gas concentration using multi-layer perceptrons, such that researchers and engineers can use in real application.

Chapter 6 introduces a hybrid support vector machine model based on linear discriminant analysis, such that the odor recognition accuracy can be improved.

Chapter 7 introduces a new kernel discriminant analysis method for feature extraction that benefits to the odor recognition accuracy.

Chapter 8 introduces a classifier ensemble method for classification of multiple kinds of odors.

Chapter 9 presents a novel sensor drift prediction approach based on chaotic time series, which can show readers that drift can be recognized to be a kind of chaotic signal.

Chapter 10 presents a very new perspective of machine learning for effectively addressing the drift compensation issue, by using classifier transfer learning idea.

Chapter 11 presents a cross-domain subspace projection approach, for addressing drift reduction issue, by using feature transfer learning idea.

Chapter 12 presents a common subspace learning model for drifted data classification, such that the drifted data can share a common feature space with the non-drifted data.

Chapter 13 introduces a transfer learning-guided domain correction method for drift compensation.

Chapter 14 introduces a multi-feature jointly semi-supervised learning approach, which aims to improve the robustness of electronic nose to outliers and drift.

Chapter 15 presents a pattern recognition-based background interference reduction, which describes a simple idea for interference recognition and then achieves interference reduction.

Chapter 16 presents a pattern mismatch-based interference elimination method by sensor reconstruction.

Chapter 17 introduces a new self-expression model for interference recognition and reduction, which is a progress of Chap. 14.

Chapter 18 introduces an affine transformation-based discreteness calibration model, which is simple but effective, and benefits to large-scale instrument calibration.

Chapter 19 presents a very effective scheme for large-scale instrument standardization by using the affine calibration shown in Chap. 18.

Chapter 20 gives the book review and future work in electronic nose topic.

References

1. K. Persaud, G. Dodd, Analysis of discrimination mechanisms in the mammalian olfactory system using a model nose. Nature **299**, 352–355 (1982)
2. J.W. Gardner, P.N. Bartlett, A brief history of electronic noses. Sens. Actuators B: Chem. **18–19**(1), 210–211 (1994)
3. F. Röck, N. Barsan, U. Weimar, Electronic nose: current status and future trends. Chem. Rev. **108**, 705–725 (2008)
4. K. Brudzewski, S. Osowski, T. Markiewicz, Classification of milk by means of an electronic nose and SVM neural network. Sens. Actuators B: Chem. **98**, 291–298 (2004)
5. P.C. Lorenzen, H.G. Walte, B. Bosse, Development of a method for butter type differentiation by electronic nose technology. Sens. Actuators B: Chem. **181**, 690–693 (2013)
6. N. Bhattacharyya, R. Bandyopadhyay, M. Bhuyan, B. Tudu, D. Ghosh, A. Jana, Electronic nose for black tea classification and correlation of measurement with "Tea Taster" marks. IEEE Trans. Instrum. Measure. **57**(7), 1313–1321 (2008)
7. Q. Chen, J. Zhao, Z. Chen, H. Lin, D.A. Zhao, Discrimination of green tea quality using the electronic nose technique and the human panel test, comparison of linear and nonlinear classification tools. Sens. Actuators B: Chem. **159**(1), 294–300 (2011)
8. R. Dutta, E.L. Hines, J.W. Gardner, K.R. Kashwan, M. Bhuyan, Tea quality prediction using a tin oxide-based electronic nose: an artificial intelligent approach. Sens. Actuators B: Chem. **94**, 228–237 (2003)

9. G. Hui, Y. Wu, D. Ye, W. Ding, Fuji apple storage time predictive method using electronic nose. Food Anal. Methods **6**, 82–88 (2013)
10. M.G. Varnamkhasti, S.S. Mohtasebi, M. Siadat, J. Lozano, H. Ahmadi, S.H. Razavi, A. Dicko, Aging fingerprint characterization of beer using electronic nose. Sens. Actuators B: Chem. **159**(1), 51–59 (2011)
11. M. Peris, L.E. Gilabert, A 21st century technique for food control: electronic noses. Anal. Chim. Acta **638**, 1–15 (2009)
12. A. Berna, Metal oxide sensors for electronic noses and their application to food analysis. Sensors **10**, 3882–3910 (2010)
13. E.A. Baldwin, J. Bai, A. Plotto, S. Dea, Electronic noses and tongues: applications for the food and pharmaceutical industries. Sensors **11**, 4744–4766 (2011)
14. A. D'Amico, C. Di Natale, R. Paolesse, A. Macagnano, E. Martinelli, G. Pennazza, M. Santonico, M. Bernabei, C. Roscioni, G. Galluccio, Olfactory systems for medical applications. Sens. Actuators B: Chem. **130**, 458–465 (2008)
15. K. Yan, D. Zhang, Feature selection and analysis on correlated gas sensor data with recursive feature elimination. Sens. Actuators B: Chem. **212**, 353–363 (2015)
16. C. Di Natale, A. Macagnano, E. Martinelli, R. Paolesse, G. D'Arcangelo, C. Roscioni, A.F. Agro, A. D'Amico, Lung cancer identification by the analysis of breath by means of an array of non-selective gas sensosrs. Biosens. Bioelectron. **18**, 1209–1218 (2003)
17. A.K. Pavlou, N. Magan, C. McNulty, J.M. Jones, D. Sharp, J. Brown, A.P.F. Turner, Use of an electronic nose system for diagnoses of urinary tract infections. Biosens. Bioelectron. **17**, 893–899 (2002)
18. J. Getino, M.C. Horrillo, J. Gutiérrez, L. Arés, J.I. Robla, C. Garcia, I. Sayago, Analysis of VOCs with a tin oxide sensor array. Sens. Actuators B: Chem. **43**, 200–205 (1997)
19. E.J. Wolfrum, R.M. Meglen, D. Peterson, J. Sluiter, Metal oxide sensor arrays for the detection, differentiation, and quantification of volatile organic compounds at sub-part-per-million concentration levels. Sens. Actuators B: Chem. **115**, 322–329 (2006)
20. L. Zhang, F. Tian, C. Kadri, G. Pei, H. Li, L. Pan, Gases concentration estimation using heuristics and bio-inspired optimization models for experimental chemical electronic nose. Sens. Actuators B: Chem. **160**(1), 760–770 (2011)
21. L. Zhang, F. Tian, S. Liu, J. Guo, B. Hu, Q. Ye, L. Dang, X. Peng, C. Kadri, J. Feng, Chaos based neural network optimization for concentration estimation of indoor air contaminants by an electronic nose. Sens. Actuators, A **189**, 161–167 (2013)
22. L. Dentoni, L. Capelli, S. Sironi, R.D. Rosso, S. Zanetti, M.D. Torre, Development of an electronic nose for environmental odour monitoring. Sensors **12**, 14363–14381 (2012)
23. R.E. Baby, M. Cabezas, E.N.W. de Reca, Electronic nose: a useful tool for monitoring environmental contamination. Sens. Actuators B: Chem. **69**, 214–218 (2000)
24. A. Fort, N. Machetti, S. Rocchi, M.B.S. Santos, L. Tondi, N. Ulivieri, V. Vignoli, G. Sberveglieri, Tin oxide gas sensing: comparison among different measurement techniques for gas mixture classification. IEEE Trans. Instrum. Measure. **52**(3), 921–926 (2003)
25. J.W. Gardner, H.W. Shin, E.L. Hines, C.S. Dow, An electronic nose system for monitoring the quality of potable water. Sens. Actuators B: Chem. **69**, 336–341 (2000)
26. M. Cano, V. Borrego, J. Roales, J. Idígoras, T.L. Costa, P. Mendoza, J.M. Pedrosa, Rapid discrimination and counterfeit detection of perfumes by an electronic olfactory system. Sens. Actuators B: Chem. **156**, 319–324 (2011)
27. K. Brudzewski, S. Osowski, A. Golembiecka, Differential electronic nose and support vector machine for fast recognition of tobacco. Expert Syst. Appl. **39**, 9886–9891 (2012)
28. K. Brudzewski, S. Osowski, A. Dwulit, Recognition of coffee using differential electronic nose. IEEE Trans. Instrum. Measure. **61**(6), 1803–1810 (2012)
29. P. Ciosek, Z. Brzózka, W. Wróblewski, Classification of beverages using a reduced sensor array. Sens. Actuators B: Chem. **103**, 76–83 (2004)
30. K. Brudzewski, S. Osowski, W. Pawlowski, Metal oxide sensor arrays for detection of explosives at sub-parts-per million concentration levels by the differential electronic nose. Sens. Actuators B: Chem. **161**, 528–533 (2012)

31. A.D. Wilson, M. Baietto, Applications and advances in electronic-nose technologies. Sensors **9**, 5099–5148 (2009)
32. L. Zhang, F. Tian, Performance study of multilayer perceptrons in a low-cost electronic nose. IEEE Trans. Instrum. Measure. **63**(7), 1670–1679 (2014)
33. L. Zhang, F. Tian, X. Peng, X. Yin, G. Li, L. Dang, Concentration estimation using metal oxide semi-conductor gas sensor array based e-noses. Sens. Rev. **34**, 284–290 (2014)
34. H.K. Hong, C.H. Kwon, S.R. Kim, D.H. Yun, K. Lee, Y.K. Sung, Portable electronic nose system with gas sensor array and artificial neural network. Sens. Actuators B: Chem. **66**, 49–52 (2000)
35. I.R. Lujan, J. Fonollosa, A. Vergara, M. Homer, R. Huerta, On the calibration of sensor arrays for pattern recognition using the minimal number of experiments. Chemometr. Intell. Lab. Syst. **130**, 123–134 (2014)
36. A.P. Lee, B.J. Reedy, Temperature modulation in semiconductor gas sensing. Sens. Actuators B: Chem. **60**, 35–42 (1999)
37. E. Llobet, R. Ionescu, S.A. Khalifa, J. Brezmes, X. Vilanova, X. Correig, N. Barsan, J.W. Gardner, Multicomponent gas mixture analysis using a single tin oxide sensor and dynamic pattern recognition. IEEE Sens. J. **1**(3), 207–213 (2001)
38. E. Martinelli, D. Polese, A. Catini, A. D'Amico, C. Di Natale, Self-adapted temperature modulation in metal-oxide semiconductor gas sensors. Sens. Actuators B: Chem. **161**, 534–541 (2012)
39. F. Hossein-Babaei, A. Amini, A breakthrough in gas diagnosis with a temperature-modulated generic metal oxide gas sensor. Sens. Actuators B: Chem. **166–167**, 419–425 (2012)
40. F. Hossein-Babaei, A. Amini, Recognition of complex odors with a single generic tin oxide gas sensor. Sens. Actuators B: Chem. **194**, 156–163 (2014)
41. X. Yin, L. Zhang, F. Tian, D. Zhang, Temperature modulated gas sensing e-nose system for low-cost and fast detection. IEEE Sens. J. (2015). https://doi.org/10.1109/JSEN.2015.2483901
42. R. Gosangi, R. Gutierrez-Osuna, Active temperature programming for metal-oxide chemoresistors. IEEE Sens. J. **10**(6), 1075–1082 (2010)
43. R. Gosangi, R. Gutierrez-Osuna, Active temperature modulation of metal-oxide sensors for quantitative analysis of gas mixtures. Sens. Actuators B: Chem. **185**, 201–210 (2013)
44. F. Herrero-Carrón, D.J. Yáñez, F.D.B. Rodríguez, P. Varona, An active, inverse temperature modulation strategy for single sensor odorant classification. Sens. Actuators B: Chem. **206**, 555–563 (2015)
45. M. Imahashi, K. Hayashi, Odor clustering and discrimination using an odor separating system. Sens. Actuators B: Chem. **166–167**, 685–694 (2012)
46. S.K. Jha, K. Hayashi, A novel odor filtering and sensing system combined with regression analysis for chemical vapor quantification. Sens. Actuators B: Chem. **200**, 269–287 (2014)

Chapter 2
E-Nose Algorithms and Challenges

Abstract This chapter focuses on the up-to-date progress and development in algorithm level for E-nose. We discuss from the viewpoint of feature extraction algorithms, signal de-noising algorithms, pattern recognition algorithms, and drift compensation algorithms that have been fully studied in electronic noses. Then, the challenges of E-nose technology are defined and described, including drift compensation, disturbance elimination, and discreteness correction.

Keywords Feature extract · Signal de-noising · Pattern recognition
Drift compensation

2.1 Feature Extraction and De-noising Algorithms

Feature extraction is the first step of a recognition system. Without exception, multi-dimensional feature extraction is also the key part of E-nose system. Generally, handcrafted feature extraction includes normalization, feature selection, and feature enhancement [1]. *Normalization* is used to remove the scaling effect caused by odorant concentration, such that the interrelation among patterns can be better shown. Suppose the original feature matrix $\mathbf{X} = [\mathbf{x}_1, \ldots, \mathbf{x}_n] \in \Re^{d \times n}$, then the normalization is formulated as follows

$$x_{ij} = x_{ij}/\max(\mathbf{x}_i), \quad i = 1, \ldots, d; \quad j = 1, \ldots, n \tag{2.1}$$

where d and n denote the feature dimension and the number of samples, respectively. More normalization techniques such as baseline subtraction, centralization, and scaling can be found in [1]. *Feature selection* aims at identifying the most informative and discriminative subset that leads to the best classification performance [2]. Principal component analysis (PCA) and linear discriminant analysis (LDA) are used for extracting the most informative and discriminative features, respectively. Suppose $\mathbf{W} \in \Re^{d \times k}$ to be the linear transformation (basis), the extracted features $\widehat{\mathbf{X}}$ are as follows

L. Zhang et al., *Electronic Nose: Algorithmic Challenges*,
https://doi.org/10.1007/978-981-13-2167-2_2

$$\widehat{\mathbf{X}} = \mathbf{W}^{\mathrm{T}}\mathbf{X} \tag{2.2}$$

where k denotes the number of selected components (eigenvectors). Further, Peng et al. [3] proposed a KECA method by considering the components w.r.t. maximum entropy, rather than maximum eigenvalue. Martinelli et al. [4] proposed a phase space-based feature extraction with temporal evolution of sensor response. Leone et al. [5] proposed a representation method, which solves a dictionary algorithm by minimizing reconstruction error. *Feature enhancement* targets at the best features in another domain (e.g., frequency domain). Ehret et al. [6] proposed a Fourier transform method without information loss, and Kaur et al. [7] proposed a dynamic social impact theory and moving window time slicing method.

De-noising algorithms pursue noise removal (i.e., de-noising) on data mixed with some unknown noise. First, in preprocessing, smooth filter (window moving average) in real-time sensing was used [8]. Suppose the response sequence to be $\{S(i), \quad i = 1, \ldots, Q\}$, the filtered response is

$$\hat{S}(j) = \frac{\sum_{i=j}^{j+q-1} S(i) - \max\{S(j), \ldots, S(j+q-1)\} - \min\{S(j), \ldots, S(j+q-1)\}}{q-2} \tag{2.3}$$

where Q and q represent the length of signal and smooth filter, $j = 1, \ldots, Q-q+1$. Moreover, Kalman filter was used [9]. Second, in feature extraction, Jha and Yadava [10] proposed singular value decomposition (SVD)-based de-noising. The noise removal is done by truncating the components w.r.t. a few largest singular values and reconstructing the noiseless data as

$$\widehat{\mathbf{X}} = \sum_{i=1}^{k} \hat{\sigma}_i \mathbf{u}_i \mathbf{v}_i^{\mathrm{T}} \tag{2.4}$$

where $\hat{\sigma}_i$ represents the truncated singular value, $\mathbf{u}_i$ and $\mathbf{v}_i$ represent the singular vectors w.r.t. $\hat{\sigma}_i$, and k denotes the rank of noiseless data (i.e., $\hat{\sigma}_i = 0 \quad \text{for} \quad i > k$). The principle of SVD can be referred as [11]. Furthermore, following component analysis-based methods, Di Natale et al. [12], Kermit, and Tomic [13] proposed higher order statistical method, i.e., independence component analysis (ICA)-based de-noising, which shows better representation for non-Gaussian data than PCA. The ICs that are highly correlated with disturbance will be recognized as noise. Suppose that $\mathbf{X} = [\mathbf{x}_1, \ldots, \mathbf{x}_n]^{\mathrm{T}} \in \Re^{n \times d}$ is mixed by source signal $\mathbf{S} = [\mathbf{s}_1, \ldots, \mathbf{s}_m]^{\mathrm{T}} \in \Re^{m \times d}$ through a linearly mixed system $\mathbf{A} \in \Re^{n \times m}$, then there is

$$\mathbf{X} = \mathbf{AS} \tag{2.5}$$

ICA aims at solving a linear transform or un-mixing system $\mathbf{W} \in \Re^{m \times n}$, and the estimated source signal $\widehat{\mathbf{X}}$ is formulated as

$$\widehat{\mathbf{X}} = \mathbf{WX} = \mathbf{WAS} \approx \mathbf{S} \tag{2.6}$$

where $\widehat{\mathbf{X}} \approx \widehat{\mathbf{S}}$ and $\mathbf{W} \approx \widehat{\mathbf{A}}^{-1}$. By solving $\mathbf{W}$, the unmixed source signal can be recovered. Tian et al. [14] also proposed a hybrid PCA plus ICA de-nosing. We refer to as [1, 15, 16] for more knowledge about signal processing in E-nose.

2.2 Pattern Recognition Algorithms

Pattern recognition methods are powerful tools to endow E-nose with "intelligence." Therefore, conventional pattern recognition algorithms such as nearest neighbors [17, 18], neural networks [19, 20], support vector machines (SVMs) [17, 21, 22], and decision tree [23] have been proposed for E-nose applications. Some improved SVMs are also proposed. For example, Wang et al. [24] proposed a relevance vector machine (RVM) with fewer kernel functions, Zhang et al. [25] proposed a hybrid support vector machine (HSVM) with fisher linear discriminant analysis, and Vergara et al. [26] proposed an inhibitory support vector machine (ISVM), inspired by the inhibition process in animal neural system. Recently, *committee classifiers* (ensemble models) are used in E-nose for classification. For example, Shi et al. [27] proposed a committee machine (GIEM) combined with five algorithms (experts) in decision level, Szczurek et al. [28] proposed a multiple classifiers system (MCS) of four sensor-specific base classifiers, and Dang et al. [29] proposed an improved support vector machine ensemble (ISVMEN) method with base classifier weighting. *Fuzzy algorithms* have also been proposed in E-nose. For example, Tudu et al. [30] proposed an incremental learning fuzzy approach for black tea classification, and Jha et al. [31] proposed an adaptive neuro-fuzzy inference system (ANFIS). Recently, *semi-supervised learning* has been used in E-nose for dealing with more challenging problem (i.e., insufficient labeled data). Specifically, De Vito et al. [32] proposed a semi-supervised learning technique (COREG) based on cluster assumption, and Hong et al. [33] also proposed to use cluster-based semi-supervised approach for E-nose data. From the analysis above, E-nose has observed a progress in pattern recognition algorithms. We refer to readers as [34] for an insight of pattern analysis methods in machine olfaction.

2.3 Drift Compensation Algorithms

Though much endeavor has been made on algorithms, sensor drift caused by unknown dynamic processes (aging, poisoning, etc.) is seriously deteriorating classification [35]. From the viewpoint of machine learning, the drifted data cannot well fit the training data due to their differences in probability distribution, and classifier retraining/recalibration with new samples is required. In the past 10 years, a progress is observed in drift compensation. Zuppa et al. [36] proposed a multiple self-organizing maps method, by adapting each map to the changes of input probability distribution. Ding et al. [37] proposed a hybrid method of PCA and wavelet for detecting drift on-line, and then compensate it using an adaptive dynamic drift compensation algorithm (ADDC). The drift model of each sensor can be dynamically updated on-line, which assumes that there exists some linear/nonlinear relation between drift and time. Artursson et al. [38] proposed a component correction-based principal component analysis (CCPCA) method, by finding the drift direction and correct the data. The model of component correction is shown as follows

$$\hat{\mathbf{X}} = \mathbf{X} - (\mathbf{X}\mathbf{p})\mathbf{p}^{\mathrm{T}} \tag{2.7}$$

where $\mathbf{p}$ is the loading vector (drift direction) calculated by PCA on the reference gas data with serious drift, $\mathbf{X}$ is the measurement data and $\hat{\mathbf{X}}$ is the drift component corrected data.

Ziyatdinov et al. [39] proposed a common principal component analysis (CPCA) method, which explicitly computes the drift direction for all classes. Padilla et al. [40] proposed a linear orthogonal signal correction (OSC) by removing the components (e.g., drift components) orthogonal to the data. The correction model is shown as

$$\hat{\mathbf{X}} = \mathbf{X} - \sum_{i=1}^{n} \mathbf{t}_i \mathbf{p}_i^{\mathrm{T}} \tag{2.8}$$

where n denotes the number of OSC factors, $\mathbf{p}_i$ is the loading vector, and $\mathbf{t}_i$ is the score vector.

Additionally, Di Carlo et al. [41] proposed an additive factor correction method, in which the correction matrix is optimized by using evolutionary algorithm (EA), described as

$$\hat{\mathbf{X}} = \mathbf{X} + \mathbf{X}\mathbf{M} \tag{2.9}$$

where $\mathbf{M}$ is the optimized correction factor by using EA.

These methods suppose that drift can be corrected additively, but capability restricted due to its nonlinear dynamic behavior [42]. To this end, Vergara et al. [43] proposed a classifier ensemble method for enhancing the classifier generalization to drift and also provided a long-term drift dataset for validation. Liu et al. [44] also proposed a fitting-based dynamic classifier ensemble. Martinelli et al. [45] proposed an AIS-based adaptive classification method. Recently, transfer learning-based models have been proposed by Liu et al. [46] and Zhang et al. [47] for drift adaptation and the classification accuracy on drifted data has been much improved. The DAELM transfer learning model proposed in [47] for drift adaptation is formulated as follows

$$\min_{\boldsymbol{\beta}_S,\,\xi_S^i,\,\xi_T^j} \frac{1}{2}\|\boldsymbol{\beta}_S\|^2 + C_S\frac{1}{2}\sum_{i=1}^{N_S}\|\xi_S^i\|^2 + C_T\frac{1}{2}\sum_{j=1}^{N_T}\|\xi_T^j\|^2 \tag{2.10}$$

$$\text{s.t.}\begin{cases} \mathbf{H}_S^i\boldsymbol{\beta}_S = \mathbf{t}_S^i - \xi_S^i, & i = 1,\ldots,N_S \\ \mathbf{H}_T^j\boldsymbol{\beta}_S = \mathbf{t}_T^j - \xi_T^j, & j = 1,\ldots,N_T \end{cases} \tag{2.11}$$

where $\boldsymbol{\beta}_S$ is the learned classifier with drift adaptation, $\mathbf{H}_S$ and $\mathbf{H}_T$ denote the handcrafted data matrix for clean and drifted sensor data, $\mathbf{t}_S$ and $\mathbf{t}_T$ denote the label matrix, ξ_S and ξ_T denote the prediction error on source and target data, respectively.

From the discussions in the previous sections, E-nose has witnessed significant progress in *applications*, *systems*, *feature extraction*, *pattern analysis,* and *drift compensation* aspects. However, for industrialization, commercialized E-noses still face with several novel challenges, such as *discreteness* (reproducibility), *drift* recovery, and non-target *disturbance* counteraction. These challenges are closely related with the fate of E-nose technology in large-scale industrialization. Therefore, in this chapter, we aim at proposing these challenges in terms of the current E-noses, and also proposing the highly efficient solutions for dealing with the key problems.

2.4 Current E-Nose Challenges

2.4.1 Discreteness

Reproducibility represents the signal discrepancy of multiple E-nose systems with identical sensor array. It is known that in the sensor manufacturing process, the inherent variability may cause slight difference in the reactivity of the tin oxide substrate, such that the response of two identical sensors under the same condition is different [48]. This discrepancy between two identical sensors can be termed as sensor discreteness [49], which results in the worse reproducibility problem of E-noses. From the viewpoint of E-nose system, sensor discreteness reflects the output differences (signal shift) among completely the same E-nose systems. Specifically, the discreteness can be described as two facets [49]: (1) baseline difference—The sensitive resistance R_o of two identical sensors in clear air under the same ambient temperature and relative humidity is different, which results in that the response (output voltage) of the identical sensors is also different; (2) sensitivity difference: Two identical sensors that exposed to some odorant have different sensitivity R_s/R_o, where R_s is the sensitive resistance in odorant and R_o in clean air, such that the responses of two sensors are also different, as shown in Fig. 2.1.

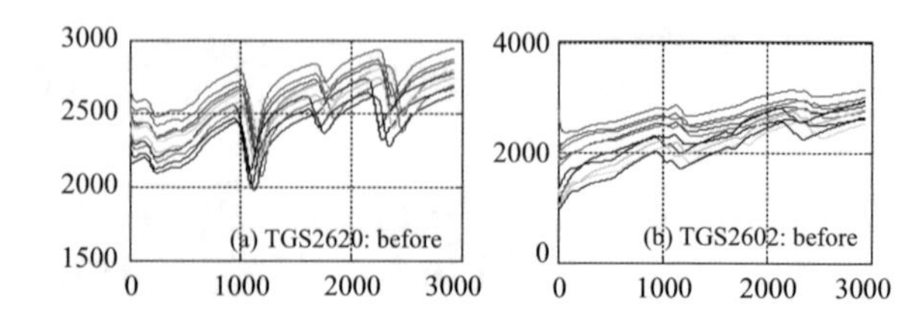

Fig. 2.1 Sensor discreteness (signal uniqueness is bad) of 10 identical E-nose instruments under the same scene

2.4.2 Drift

Drift is another challenge, due to that the sensor array is the most important part in E-noses which provides source signal of odorants. In the past two decades, drift has been paid more attention with different drift counteraction and compensation techniques. However, drift still plagued researchers and prohibit the long-term and stable usage of E-noses. In principle, drift has been caused by a number of factors, such as aging, ambient temperature, humidity, pressure, and poisoning, such that the chemical reaction inside the sensors when exposed to some odorant will be broken. Therefore, it is difficult to establish a drift model once and for all, especially that the sensor with identical type shows different drift effects. In our opinion, drift cannot be explicitly shown but may be implicitly learned through rich prior knowledge.

2.4.3 Disturbance

Non-target disturbance has seriously caused a failure of E-nose in real-time applications. For better understanding, the "non-target disturbance," we call the odorants that will be measured as target gases. Therefore, the non-target gases other than the several target odorants being detected are uniformly recognized to be the "disturbances." For our E-nose system, six odorants such as formaldehyde, benzene, toluene, carbon monoxide, ammonia, and nitrogen dioxide are the target gases being tested. Specifically, any unknown odorant except the six target gases would be viewed as disturbance, such as alcohol, perfume, smoke smell and fruit smell because the metal oxide semiconductor gas sensors produce much stronger response to these disturbances that are undesired, as shown in Fig. 2.2. The sampling process in Fig. 2.2 is elaborated as follows. First, the baseline is collected in the first 2 min. Second, the disturbance is within the next 5 min. Third, the disturbance exhaust continues for 10 min. Fourth, repeat the three steps four times. Although the non-target disturbance is serious, research on this specific problem has never been reported in E-nose community except [8], in which the non-target disturbances were treated by using *disturbance recognition* and *disturbance elimination*-based route. The disturbance

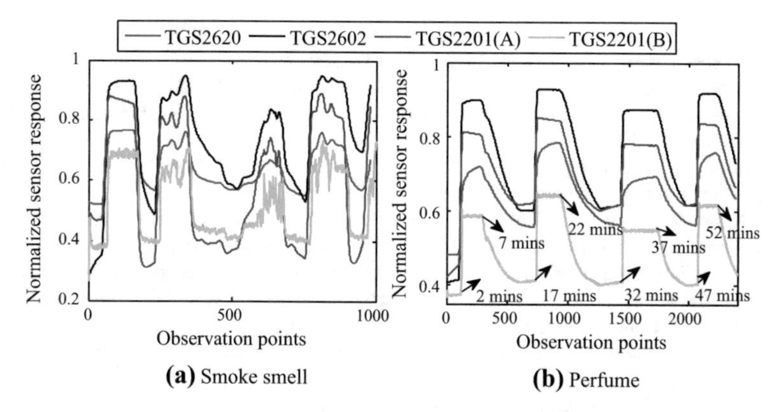

Fig. 2.2 Sensor response to disturbance, i.e., smoke smell and perfume; **a** and **b** are experimented with different sampling rate

detection step is the most important part for non-targets disturbance counteraction. The disadvantage of [8] is that the proposed method can only treat very few kinds of disturbances. However, it is known that there are thousands of "non-target" disturbances in real world. Therefore, we claim that non-target disturbances counteraction is another huge challenge of E-nose for real applications.

2.5 Summary

In this chapter, we have summarized the progress in algorithm level and new challenges in current E-nose research. *First*, the progress is described in algorithm level. Specifically, the state-of-the-art algorithms in feature extraction, pattern recognition, and drift compensation are overviewed. *Second*, with a deep insight of the observed achievements in E-nose, we present new challenges that have not been properly handled and treated, i.e., 3D issues including: (I) **D**iscreteness: sensor reproducibility enhancement; (II) **D**rift: sensor drift recovery; (III) **D**isturbance: non-targets disturbances counteraction.

References

1. S.M. Scott, D. James, Z. Ali, Data analysis for electronic nose systems. Microchim. Acta **156**, 183–207 (2007)
2. L. Zhang, F. Tian, G. Pei, A novel sensor selection using pattern recognition technique in electronic nose. Measurement **54**, 31–39 (2014)
3. X. Peng, L. Zhang, F. Tian, D. Zhang, A novel sensor feature extraction based on kernel entropy component analysis for discrimination of indoor air contaminants. Sens. Actuators, A **234**, 143–149 (2015)

4. E. Martinelli, C. Falconi, A. D'Amico, C. Di Natale, Feature extraction of chemical sensors in phase space. Sens. Actuators B: Chem. **95**, 132–139 (2003)
5. A. Leone, C. Distante, N. Ancona, K.C. Persaud, E. Stella, P. Siciliano, A powerful method for feature extraction and compression of electronic nose responses. Sens. Actuators B: Chem. **105**, 378–392 (2005)
6. B. Ehret, K. Safenreiter, F. Lorenz, J. Biermann, A new feature extraction method for odour classification. Sens. Actuators B: Chem. **158**, 75–88 (2011)
7. R. Kaur, R. Kumar, A. Gulati, C. Ghanshyam, P. Kapur, A.P. Bhondekar, Enhancing electronic nose performance: a novel feature selection approach using dynamic social impact theory and moving window time slicing for classification of Kangra orthodox black tea (*Camellia sinensis* (L.) O. Kuntze). Sens. Actuators B: Chem **166–167**, 309–319 (2012)
8. L. Zhang, F. Tian, L. Dang, G. Li, X. Peng, X. Yin, S. Liu, A novel background interferences elimination method in electronic nose using pattern recognition. Sens. Actuators, A **201**, 254–263 (2013)
9. L. Zhang, F. Tian, S. Liu, H. Li, C. Kadri, L. Pan, Applications of adaptive kalman filter coupled with multilayer perceptron for quantification purposes in electronic nose. J. Comput. Inf. Syst. **8**, 275–282 (2012)
10. S.K. Jha, R.D.S. Yadava, Denosing by singular value decomposition and its application to electronic nose data processing. IEEE Sens. J. **11**(1), 35–44 (2011)
11. A. Van der Veen, E.F. Deprettere, A.L. Swindlehurst, Subspace based signal analysis using singular value decomposition. Proc. IEEE **81**(9), 1277–1308 (1993)
12. C. Di Natale, E. Martinelli, A. D'Amico, Counteraction of environmental disturbances of electronic nose data by independent component analysis. Sens. Actuators B: Chem. **82**, 158–165 (2002)
13. M. Kermit, O. Tomic, Independent component analysis applied on gas sensor array measurement data. IEEE Sens. J. **3**(2), 218–228 (2003)
14. F. Tian, H. Li, L. Zhang, S. Liu, Q. Ye, B. Hu, B. Xiao, A denoising method based on PCA and ICA in electronic nose for gases quantification. J. Comput. Inf. Syst. **8**, 5005–5015 (2012)
15. A. Hyvärinen, E. Oja, Independent component analysis: algorithms and applications. Neural Networks **13**(4–5), 411–430 (2000)
16. S. Marco, A.G. Gálvez, Signal and Data processing for machine olfaction and chemical sensing: a review. IEEE Sens. J. **12**(11) (2012)
17. S. Güney, A. Atasoy, Multiclass classification of n-butanol concentrations with k-nearest neighbor algorithm and support vector machine in an electronic nose. Sens. Actuators B: Chem. **166–167**, 721–725 (2012)
18. K.T. Tang, Y.S. Lin, J.M. Shyu, A local weighted nearest neighbor algorithm and a weighted and constrained least-squared method for mixed odor analysis by electronic nose systems. Sensors **10**, 10467–10483 (2010)
19. S. Omatu, M. Yano, E-nose system by using neural networks. Neurocomputing **172**, 394–398 (2016)
20. Z. Xu, X. Shi, L. Wang, J. Luo, C.J. Zhong, S. Lu, Pattern recognition for sensor array signals using Fuzzy ARTMAP. Sens. Actuators B: Chem. **141**, 458–464 (2009)
21. D. Gao, F. Liu, J. Wang, Quantitative analysis of multiple kinds of volatile organic compounds using hierarchical models with an electronic nose. Sens. Actuators B: Chem. **161**, 578–586 (2012)
22. S.J. Dixon, R.G. Brereton, Comparison of performance of five common classifiers represented as boundary methods: euclidean distance to centroids, linear discriminant analysis, quadratic discriminant analysis, learning vector quantization and support vector machines, as dependent on data structure. Chemometr. Intell. Lab. Syst. **95**, 1–17 (2009)
23. J.H. Cho, P.U. Kurup, Decision tree approach for classification and dimensionality reduction of electronic nose data. Sens. Actuators B: Chem. **160**(1), 542–548 (2011)
24. X. Wang, M. Ye, C.J. Duanmu, Classification of data from electronic nose using relevance vector machines. Sens. Actuators B: Chem. **140**, 143–148 (2009)

25. L. Zhang, F. Tian, H. Nie, L. Dang, G. Li, Q. Ye, C. Kadri, Classification of multiple indoor air contaminants by an electronic nose and a hybrid support vector machine. Sens. Actuators B: Chem. **174**, 114–125 (2012)
26. A. Vergara, J. Fonollosa, J. Mahiques, M. Trincavelli, N. Rulkov, R. Huerta, On the performance of gas sensor arrays in open sampling systems using inhibitory support vector machines. Sens. Actuators B: Chem. **85**, 462–477 (2013)
27. M. Shi, A. Bermak, S.B. Belhouari, P.C.H. Chan, Gas identification based on committee machine for microelectronic gas sensor. IEEE Trans. Instrum. Measure. **55**(5), 1786–1793 (2006)
28. A. Szczurek, B. Krawczyk, M. Maciejewska, VOCs classification based on the committee of classifiers coupled with single sensor signals. Chemometr. Intell. Lab. Syst. **125**, 1–10 (2013)
29. L. Dang, F. Tian, L. Zhang, C. Kadri, X. Yin, X. Peng, S. Liu, A novel classifier ensemble for recognition of multiple indoor air contaminants by an electronic nose. Sens. Actuators, A **207**, 67–74 (2014)
30. B. Tudu, A. Metla, B. Das, N. Bhattacharyya, A. Jana, D. Ghosh, R. Bandyopadhyay, Towards versatile electronic nose pattern classifier for black tea quality evaluation: an incremental fuzzy approach. IEEE Trans. Instrum. Measure. **58**(9), 3069–3078 (2009)
31. S.K. Jha, K. Hayashi, R.D.S. Yadava, Neural, fuzzy and neuro-fuzzy approach for concentration estimation of volatile organic compounds by surface acoustic wave sensor array. Measurement **55**, 186–195 (2014)
32. S. De Vito, G. Fattoruso, M. Pardo, F. Tortorella, G. Di Francia, Semi-supervised learning techniques in artificial olfaction: a novel approach to classification problems and drift counteraction. IEEE Sens. J. **12**(11), 3215–3224 (2012)
33. X. Hong, J. Wang, G. Qi, Comparison of semi-supervised and supervised approaches for classification of e-nose datasets: case studies of tomato juices. Chemometr. Intell. Lab. Syst. **146**, 457–463 (2015)
34. R. Gutierrez-Osuna, Pattern analysis for machine olfaction: a review. IEEE Sens. J. **2**(3), 189–202 (2002)
35. M. Holmberg, F.A.M. Davide, C.D. Natale, A. D'Amico, F. Winquist, I. Lundström, Drift counteraction in odour recognition applications: lifelong calibration method. Sens. Actuators B: Chem. **42**(3), 185–194 (1997)
36. M. Zuppa, C. Distante, P. Siciliano, K.C. Persaud, Drift counteraction with multiple self-organising maps for an electronic nose. Sens. Actuators B: Chem. **98**, 305–317 (2004)
37. H. Ding, J.H. Liu, Z.R. Shen, Drift reduction of gas sensor by wavelet and principal component analysis. Sens. Actuators B: Chem. **96**, 354–363 (2003)
38. T. Artursson, T. Eklov, I. Lundstrom, P. Martensson, M. Sjostrom, M. Holmberg, Drift correction for gas sensors using multivariate methods. J. Chemometr. **14**(5–6), 711–723 (2000)
39. A. Ziyatdinov, S. Marco, A. Chaudry, K. Persaud, P. Caminal, A. Perera, Drift compensation of gas sensor array data by common principal component analysis. Sens. Actuators B: Chem. **146**, 460–465 (2010)
40. M. Padilla, A. Perera, I. Montoliu, A. Chaudry, K. Persaud, S. Marco, Drift compensation of gas sensor array data by orthogonal signal correction. Chemometr. Intell. Lab. Syst. **100**, 28–35 (2010)
41. S. Di Carlo, M. Falasconi, E. Sanchez, A. Scionti, G. Squillero, A. Tonda, Increasing pattern recognition accuracy for chemical sensing by evolutionary based drift compensation. Pattern Recogn. Lett. **32**, 1594–1603 (2011)
42. L. Zhang, F. Tian, S. Liu, L. Dang, X. Peng, X. Yin, Chaotic time series prediction of e-nose sensor drift in embedded phase space. Sens. Actuators B: Chem. **182**, 71–79 (2013)
43. A. Vergara, S. Vembu, T. Ayhan, M.A. Ryan, M.L. Homer, R. Huerta, Chemical gas sensor drift compensation using classifier ensembles. Sens. Actuators B: Chem. **166–167**, 320–329 (2012)
44. H. Liu, Z. Tang, Metal oxide gas sensor drift compensation using a dynamic classifier ensemble based on fitting. Sensors **13**, 9160–9173 (2013)

45. E. Martinelli, G. Magna, S. DeVito, R. Di Fuccio, G. Di Francia, A. Vergara, C. Di Natale, An adaptive classification model based on the artificial immune system for chemical sensor drift mitigation. Sens. Actuators B: Chem. **177**, 1017–1026 (2013)
46. Q. Liu, X. Li, M. Ye, S.S. Ge, X. Du, Drift compensation for electronic nose by semi-supervised domain adaptation. IEEE Sens. J. **14**(3), 657
47. L. Zhang, D. Zhang, Domain adaptation extreme learning machines for drift compensation in E-nose systems. IEEE Trans. Instrum. Measure. **64**(7), 1790–1801 (2015)
48. E.J. Wolfrum, R.M. Meglen, D. Peterson, J. Sluiter, Calibration transfer among sensor arrays designed for monitoring volatile organic compounds in indoor air quality. IEEE Sens. J. **6**(6), 1638–1643 (2006)
49. L. Zhang, F.C. Tian, X.W. Peng, X. Yin, A rapid discreteness correction scheme for reproducibility enhancement among a batch of MOS gas sensors. Sens. Actuators, A **205**, 170–176 (2014)

Part II
E-Nose Odor Recognition and Prediction: Challenge I

Chapter 3
Heuristic and Bio-inspired Neural Network Model

Abstract E-nose technology for detecting indoor harmful gases and concentration estimation of harmful gases and estimating the concentration become feasible by using a multi-sensor system. The estimation accuracy in actual application is concerned too much by manufacturers and researchers. This chapter analyzes the application of different bio-inspired and heuristic techniques to improve the concentration estimation in experimental electronic nose application. In this chapter, seven different particle swarm optimization models are studied including six models used for numerical function optimization, and a novel hybrid model of particle swarm optimization and adaptive genetic algorithm, for optimizing back-propagation multilayer perceptron neural network. We present the performance of a particle swarm optimization technique, an adaptive genetic strategy, and a back-propagation artificial neural network approach to perform concentration estimation of chemical gases and improve the intelligence of an E-nose.

Keywords Particle swarm optimization · Adaptive genetic strategy
Back-propagation multilayer perceptron neural network · Electronic nose
Concentration estimation

3.1 Introduction

Electronic noses (E-nose) employ an array of chemical gas sensors and have been widely used for the analysis of volatile organic compounds [1] and vapor chemicals [2]. Pattern recognition provides a higher degree of selectivity and reversibility to the system, leading to an extensive range of applications. An E-nose is an instrument consisting of an array of reversible, but only semi-selective gas sensors coupled to a pattern recognition algorithm [3], and the formal structure of an electronic nose system has also been introduced. An excellent overview of the electronic nose technology is contained in Gardner and Bartlett [4], and techniques for processing the sensor responses were reviewed by Jurs [5] and Gutierrez-Osuna [6]. Data acquisition is the first step for data analysis. Sensors collect the data and

L. Zhang et al., *Electronic Nose: Algorithmic Challenges*,
https://doi.org/10.1007/978-981-13-2167-2_3

convert it into an electrical signal pattern that is more suitable for computer analysis [3]. The output of each sensor is a pattern vector in the pattern space. Then, the pattern vector is passed into the second stage, feature selection. Feature selection is the process of identifying the most effective subset of the original features for obtaining the smallest classification error. The data produced by an electronic nose can be classified into two kinds: a set of semi-independent variables (the sensor array outputs) and a set of dependent variables (gases concentrations). Related applications of electronic nose technique have been researched [7–10]. The selectivity of the instrument is achieved through the application of pattern recognition methods to responses of the sensor array [11, 12]. Regarding the neural network training, back-propagation neural network (BP-MLP-NN) which is a gradient-based method has been widely used to classify nonlinearly separable patterns in the real application for its strong ability in recognition [13]. However, BP neural network still posses some inherent problems. First, the BP model can easily get trapped in local minima for the problems of pattern recognition and complex functions approximation [14], so that a local optimal solution is obtained. Second, the solutions are different for every train with the random initial weights. Thus, neural network optimization algorithms have been proposed by researchers to improve the ability of finding global minima of BP by using genetic algorithm (GA) [15–18].

In this chapter, we describe the application of several heuristics and bio-inspired optimization models for multi-dimensional nonlinear concentration estimation problems in experimental electronic nose applications. Similar heuristics and bio-inspired optimization model was also used for curve fitting in chemistry [19]. Particle swarm optimization (PSO) algorithm, which was originally developed by Kennedy and Eberhart in 1995, is an optimization method based on social behavior simulations [20], used to visualize the movements of a flock of birds which has the superior dynamic characteristics. Currently, a number of improved PSO models using different strategies have been investigated by researchers, such as inertia weight approach (IWA) [21], adaptive PSO (APSO) [13], attractive and repulsive PSO (ARPSO) [22], particle swarm optimization based on diffusion and repulsion (DRPSO) [23], and PSO hybrid model based on bacterial chemotaxis (PSOBC) [24]. PSO has been widely used in many fields [25–28]. It has attracted attention in recent years because of its simplicity and high performance in searching global optima. Similar to genetic algorithm, the PSO algorithm is also an optimization tool based on population which is initialized with a population of random solutions and search for optimum by continuous updating of generations. With the introduction of inertia weights, we first proposed the performance of the PSO algorithm that has been improved significantly.

The genetic algorithm (GA) is a global search procedure that searches from one population of points to another. Sexton et al. demonstrated that the GA significantly outperforms BP in optimizing neural network for a set of computer-generated data [18]. Genetic algorithm is first used by choosing an objective function for optimizing the network. The objective function values are then used to assign probabilities for each point of the population, and points generating the lowest error are the most likely to be represented in the new population. The points forming this new population are then randomly paired for the crossover operation. This crossover operation results in each new point having information from both parent points. In addition, each point has a small probability of being replaced with a value randomly chosen from the parameter space. This operation is named as mutation. The genetic algorithm has been applied to particle swarm optimization for radial basic function (RBF) network [17] and order clustering [28].

In this chapter, considering the fast convergence velocity with strong ability to find the global best solution of PSO and so many directions that genetic algorithm can also search in, an improved PSO optimization neural network based on adaptive genetic strategy (PSOAGS) has been proposed. We take three harmful gases (formaldehyde, CO, and NO_2) for experiments, through analysis of simple approaches such as principle component regression (PCR), partial least square (PLS), third-order polynomial fitting-based nonlinear least square (NLS), and seven optimized neural network models, results demonstrate that the proposed heuristics and bio-inspired optimization neural network model can significantly improve the ability of global search for optimum, and it can more accurately estimate the concentrations of gases by an electronic nose.

3.2 Particle Swarm Optimization Models

3.2.1 Standard Particle Swarm Optimization (SPSO)

In this standard PSO system, a number of particles cooperate to search for the best solutions by simulating the movement and flocking of birds. These particles fly with a certain velocity and find the global best position after certain generations. At each generation t, the velocity is updated and the particle is moved to a new position. This new position is simply calculated as the sum of the previous position and the new velocity. The mathematical notation of PSO is defined as follows.

Supposing the dimension for a searching space is D, the total number of particles is N, the position of the ith particle can be expressed as vector $X_i = (x_{i1}, x_{i2},\ldots, x_{iD})$, the best position of the ith particle searching until now is $P_i = (p_{i1}, p_{i2},\ldots, p_{iD})$, the best position of all the particles searching until now is $P_g = (p_{g1}, p_{g2},\ldots, p_{gD})$, the velocity of the ith particle is represented as $V_i = (v_{i1}, v_{i2},\ldots,v_{iD})$, then the standard PSO can be illustrated as

$$v_{id}(t+1) = w \cdot v_{id}(t) + c_1 \cdot r_{1i}^p(t) \cdot [p_{id}(t) - x_{id}(t)] + c_2 \cdot r_{2i}^p(t) \cdot [p_{gd}(t) - x_{id}(t)] \tag{3.1}$$

$$x_{id}(t+1) = x_{id}(t) + v_{id}(t+1), \quad 1 \le i \le N, \quad 1 \le d \le D \tag{3.2}$$

where c_1, c_2 are the acceleration constants with positive values and w is called inertia factor;

$$r_{1i}^{p}(t) \leftarrow U(0,\ 1),\ r_{2i}^{p}(t) \leftarrow U(0,\ 1) \tag{3.3}$$

Noteworthy is that, placing a limit on the velocity $v_{\max}$ and adjusting the inertia weight w, the PSO can achieve better search performance.

$$v_p^i(t+1) = \begin{cases} v_{\max}^i, & v_p^i(t+1) > v_{\max}^i \\ -v_{\max}^i, & v_p^i(t+1) < -v_{\max}^i \\ v_p^i(k+1), & \text{otherwise} \end{cases} \tag{3.4}$$

Further, the inertia weight approach (IWA-PSO), in which a linear reduction from a large value to a small value during the search has been proposed [21]. The w is updated by

$$w(t) = w_{\text{start}} - (w_{\text{start}} - w_{\text{end}}) \cdot t/t_{\max} \tag{3.5}$$

where $t_{\max}$ is the maximum generation, w_{start} is the initial value, and w_{end} is the terminal value.

3.2.2 *Adaptive Particle Swarm Optimization (APSO)*

The adaptive particle swarm optimization (APSO) algorithm which is based on the standard PSO was firstly proposed [21]. The APSO is called adaptive PSO for its new inertial weight improved by [13]

$$w = \begin{cases} w_{\text{start}} - (w_{\text{end}}/\max \text{gen1}) \cdot \text{gen}, & 1 \le \text{gen} \le \max \text{gen1} \\ (w_{\text{start}} - w_{\text{end}}) \cdot \exp[(\max \text{gen1} - \text{gen})/k], & \max \text{gen1} \le \text{gen} \le \max \text{gen2} \end{cases} \tag{3.6}$$

where w_{start} is the initial inertial weight, w_{end} is the ending inertial weight of linear section, max gen2 is the total searching generations, max gen1 is the used generations that inertial weight reduced linearly, and k should be adjusted for the best solution. In the experiment, we set k to 2.

3.2.3 *Attractive and Repulsive Particle Swarm Optimization (ARPSO)*

The attractive and repulsive PSO (ARPSO) was introduced to overcome the problem of premature convergence [22]. It uses a diversity measure to control the swarm. This algorithm is based on the phases between attraction and repulsion. The velocity update formula of the particles based on the SPSO was described as:

$$v_{id}(t+1) = w \cdot v_{id}(t) + \text{dir} \cdot c_1 \cdot [p_{id}(t) - x_{id}(t)] + \text{dir} \cdot c_2 \cdot [p_{gd}(t) - x_{id}(t)] \quad (3.7)$$

where dir is the coefficient (with value of 1 or −1) which decides whether the particles attract or repel each other, and c_1 and c_2 are random number in the range of [0, 2]. The dir can be determined by

$$\text{if}(\text{dir} > 0 \text{ and diversity} < \mathrm{d_{low}}), \text{then} \quad \text{dir} = -1; \quad (3.8a)$$

$$\text{if}\left(\text{dir} < 0 \text{ and diversity} > \mathrm{d_{high}}\right), \text{then} \quad \text{dir} = 1; \quad (3.8b)$$

where $\mathrm{d_{low}}$ denotes the low diversity and $\mathrm{d_{high}}$ denotes the high diversity. The diversity can be calculated using the following formula [24]

$$\text{diversity}(S) = \frac{1}{|S| \cdot |L|} \cdot \sum_{i=1}^{|S|} \sqrt{\sum_{j=1}^{n} (p_{ij} - \overline{p_j})^2} \quad (3.9)$$

where S is the swarm, |S| is the size of the swarm, |L| is the length of the longest diagonal in the search space S, n is the dimensionality of the problem, and p_{ij} is the jth value of the ith particle. $\overline{p_j}$ is the jth value of the average points $\overline{p}$. The attractive phase is also defined as the SPSO, while the individual particle is repelled by the best-known particle position in the repulsion phase.

3.2.4 *Diffusion and Repulsion Particle Swarm Optimization (DRPSO)*

The diffusion repellent PSO model, in which particle swarm is not only attracted by the region of the best solution but also carried out diffusion repellent movement by the region of the worst solution for the potentially dangerous property premature of PSO, was proposed by [23].

The velocity updating formula of DRPSO based on the SPSO model is given by:

$$v_{id}(t+1) = -w \cdot f(v_{id}(t), \theta) - c_1 \cdot \text{rand} \cdot [W_{id}(t) - x_{id}(t)] - c_2 \cdot \text{rand} \cdot [W_{gd}(t) - x_{id}(t)] \quad (3.10)$$

where W_{id} is the worst previous particle, W_{gd} is the worst particle among all the particles, and $f(v_{id}(t), \theta)$ is $v_{id}(t)$ after clockwise rotation in a certain random angle θ [0°, 360°].

3.2.5 Bacterial Chemotaxis Particle Swarm Optimization (PSOBC)

This PSO is a hybrid model based on bacterial chemotaxis and proposed by [24]. In this model, the particle swarm is not only attracted by the regions where the best results were found, but also repelled by the regions where the worst results were found. In PSOBC model, the location of the worst point found so far and location of the worst point found by the total particles are added to the SPSO algorithm, and the velocity updating formula is illustrated as follows.

$$v_{id}(t+1) = w \cdot v_{id}(t) - c_1 \cdot \text{rand}(\cdot) \cdot [W_{id}(t) - x_{id}(t)] - c_2 \cdot \text{rand}(\cdot) \cdot [W_{gd}(t) - x_{id}(t)] \quad (3.11)$$

where W_{id} is the worst previous particle and W_{gd} is the worst particle among the total particles. In this model, the same diversity-guided method as the ARPSO and the DRPSO model was also employed and the individual particle was no longer attracted, but repelled by the worst known particle position and its own previous worst position.

3.3 Hybrid Evolutionary Algorithm

In this section, considering the fact that PSO has potentially dangerous properties such as premature convergence and stagnation, a hybrid evolutionary algorithm based on an improved PSO and adaptive genetic strategy (PSOAGS) is presented. In actual applications, a large number of samples and the characteristic of real time are necessary; thus, it is meaningful and important to find the best global solution in a fast converging velocity as soon as possible, aiming at reducing the search generations. This section describes the strategies of inertial weight, acceleration constant based on cosine mechanism. Also, an improved GA with adaptive probability variation is presented.

3.3.1 PSO with Cosine Mechanism

Through the above analysis, the inertial weight and acceleration constants influence the convergence of PSO in searching the global best solution. The linear reduction and exponent reduction of inertial weight can play a good role in finding the optimum, but it is easy to trap into the local minimum [13]; thus, a regular form of cosine wave is presented in this PSOAGS model. The strategies of inertial weight and acceleration constants based on the cosine mechanism are illustrated as follows

$$w(t) = a + b \cdot \cos\left(\frac{t}{A} \times \pi\right) \tag{3.12}$$

where parameters a and b can be adjusted according to the boundary of w_{max} and w_{min}.

For parameters setup of inertial weight w, a suitable value for the inertia weight w usually provides balance between global and local exploration abilities and consequently results in a reduction of the number of generations required to locate the optimum solution. We determine the w_{min} as 0.4 and the w_{max} as 0.9 which are recommended in PSO literature [13, 21–24]. After the fixed minimum and maximum of w, we can derive that $a = 0.65$ and $b = 0.25$ through Eq. 3.12. For the acceleration coefficients, $c_1 = c_2 = 2$ were proposed as default values in the previous literature, but experimental results indicate that it can provide better results when acceleration coefficients c_1 and c_2 were described in a similar form with Eq. 3.12. We set the c_{min} to 0.6 and the c_{max} to 1.5 for both c_1 and c_2, so that $a = 1.05$ and $b = 0.45$. The value of A controls the frequency of cosine function, which also denotes the velocity of changing w; through experimental simulations, we finally provide A with 20 which can be a good choice.

3.3.2 *Adaptive Genetic Strategy (AGS)*

The crossover probability and mutation probability are the important parameters in genetic algorithm, for that the two probabilities decide the quality of population. It has been well established in the GA literature that moderately large values of the crossover probability p_c ($0.5 < p_c < 1$) and small values of the mutation probability p_m ($0.001 < p_m < 0.05$) are essential for the successful working of GAs [29–31]. And GA has also been used in architecture optimization of neural network [32]. In this chapter, we make the crossover and mutation probability varied adaptively with the Euclidean distances between individuals of the whole population. It is known that the Euclidean distances will be smaller and smaller with the continuous search evolutions. We define the distance vector between individuals of each generation t as follows

$$\vec{d} = (d_1^{(t)}, d_2^{(t)}, \ldots, d_J^{(t)}) \tag{3.13}$$

where J denotes the possible combinations among individuals, which can be calculated as

$$J = \text{PopSize} \times (\text{PopSize} - 1)/2 \tag{3.14}$$

where PopSize denotes the number of the total particles (PopSize = 50, in this chapter). Then, the adaptive crossover probability of each generation t is shown as

$$\text{probability}_{\text{crossover}} = (\vec{d}_{\max}^{(t)} - \vec{d}_{\text{average}}^{(t)})/(\vec{d}_{\max}^{(t)} - \vec{d}_{\min}^{(t)}) \tag{3.15}$$

The adaptive mutation probability of each generation is shown as

$$\text{probability}_{\text{mutation}} = \vec{d}_{\text{average}}^{(t)}/\vec{d}_{\max}^{(t)} \tag{3.16}$$

Through Eqs. 3.15 and 3.16, we can see that the probability can be adaptively changed according to the whole population; when the GA converges to a local optimum, p_c and p_m have to be increased.

Combined with the adaptive GA operators such as selection, crossover, and mutation, the repetitive search guideline is used for improving the PSO. For each particle, the fitness values in solution searching are evaluated from the PSO and AGS, respectively. Note that, the genetic operators only perform on the positions of each particle. If the best fitness of PSO is superior to that of AGS, keep the best particle of PSO; otherwise, the best particle of AGS is used in the next generation instead. In addition, to prevent the best solution from trapping into the local optimal, we present the adaptive mutation of the best particle according to the best fitness value. If the best fitness is invariable with the maximum number of tolerated generations, the best particle so far will be randomly mutated according to the following formula

$$X_{\text{best}} = X_{\text{best}} + R \times (X_{\text{worst}} - X_{\text{best}}) \tag{3.17}$$

where $R \leftarrow U(0, 1)$, X_{best} is the best particle so far, and X_{worst} is the worst particle so far; the purpose is to change X_{best} in the range of X_{best} and X_{worst};

3.4 Concentration Estimation Algorithm

Through the analysis above, we have made a significant investigation on the heuristics and bio-inspired optimization algorithms considered. In this section, we aim to apply the PSO-based techniques and genetic algorithm to optimize feed-forward artificial neural network based on error back-propagation for gases concentration estimation by an electronic nose and verify the effectiveness of these evolutionary algorithms in concentration estimation.

3.4.1 Multilayer Perceptron

The back-propagation (BP) algorithm based on gradient descending was proposed by [33]. Rumelhart further formulated the standard back-propagation (BPA) algorithm for multilayer perceptrons. The architecture of simple multilayer

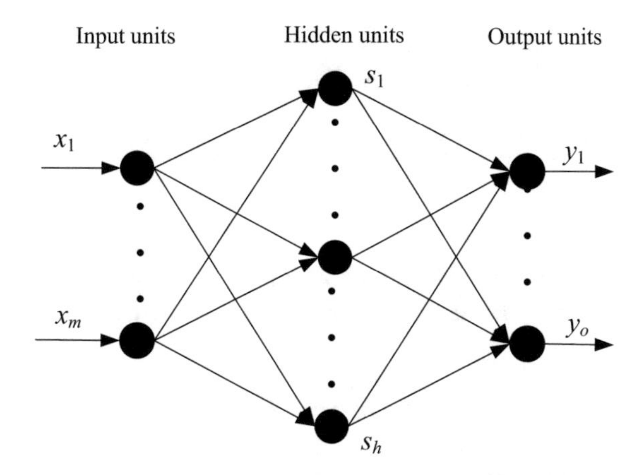

Fig. 3.1 Structure with two-layered feed-forward ANN

perceptron neural network is shown in Fig. 3.1 which presents a two-layered neural network. BP algorithm has been improved by several adaptive back-propagation algorithms for its inherent disadvantages [13, 17, 18, 34]. BP network based on a gradient descent has been widely used to classify nonlinearly separable patterns in real application for its strong ability in recognition [13]. However, some inherent problems still exist in BP neural network. First, the BP model can easily get trapped in local minima for the problems of pattern recognition [33] and fail to find the global optimal solution. Second, the initial weight matrix **W** and bias vectors **B** of back-propagation neural network are randomly produced for training so that different weights and biases would produce different trained neural networks. Thus, obtaining the global minimum of regression error by only using back-propagation neural network becomes little impractical.

3.4.2 Network Optimization

During weights optimization, the first step is to determine the encoding strategy. In this chapter, each particle is encoded as a vector with real values. For the neural network structure, each particle should represent all weights of the whole network structure. According to the subsection mentioned above, the structure of BP-MLP-NN is *m*-*h*-*o* consisted of one hidden layer, and then the total number of weights can be calculated as $m \times h + h + h \times o + o$; thus, the dimension of each particle is shown as

$$n = m \times h + h + h \times o + o \tag{3.18}$$

where m, h, and o denote the number of input neurons, hidden neurons, and output neurons, respectively. To the BP-MLP-NN with three layers (one input layer, one hidden layer, and one output layer), the weight matrix $\mathbf{W_1}$ ($m \times h$), $\mathbf{W_2}$ ($h \times o$) and the bias vectors $\mathbf{B_1}$ ($h \times 1$), $\mathbf{B_2}$ ($o \times 1$) can be obtained from each particle $\boldsymbol{x}$ using the following decoding rule

$$\underbrace{x_1, x_2, \ldots, x_{m\times h}}_{W_1}, \underbrace{x_{m\times h+1}, \ldots, x_{m\times h+h}}_{B_1}, \underbrace{x_{m\times h+h+1}, \ldots, x_{m\times h+h+h\times o}}_{W_2}, \underbrace{x_{m\times h+h+h\times o+1}, \ldots, x_n}_{B_2} \tag{3.19}$$

The active functions of the hidden layer and output layer are selected as log-sigmoid and pure linear function. The output of the jth hidden node is

$$f(\text{node}_j) = 1/(1 + \mathrm{e}^{-(\sum_{i=1}^{m} w_{ij}\cdot s_i - b_j)}), \quad j = 1, \ldots, h \tag{3.20}$$

where w_{ij} (one element of $\mathbf{W_1}$) is the connection weight from the ith node of input layer to the jth node of hidden layer and b_j (one element of $\mathbf{B_1}$) is the bias of the jth hidden layer.

The output of the kth output node is

$$y_k = \sum_{j=1}^{h} w_{kj} \cdot f(node_j) - b_k, \quad k = 1, \ldots, o \tag{3.21}$$

where w_{kj} (one element of $\mathbf{W_2}$) is the connection weight from the jth hidden node to the kth output node and b_k (one element of $\mathbf{B_2}$) is the bias of the kth output layer.

Considering the robustness of BP-MLP-NN, to avoid big difference between train error and test error and overfitting in training, the objective function of PSO is selected as the maximum between the relative validation errors and the relative training error to improve the robustness of NN, which is shown by

$$F = \max\left\{ 1/N_1 \cdot \sum_{i=1}^{N_1} \left|(Y_i^{\mathrm{tr}} - T_i^{\mathrm{tr}})/T_i^{\mathrm{tr}}\right|, 1/N_2 \cdot \sum_{j=1}^{N_2} \left|(Y_j^{\mathrm{cv}} - T_j^{\mathrm{cv}})/T_j^{\mathrm{cv}}\right| \right\} \tag{3.22}$$

where N_1 denotes the number of training samples, N_2 denotes the number of cross-validation samples, Y^{tr} and T^{tr} represent the estimated and target concentrations of training samples, and Y^{cv} and T^{cv} represent the estimated and target concentrations of cross-validation samples. The detailed evolutionary pseudo-codes of the PSOAGS model are illustrated as Algorithm 3.1.

Algorithm 3.1. PSOAGS

Algorithm 3.1 . PSOAGS	
1,	randomly initialize the positions and velocities of the total particles;
2,	**for** i = 1 to number of particles
3,	Evaluates each initialized particles using BP-MLP-NN train and test, through Eq. 3.22;
4,	**end for**;
5,	set *Generation* = 0;
6,	**while** *Generation* <Maximum generations
7,	*Generation* = *Generation*+1;
8,	Update the best fitness, the positions, and the velocities using Eq. 3.1 and Eq. 3.2 under the strategies based on the cosine mechanism;
9,	**for** $i = 1$ to number of particles
10,	evaluates each updated particle using BP-MLP-NN and obtain the best fitness PSO_{best};
11,	update the positions using AGS;
12,	evaluates each updated particle from AGS using BP-MLP-NN and obtain the best fitness AGS_{best};
13,	**end for**;
14,	**if** $PSO_{best} < AGS_{best}$
15,	keeps the best particle from PSO and the worst particle from AGS, go to the next generation;
16,	**else** the best particle from AGS and the worst particle from PSO are kept to the next generation;
17,	**end if**;
18,	**if** the best fitness is unchanged during the maximum tolerated generations
19,	Mutated the best particle using Eq. 3.17;
20,	**end if**;
21,	**end while**;

3.5 Experiments

In this section, the presented heuristics and bio-inspired algorithms are applied to concentration estimation problems for an electronic nose. The experimental design by an electronic nose system is illustrated including the experimental platform, sensor array on the printed circuit board of an electronic nose. Three chemical gases are measured using a devised electronic nose system to estimate the unknown concentration of chemical analytes through the extrapolation ability of optimized BP-MLP-NN.

3.5.1 Experimental Setup

Our sensor array in E-nose system consists of six different types of gas sensors: four from the TGS series (TGS2602, TGS2620, TGS2201A, and TGS2201B), an oxygen sensor (O2A2, type of electrochemical), and one GSBT11 sensor of Ogam Technology in Korea. In addition, a module (SHT2230 of Sensirion in Switzerland) with two auxiliary sensors for the temperature (T) and humidity (H) is also used for compensation. The sensors were mounted on a custom-designed printed circuit board (PCB), along with associated electrical components. An analog–digital converter is used as an interface between the field-programmable gate array (FPGA) processor and the sensors. FPGA can be used for data collection, storage, and processing. The E-nose system is connected to a personal computer (PC) via a Joint Test Action Group (JTAG) port which can be used to transfer data and debug programs. An additional flash memory is used to save the embedded neural network as well as the weights and biases of the neural network trained on the PC. The datasets for these gases are made up of samples in R8 space, and it just means that an input vector with eight variables can be obtained in each observation; the multi-dimensional response dataset denotes the nonlinear relation with the gas concentration. The gases measurements are implemented by an E-nose in the constant temperature and humidity chamber whose type is LRH-150S in which the temperature and humidity can be effectively controlled in terms of the target temperatures and humidity.

3.5.2 Datasets

In this chapter, for validation of the proposed model, we measured three harmful gases: formaldehyde, carbon monoxide (CO), and nitrogen dioxide (NO_2). Each of the three candidates should be presented in the E-nose in many different concentrations, respectively, and the responses of the sensor array are saved on PC. Totally, 186 samples (dataset) include 68 formaldehyde samples, 47 CO samples, and 71 NO_2 samples that are measured within one month, and each sample corresponds to a different concentration and environment; the total measurement cycle time for one single measurement was set to 20 min, i.e., 2 min for reference air (baseline), 8 min for gas sampling, and 10 min for cleaning the chamber through injecting clean air before the next experiment begins. These samples are measured at the target temperatures of 15, 25, 30, and 35 °C and target humidity of 40, 60, 80% RH, through different combinations of these target temperatures and humidity. The precisions of the chamber for temperature and humidity are ±0.1 °C and ±5% RH. And in each group of fixed temperature and humidity (e.g., 25 °C, 60% RH), the samples were measured at growing concentrations within desired range. In our experiments, there was little sensor drift in a short term of one month; thus, drift was not considered. Besides, we can find good reproducibility of the sensors

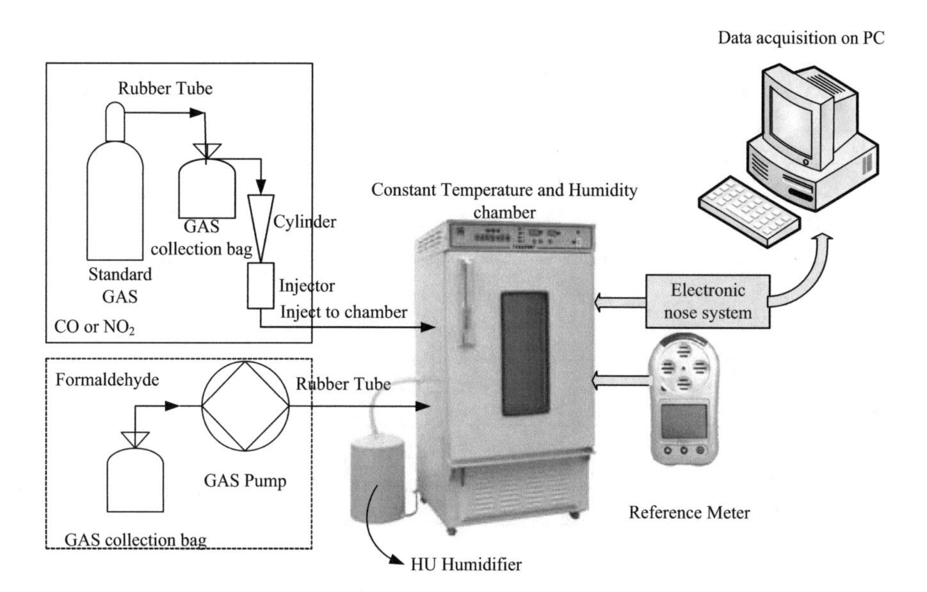

Fig. 3.2 Experimental platform for gases samples collection by an electronic nose system

through the repeated experiments under almost the same environment; the very little response difference of each sensor between almost the same two experiments can be neglected. The experimental sample collections were developed in the chamber. The basic experimental platform illustrated in Fig. 3.2 is the whole experimental process in this work. The part within the dashed lines shows the preparation for formaldehyde gas, and the part within the solid lines illustrates in preparation for CO or NO_2 gas. For the formaldehyde experiments, a gas header is necessary to store the formaldehyde gas and inject into the chamber by a pump, and the rough concentrations are controlled through the time interval of gas exhaust. For CO and NO_2 experiments, standard CO gas and NO_2 gas are prepared for the chamber with desired concentration range, and their concentrations depend on the volume of injected target gas. We can see that a reference meter which can measure the concentration of gases roughly is needed in the process for that we can have a concentration comparison with the reference meter. Note also that the true concentrations of formaldehyde were obtained by GC analysis using spectrophotometer.

To validate the effectiveness of all the models, the training and testing samples are randomly classified; 2/3 of the total samples are used as training and cross-validation samples and the remaining 1/3 as test samples. The training dataset including cross-validation was constructed of 124 measurements. There are 47 formaldehyde samples, 27 CO samples, and 50 NO_2 samples in which 10 formaldehyde samples, 10 CO samples, and 10 NO_2 samples are recognized as cross-validation samples. The target concentrations of the training samples are shown in Table 3.1. The remaining 62 measurements including 21 formaldehyde

Table 3.1 Target concentrations of 124 training samples

Formaldehyde (ppm)			CO (ppm)			NO_2 (ppm)		
0.01	0.08	0.02	1	6	5	45	45	49
0.11	0.12	0.15	5.6	1	15	63	65	71
0.13	0.48	0.23	1	5	25	5	5	
0.9	1.03	0.55	5	1	45	30	15	
0.04	0.03	0.9	1	5	65	47	32	
0.05	0.13	0.08	5	1.4	5	65	66	
0.09	0.4	0.16	1	6	15	6	5	
0.92	0.53	1.5	1.5	1.6	30	15	15	
0.04	0.9	0.02	5	6	60	46	49	
0.1	0.04	0.1	1	0.8	3	66	71	
0.25	0.39	0.4	2	5	6	5	6	
0.79	0.9	1.01	3		23	32	17	
0.02	0.03	0.05	5		45	46	50	
0.1	0.15	0.13	1.3		65	61	68	
0.5	0.3	0.65	5		5	5	18	
0.64	0.99		1.3		15	15	34	

samples, 20 CO samples, and 21 NO_2 samples are used as test dataset, to evaluate the robust performance of the neural network optimization model. Table 3.1 presents the target concentrations of the testing samples (Table 3.2).

Table 3.2 Target concentrations of 62 testing samples

Formaldehyde (ppm)		CO (ppm)		NO_2 (ppm)	
0.31	0.42	3	2.5	30	15
0.32	0.15	3	4	45	17
0.52	0.33	2.5	2.8	15	50
0.48	0.30	3.5	6	35	45
0.18	0.17	4	4	30	30
0.10	0.15	2.8	3.5	15	32
0.15	0.22	4	2.8	32	16
0.4	0.35	3	2.5	15	6
0.2	0.34	3		30	15
0.36		3		46	
0.49		3.5		30	
0.15		2.8		34	
0.34		3.5		6	

3.5.3 Concentration Estimation

In this work, the steady state values in response to each sample as our features for algorithm analysis. Thus, for each sample, a vector of eight feature values (eight sensors) is selected as one observation. Assuming the number of samples is S, we will get a matrix **X** with eight row vectors having S columns. This feature matrix **X** is the input matrix of BP-MLP-NN which will be analyzed using our proposed model. The targets of the BP-MLP-NN are the target concentrations of the corresponding gas samples. The matrix of sensor responses can be normalized as [0, 1] using the following formula

$$X'_{ij} = X_{ij}/X_{\max}, \quad i = 1, \ldots, 8; \quad j = 1, \ldots, S \tag{3.23}$$

where $X_{\max}$ is the maximum saturation value 4095 of each sensor and S is the number of total samples. Based on the input responses X and the target concentrations T, the nonlinear relation between sensors responses and the concentrations can be determined through $\mathbf{W}_1$, $\mathbf{W}_2$, $\mathbf{B}_1$, and $\mathbf{B}_2$ which are obtained by back-propagation algorithm learning based on heuristic and bio-inspired optimization models. Assume the number of samples for each gas is set to n_1, n_2, and n_3, respectively. The training input data matrix **P** and training targets matrix **T** of neural network are described as the following forms

$$P = \left[\begin{array}{l} \overbrace{\begin{matrix} S_{1,1} & S_{1,2} & \cdots & S_{1,n_1} \\ S_{2,1} & S_{2,2} & \cdots & S_{2,n_1} \\ \vdots & \vdots & \vdots & \vdots \\ S_{8,1} & S_{8,2} & \cdots & S_{8,n_1} \end{matrix}}^{\text{gas 1}} \overbrace{\begin{matrix} S_{1,n_1+1} & S_{1,n_1+2} & \cdots & S_{1,n_1+n_2} \\ S_{2,n_1+1} & S_{2,n_1+2} & \cdots & S_{2,n_1+n_2} \\ \vdots & \vdots & \vdots & \vdots \\ S_{8,n_1+1} & S_{8,n_1+2} & \cdots & S_{8,n_1+n_2} \end{matrix}}^{\text{gas 2}} \overbrace{\begin{matrix} S_{1,n_1+n_2+1} & \cdots & S_{1,n_1+n_2+n_3} \\ S_{2,n_1+n_2+1} & \cdots & S_{2,n_1+n_2+n_3} \\ \vdots & \vdots & \vdots \\ S_{8,n_1+n_2+1} & \cdots & S_{8,n_1+n_2+n_3} \end{matrix}}^{\text{gas 3}} \end{array} \right] \tag{3.24}$$

$$T = \left[\begin{array}{l} \overbrace{\begin{matrix} t_{1,1} & t_{1,2} & \cdots & t_{1,n_1} \\ 0 & 0 & \cdots & 0 \\ 0 & 0 & \cdots & 0 \end{matrix}}^{\text{gas 1}} \overbrace{\begin{matrix} 0 & 0 & \cdots & 0 \\ t_{2,n_1+1} & t_{2,n_1+2} & \cdots & t_{2,n_1+n_2} \\ 0 & 0 & \cdots & 0 \end{matrix}}^{\text{gas 2}} \overbrace{\begin{matrix} 0 & \cdots & 0 \\ 0 & \cdots & 0 \\ t_{3,n_1+n_2+1} & \cdots & t_{3,n_1+n_2+n_3} \end{matrix}}^{\text{gas 3}} \end{array} \right] \tag{3.25}$$

where $S_{i,j}$ (i = 1,…,8, j = 1,…,n_1 + n_2 + n_3) is the selected feature at the steady state sensor response of one measurement, $t_{1,k}$ (k = 1,…,n_1) are the actual concentrations of gas 1 for the n_1 times of measurements, and $t_{2,k}$ (k = n_1 + 1,…, n_1 + n_2) are actual concentrations of gas 2 for n_2 times of measurements, the same meaning to gas 3. The values of zero concentration (0 ppm) denote the absence of the corresponding gas components. The arrangement order of gas in the input data matrix **P** is formaldehyde (gas 1), CO (gas 2), and NO_2 (gas 3), respectively. In this case, the outputs of network are just the estimated concentrations. In one observation, the output is a vector with three values corresponding to concentrations of the three known gas components. The test samples are also arranged as the similar form Eqs. 3.24 and 3.25. In this chapter, $n_1 = 37$, $n_2 = 17$, $n_3 = 40$ for training samples; $n_1 = 10$, $n_2 = 10$, $n_3 = 10$ for cross-validation samples; and $n_1 = 21$, $n_2 = 20$, $n_3 = 21$ for test samples.

Training was carried out with a sum-squared output error goal of 0.005 up to a maximum of 2000 iterations. Sometimes, the network did not reach the goal after 2000 iterations and the sum-squared error was still higher than 0.005. The objective function used in PSO optimization procedure is shown in Eq. 3.22. The output matrix Y of neural network should be the predicted concentration matrix shown as

$$Y = \begin{bmatrix} \overbrace{y_{1,1} \quad y_{1,2} \quad \cdots \quad y_{1,n_1}}^{\text{gas 1}} & \overbrace{- \quad - \quad \cdots \quad -}^{\text{gas 2}} & \overbrace{- \quad \cdots \quad -}^{\text{gas 3}} \\ - \quad - \quad \cdots \quad - & y_{2,n_1+1} \quad y_{2,n_1+2} \quad \cdots \quad y_{2,n_1+n_2} & - \quad \cdots \quad - \\ - \quad - \quad \cdots \quad - & - \quad - \quad \cdots \quad - & y_{3,n_1+n_2+1} \quad \cdots \quad y_{3,n_1+n_2+n_3} \end{bmatrix} \tag{3.26}$$

where the diagonal elements are the predicted concentrations of gas 1, gas 2, and gas 3; the values in the positions with "–" which are very close to zero can be neglected.

In this work, to evaluate the final estimation model, the relative estimation errors (train error and test error) of gas l (l = 1, 2, 3) are shown as follows

$$\text{TRE} = 1/N_l \cdot \sum_{i=1}^{N_l} \left| (Y^{\text{tr}}_{l,i+N_{l-1}+N_{l-2}} - T^{\text{tr}}_{l,i+N_{l-1}+N_{l-2}})/T^{\text{tr}}_{l,i+N_{l-1}+N_{l-2}} \right| \tag{3.27}$$

$$\text{TEE} = 1/N_l \cdot \sum_{j=1}^{M_l} \left| (Y^{\text{te}}_{l,j+M_{l-1}+M_{l-2}} - T^{\text{te}}_{l,j+M_{l-1}+M_{l-2}})/T^{\text{te}}_{l,j+M_{l-1}+M_{l-2}} \right| \tag{3.28}$$

where $\mathbf{Y}^{\text{tr}}$ and $\mathbf{T}^{\text{tr}}$ denote the training output matrix and training target matrix; $\mathbf{Y}^{\text{te}}$ and $\mathbf{T}^{\text{te}}$ denote the test output matrix and test target matrix; and N_l and M_l denote the number of training samples and test samples of gas l. Note that $N_{-1} = N_0 = 0$, $M_{-1} = M_0 = 0$. Then, the correct estimation rates (CER) can be calculated as follows

$$\text{CER} = 100 - (\text{TRE} + \text{TEE})/2 \tag{3.29}$$

Worthy noting that the zero concentration should be neglected when calculating TRE and TEE using Eqs. 3.27 and 3.28 for that infinite great would be obtained as the denominator is zero and the diagonal elements are only considered.

3.6 Results and Discussion

3.6.1 Experimental Results

The steady state responses of sensors when exposed to analytes were used as features for concentration estimation. Figure 3.3 illustrates the normalized gas sensor responses, and the selected steady state response points with humidity, temperature, and oxygen at the same position are recognized as features for each measurement. The structure of the BP-MLP-NN consisted of three layers (one input layer, one hidden layer, and one output layer) which is set as 8-16-3 (8 input neurons, 16 hidden neurons, and 3 output neurons); hence, the dimension of each particle is 303, calculated from Eq. 3.18. The eight input neurons denote the number of sensors, and the three output neurons correspond to three measured chemical gases. For description of the responses of sensors in the original dataset, a principle component analysis (PCA) result is obtained from the covariance of mean-centered data matrix. Figure 3.4 shows PCA result of the original dataset. The scores of three families are plotted for principal component 1 versus principal component 2. A PCA result can give the most information of raw response matrix. From this figure, the CO family can be separated quite well, but the other two families overlap. Besides, it can be seen that CO occupies relatively larger region of the PC space. In Fig. 3.5, the loadings of first two principal components are shown, indicating the contribution of each sensor to the analysis. We can obviously see that humidity sensor and temperature sensor, particularly the humidity, have a significant contribution to our problem and they are far from the gas sensors. And the distribution of the five gas sensors except the oxygen is dispersive so that they have no collinearity in the response matrix. Apparently, the oxygen sensor plays a little role. Thus, we can speculate that the linear approaches would fail to solve the problem of concentration estimation effectively in our project because the responses

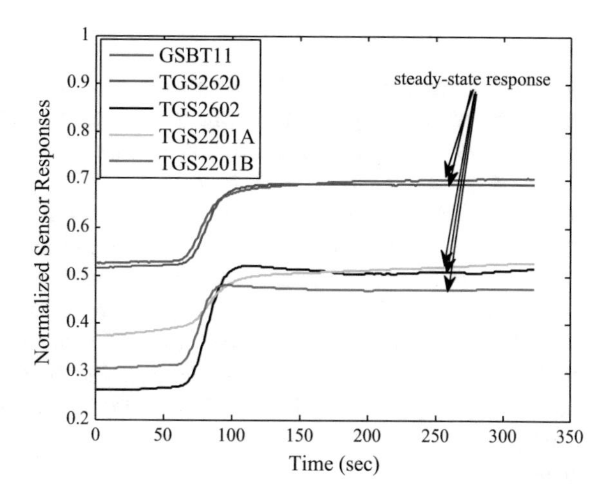

Fig. 3.3 Normalized gas sensor responses when exposed to formaldehyde with concentration of 0.42 ppm under the condition of temperature = 15 °C and humidity = 60% RH

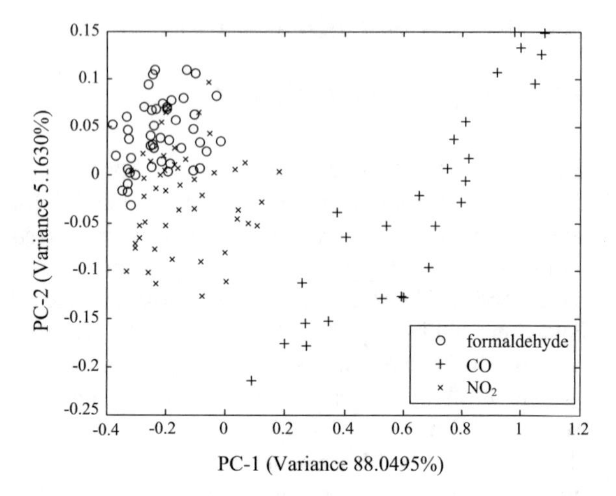

Fig. 3.4 Principal component score plots of original dataset

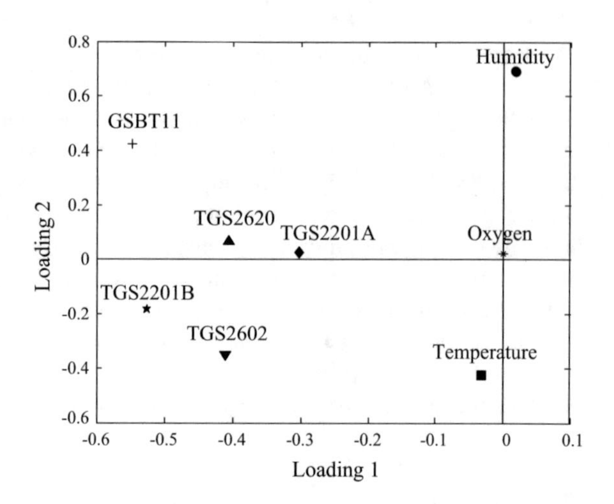

Fig. 3.5 Results of PCA for the eight-element sensor array showing the loadings of each sensor for principal components 1 and 2

of harmful gas sensors would be affected by both concentration and other environmental factors (e.g., temperature and humidity). However, for quantification comparisons of the problem, we first attempt to use two simple linear methods including principle component regression (PCR) [35] and partial least square (PLS) [35, 36] to estimate concentrations of gases. Besides, a nonlinear least square (NLS) [37] third-order polynomial fitting is also used to estimate gas concentration.

Then, PSO optimization multilayer perceptron neural networks with back-propagation algorithm learning are used to validate the improved PSO model. Considering the instability of BP-MLP-NN resulted from randomly initialized particles, we run each PSO-based concentration estimation program for 20 times and 50 generations for each time. The average best fitness values including train error (TRE) and test error (TEE) of the 20 runs are recognized to evaluate each model.

Table 3.3 Relative estimation error of chemical gases concentration using PCR, PLS, NLS, and BP-MLP

Concentration estimation methods	Relative estimation error of chemical gases (%)							
	Formaldehyde		CO		NO_2		Average	
	TRE	TEE	TRE	TEE	TRE	TEE	TRE	TEE
PCR	73.1578	105.4032	52.1022	66.3210	53.0121	62.2314	59.4240	77.9852
PLS	61.5431	79.5882	42.4425	50.5120	40.4178	51.2125	48.1345	60.4376
NLS	50.2152	58.0233	29.2412	38.5022	28.1586	36.5401	35.8717	44.3552
Single BP	35.7356	28.3328	10.7721	20.0285	12.9313	13.4460	19.813	20.6024
SPSO-BP	25.3224	19.7514	7.5381	9.4957	8.8106	11.8477	13.8904	13.6982
IWAPSO-BP	19.2493	22.8549	6.8687	10.1758	7.0175	7.7730	11.0602	13.6012
APSO-BP	37.8276	25.2891	7.2265	8.7764	6.3244	5.9902	17.1262	13.3519
DRPSO-BP	25.4639	22.1065	6.2033	9.1402	10.1575	10.3230	13.9416	13.8566
ARPSO-BP	20.5874	20.4023	5.8827	11.1103	11.3025	6.6240	12.5909	12.7122
SGA-BP	32.5937	18.8185	6.5593	8.4941	10.0178	6.6698	16.3903	11.3275
PSOBC-BP	31.3325	21.5227	6.1899	8.5397	8.2748	9.5581	15.2657	13.2068
PSOAGS-BP	**16.3536**	**17.5681**	**5.0970**	**7.8017**	**5.1787**	**5.4962**	**8.8764**	**10.2887**

The boldface type denotes the best performance.

Table 3.3 demonstrates the best train error and test error of each measured gas calculated by Eqs. 3.26 and 3.27. Table 3.4 presents the correct estimation rates (CER) calculated as Eq. 3.28.

Through Table 3.3, we can find that the simple linear methods including PCR and PLS obtain very big TREs and TEEs for each gas. Obviously, their correct estimation rates are also low with only average 31.2954 and 45.714% (see Table 3.4). When the NLS method by a third-order polynomial fitting is used, the concentration estimation performance has been improved a little and average 59.8886% of correct estimation rate is obtained.

When compared with neural network methods, we can see that the linear methods are far inferior to single BP approach. This also demonstrates that the responses of multi-dimensional MOS sensor array are strongly nonlinear with gas concentrations, because of their high sensitivities to temperature and humidity. For

Table 3.4 Correct estimation rates of chemical gases by using PCR, PLS, NLS, and BP-MLP

Concentration estimation methods	Correct estimation rates (CER) of chemical gases (%)			
	Formaldehyde	CO	NO_2	Average
PCR	10.7195	40.7884	42.3783	31.2954
PLS	29.4343	53.5228	54.1848	45.7140
NLS	45.8870	66.1283	67.6506	59.8886
Single BP	67.9658	84.5997	86.8114	79.7923
SPSO-BP	77.4631	91.4831	89.6709	86.2057
IWAPSO-BP	78.9479	91.4778	92.6047	87.6693
APSO-BP	68.4417	91.9985	93.8427	84.7609
DRPSO-BP	76.2148	92.3282	89.7597	86.1009
ARPSO-BP	79.5052	91.5035	91.0367	87.3485
SGA-BP	74.2939	92.4733	91.6562	86.1411
PSOBC-BP	73.5724	92.6352	91.0836	85.7638
PSOAGS-BP	**83.0392**	**93.5507**	**94.6625**	**90.4175**

The boldface type denotes the best performance.

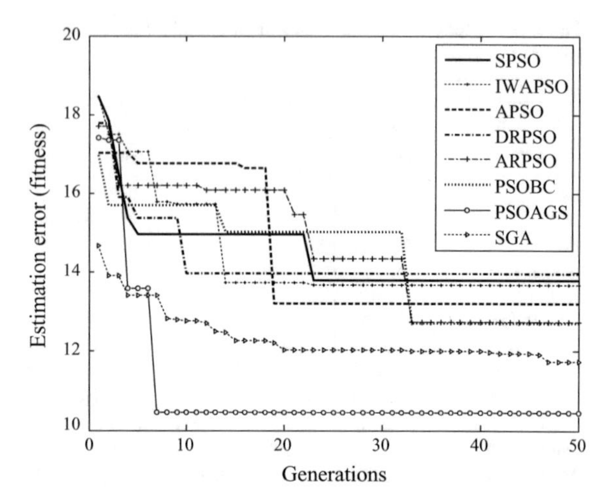

Fig. 3.6 Convergence curves of the average concentration estimation error of the three pollutant gases using BP-MLP-NN with different optimization models

comparing the PSO-based optimization models, we analyze the results for each gas separately. For formaldehyde gas, compared with single BP, the estimation errors including TRE (35.7356%) and TEE (28.3328%) are larger than the BP with the best PSOAGS optimization algorithm (16.3536% of TRE and 17.5681% of TEE). For CO gas, 10.7721% of TRE and 20.0285% TEE are with single BP, while 5.0970% and 7.8017% are obtained using the best PSOAGS. For NO_2 gas, 12.9313% train error and 13.4460% test error were obtained by single BP, while 5.1787% train error and 5.4962% test error were obtained by the best PSOAGS model. The CER of optimized methods shown in Table 3.4 are evidently higher than single BP. Further, the performance of E-nose has been improved 10% of the average CER compared to the PSOAGS model (90.4175%) and single BP algorithm (79.7923%). Other PSO-based optimization models present similar CER according to Table 3.4. IWAPSO-BP and ARPSO-BP give higher average CER (87.6693, 87.3485%) than SPSO-BP and DRPSO-BP (86.2057, 86.1009%), while APSO-BP and PSOBC-BP present relatively lower average CER (84.7609, 85.7638%). In visual, for viewing the process during searching for the best solution using optimization algorithms, we present Fig. 3.6 which describes the convergence curves with generations of the average concentration estimation error of the three chemical gases by using each optimization model. At the eighth generations, the best solution has been found out using PSOAGS model, while PSOBC and ARPSO obtain the solutions at the thirty-third generations. It can be seen from Tables 3.3 and 3.4 that the optimization algorithms are indispensable in E-nose application for accurate concentration estimation. Note that the estimation error is taken into consideration as the average best fitness (see Eq. 3.22). The proposed PSOAGS model can find the best solution in a fastest converging velocity and reduce the run time of optimization algorithm. The SGA model is inferior to the method in this work. Note also that each curve is obtained from the average results of 20 runs.

3.6.2 Computational Efficiency

We have developed these algorithms to run on programs written by us in MATLAB 2009a, operating on a laboratory computer equipped with Inter Core (TM) i3 CPU 530, 2.93 GHz processors and 2 GB of RAM. Through experimental results, we can find that the problem of concentration estimation in this chapter cannot be solved by simple linear methods or nonlinear least square polynomial fitting because the MOS gas sensors would be disturbed by temperature and humidity. This also demonstrates the strong nonlinear relation between responses of sensors and gas concentration. When we use optimized ANN to build the multi-dimensional nonlinear relation, the complexity of algorithm should be considered. First, we should note that the neural network training is employed on PC but not the E-nose itself. Second, the well-trained hyper-parameters (weight matrices and biases) should be saved on the E-nose system for real-time concentration estimation. At the first step, the neural network optimization process by heuristic and bio-inspired algorithm would increase the calculation complexity because of the maximum evolving generations and populations (*PopSize*) in optimization. In our optimization, the optimization time will depend on the maximum generations and populations. In this chapter, we set both the maximum generations and the number of populations as 50, then the total training times should be 2550 according to the pseudo-codes shown in Algorithm 3.1, and about average 4.899 s was consumed for each training. Thus, the whole neural network optimization process would consume about 3.47 h. Fast convergence velocity for the best solution would reduce the optimization time, while it will take several minutes even less than one minute to find the estimation parameters of PCR, PLS, NLS. At the second step for real-time concentration estimation using the well-trained neural network parameters obtained at the first step, it only takes 0.0138 s to calculate the concentration of one sampling and it can meet the property of real time. Similarly, 0.0020, 0.0206, and 0.0093 s would be consumed for PCR, PLS, and NLS, respectively. So, for accurate concentration estimation of harmful gases by an E-nose, the CPU time for searching a better estimation model is worthy sacrificing on PC with the fast development of computer.

3.6.3 Discussion

In PSO, the number of variables in each particle is calculated using Eq. 3.18, which is directly decided by the architecture of the BP neural network. Thus, to problems with high dimension, such as prediction in multi-sensor system (E-nose), study of nonlinear optimized models becomes necessary. Compared with genetic algorithm, we can find the superior performance of particle swarm optimization in nonlinear system. All the considered optimization algorithms adopt the same real-value encoding strategy for neural network weights and bias optimization. Also, PSO can optimize the sensor array or selection of sample subsets or the structure and related parameters of neural network for improving the performance of an E-nose. In the improvement of PSO, the convergence performance would be paid more attention to actual engineering applications. The TER, TRE, and CER analysis of each optimization model in E-nose application demonstrate the PSOAGS model can find the best optimum in a fastest convergence velocity. Therefore, the time complexity can be reduced significantly on PC.

To avoid obtaining local solution, an adaptive genetic algorithm is presented for improving the solutions searching ability of particle swarm optimization. This adaptive characteristic of crossover and mutation probability improves the performance of particles in searching according to the distances among all the particles. It also prevents the particles from being trapped into local optimal.

The apparent advantages of optimized neural network are better robustness, ability of anti-noise, and long-term stability compared with those simple approaches. As we know, the sensor drift and the reproducibility should also be considered for long-term cases when using simple methods. However, they can be neglected when using the proposed methods in this chapter for slight drifts and response variations. Even though complex algorithm would increase complexity, it is only wasted in the hyper-parameters learning of neural network on PC but not in our E-nose system. With the fast development of computer, the complexity of algorithm may be tolerated and accepted. The proposed PSOAGS model would consume a little more computer resource for considering the crossover and mutation operations of genetic algorithm. However, in current electronic nose application, especially the harmful gases detection indoor, high correct prediction rates and real-time property are necessary.

3.7 Summary

In this chapter, we addressed the issues of gases concentration estimation in electronic nose. Through the analysis of the simple linear methods (PCR, PLS) and a polynomial-based nonlinear least square (NLS), we failed to obtain an accurate concentration estimation model using approaches of low complexity for the multi-dimensional nonlinear problem in E-nose. Then, for development of the estimation problem in E-nose, we have analyzed the performance of several heuristics and bio-inspired optimization algorithms in the problems of concentration estimation. Specifically, we have studied the performance of SPSO, IWA-PSO, APSO, DRPSO, ARPSO, PSOBC, and standard genetic algorithm (SGA), optimizing the back-propagation neural network for gases quantification. Moreover, a hybrid evolutionary algorithm PSOAGS is also proposed for improving the intelligence of electronic nose further. A large number of experiments have also been carried out for traversing concentrations in different environments (temperature and humidity) by an electronic nose. We have shown that all the compared approaches provide better estimation accuracy than single BP-MLP-NN, but the PSOAGS outperforms other considered algorithms. In this work, we demonstrated that the heuristics and bio-inspired optimization algorithms are robust and effective approaches in terms of the intelligence of an electronic nose.

References

1. J.E. Haugen, K. Kvaal, Electronic nose and artificial neural network. Meat Sci. **49**, S273–S286 (1998)
2. L. Carmel, N. Sever, D. Lancet, D. Harel, An e-nose algorithm for identifying chemicals and determining their concentration. Sens. Actuators B **93**, 77–83 (2003)
3. M.S. Simon, D. James, Z. Ali, Data analysis for electronic nose systems. Microchim. Acta **156**, 183–207 (2007)
4. J.W. Gardner, P.N. Bartlett, *Electronic Noses: Principles and Applications* (Oxford University Press, Oxford, 1999)
5. P.C. Jurs, G.A. Bakken, H.F. McClelland, Computational methods for the analysis of chemical sensor array data from volatile analytes. Chem. Rev. **100**, 2649–2678 (2000)
6. R. Gutierrez-Osuna, Pattern analysis for machine olfaction: a review. IEEE Sens. J. **2**(3), 189–202 (2002)
7. G.C. Green, A.D.C. Chan, H.H. Dan, M. Lin, Using a metal oxide sensor (MOS)-based electronic nose for discrimination of bacteria based on individual colonies in suspension. Sens. Actuators B **152**, 21–28 (2011)
8. C.H. Shih, Y.J. Lin, K.F. Lee, P.Y. Chien, P. Drake, Real-time electronic nose based pathogen detection for respiratory intensive care patients. Sens. Actuators B **148**, 153–157 (2010)
9. C. Wongchoosuk, A. Wisitsoraat, A. Tuantranont, T. Kerdcharoen, Portable electronic nose based on carbon nanotube-SnO_2 gas sensors and its application for detection of methanol contamination in whiskeys. Sens. Actuators B **147**, 392–399 (2010)
10. B.A. Botre, D.C. Gharpure, A.D. Shaligram, Embedded electronic nose and supporting software tool for its parameter optimization. Sens. Actuators B **146**, 453–459 (2010)
11. P. Wang, T. Yi, A novel method for diabetes diagnosis based on electronic nose. Biosens. Bioelectron. **12**(9–10), 1031–1036 (1997)
12. P. Wang, J. Xie, A novel recognition method for electronic nose using artificial neural network and fuzzy recognition. Sens. Actuators B: Chem. **37**(3), 169–174 (1996)
13. J.R. Zhang, J. Zhang, A hybrid particle swarm optimization back-propagation algorithm for feed-forward neural network training. Appl. Math. Comput. **185**, 1026–1037 (2007)
14. M. Gori, A. Tesi, On the problem of local minima in back-propagation, in IEEE Transactions on Pattern Analysis and Machine Intelligence (1992), pp. 76–86
15. V. Maniezzo, Genetic evolution of the topology and weight distribution of neural networks neural networks. IEEE Trans. **5**(1), 39–53 (2002)
16. J.N.D. Gupta, R.S. Sexton, Comparing back-propagation with a genetic algorithm for neural network training. Omega **27**(6), 679–684 (1999)
17. C.F. Juang, Y.C. Liou, On the hybrid of genetic algorithm and particle swarm optimization for evolving recurrent neural network, in *Proceedings of IEEE International Joint Conference on Neural Networks*, 2005
18. R.S. Sexton, J.N.D. Gupta, Comparative evaluation of genetic algorithm and back-propagation for training neural networks. Inf. Sci. **129**(1–4), 45–59 (2000)
19. M.J. Polo-Corpa, S. Salcedo-Sanz, A.M. Perez-Bellido, P. Lopez-Espi, R. Benavente, E. Perez, Curve fitting using heuristics and bio-inspired optimization algorithms for experimental data processing in chemistry. Chemometr. Intell. Lab. Syst. **96**, 34–42 (2009)

20. J. Kennedy, R.C. Eberhart, Particle swarm optimization. Proc. IEEE Int. Conf. Neural Networks **4**, 1942–1948 (1995)
21. R.C. Eberhart, Y. Shi, Comparing inertia weights and constriction factors in particle swarm optimization. IEEE. Evol. Comput. **1**, 84–88 (2002)
22. J. Riget, J.S. Vesterstrm, A diversity-guided particle swarm optimizer-the ARPSO. Dept. Comput. Tech. Rep **2**, 2002 (2002)
23. W. Jiang, Y. Zhang, A Particle swarm optimization algorithm based on doffision-repulsion and application to portfolio selection. IEEE Inter. Sym. Info. Sci. Eng. (2008)
24. B. Niu, Y. Zhu, An improved particle swarm optimization based on bacterial Chemotaxis. IEEE, Intell. Control Autom. **1**, 3193–3197 (2006)
25. C.H. Hsu, W.J. Shyr, K.H. Kuo, Optimizing multiple interference cancellations of linear phase array based on particle swarm optimization. J. Inf. Hiding Multimedia Signal Process. **1**, 292–300 (2010)
26. J.F. Chang, S.C. Chu, J.F. Roddick, J.S. Pan, A parallel particle swarm optimization algorithm with communication strategies. J. Inf. Sci. Eng. **21**, 809–818 (2005)
27. M.F. Horng, Y.T. Chen, S.C. Chu, J.S. Pan, B.Y. Liao, An extensible particle swarm optimization for energy effective cluster management of underwater sensor networks, ICCCI2010. LNAI **6421**, 109–116 (2010)
28. R.J. Kuo, L.M. Lin, Application of a hybrid of genetic algorithm and particle swarm optimization algorithm for order clustering. Decis. Support Syst. **49**, 451–462 (2010)
29. J. Horn, N. Nafpliotis, An enriched Pareto genetic algorithm for multi-objective optimization, in *IEEE*, 2002
30. K.A. DeJong, An analysis of the behavior of a class of genetic adaptive system (1975)
31. D.E. Goldberg, *Genetic Algorithm in Search, Optimization and Machine Learning*, 1989
32. D. Ballabio, M. Vasighi, V. Consonni, M.K. Zareh, Genetic algorithms for architecture optimisation of counter-propagation artificial neural networks. Chemometr. Intell. Lab. Syst. **105**, 56–64 (2011)
33. W. Wu, J. Wang, M.S. Cheng, Z.X. Li, Convergence analysis of online gradient method for BP neural networks. Neural Networks **24**, 91–98 (2011)
34. O. Kisi, Multi-layer perceptrons with levenberg-marquardt training algorithm for suspended sediment concentration prediction and estimation. Hydrol. Sci. J. **49**(6), 1025–1040 (2004)
35. J. Getino, M.C. Horrillo, J. Gutiérrez, L. Arés, J.I. Robla, C. García, I. Sayago, Analysis of VOCs with a tin oxide sensor array. Sens. Actuators B **43**, 200–205 (1997)
36. S. de Jong, SIMPLS: an alternative approach to partial least squares regression. Chemometr. Intell. Lab. Syst. **18**, 251–263 (1993)
37. H. Zhou, M.L. Homer, A.V. Shevade, M.A. Ryan, Nonlinear least-squares based on method for identifying and quantifying single and mixed contaminants in air with an electronic nose. Sensors **6**, 1–18 (2006)

Chapter 4
Chaos-Based Neural Network Optimization Approach

Abstract E-nose combined with a pattern recognition module can be used for estimating gases concentration. This chapter introduces the concentration estimations of indoor contaminants using chaos-based optimization artificial neural network integrated into our E-nose instrument. Back-propagation neural network (BPNN) has been the common pattern recognition algorithm for E-nose; however, it has local optimal flaw. This chapter presents a novel chaotic sequence optimization BPNN method. Experimental results demonstrate the superiority and efficiency of the portable E-nose instrument integrated into chaos-based artificial neural network optimization algorithms in real-time monitoring of air quality in dwellings.

Keywords Electronic nose · Artificial olfactory system · Back-propagation neural network · Chaotic sequence optimization · Particle swarm optimization

4.1 Introduction

An electronic nose can be a good solution for continuous and real-time monitoring of air quality indoor in dwellings or in vehicles. Indoor air quality standard usually reports four categories of contaminants, including physical, chemical, biological, and radioactive ones [1]. Among them, chemical contaminants were recognized as harmful substances to public health indoor including sulfur dioxide, nitrogen dioxide, carbon monoxide, carbon dioxide, ammonia, ozone, formaldehyde, benzene, toluene, inhalable particle, and volatile organic compounds [1]. Formaldehyde and benzene are the contaminants we aim to detect in our E-nose technology. These odorants have been mostly recognized for their potential harms to public health as pollutants of indoor air quality from numerous studies [1–3]. Usually, they exist in the emissions from new furniture, oil paint, and building materials of residuals [4].

Chaos is a bounded unstable dynamic behavior that exhibits sensitive dependence on initial conditions and also unstable periodic motions in nonlinear systems [5]. Although it appears to be stochastic, it occurs in a deterministic nonlinear

L. Zhang et al., *Electronic Nose: Algorithmic Challenges*,
https://doi.org/10.1007/978-981-13-2167-2_4

system under deterministic conditions. Chaotic sequences generated from many chaotic maps commonly possess certainty, ergodicity, and stochastic property; therefore, they have been used instead of random sequences and somewhat good results have also been demonstrated when combined with particle swarm optimization [6] for global solutions search. In prediction, classification, and pattern recognition, hybrid forecasting models based on chaotic mapping have been presented together with Gaussian support vector machine, particle swarm optimization, and genetic algorithm in [7, 8]. Similarly, chaos optimization has also been used in prediction of silicon content in hot metal [9] and faults classification [10, 11]. Chaos search immune algorithms have been presented in [12, 13] for neuro-fuzzy controller design and pattern recognition. The choice of chaotic sequences is justified theoretically by their unpredictability including spread-spectrum characteristic, non-periodic, complex temporal behavior, and ergodic properties.

Electronic nose, as an artificial olfactory system, includes a central process unit, chemical gas sensor array, other peripheral circuits, and pattern recognition module [14]. It has been widely used for analysis of volatile organic compounds [15], vapor chemicals [16], waters [17], wine [18], and breath alcohol measurement [19]. Neural network, especially back-propagation neural network (BPNN), has been widely used for recognition and function approximation based on its strong regression ability [20]. So, in this chapter, BPNN is used for concentration estimation of formaldehyde in an electronic nose. However, BP neural network still has some inherent problems. First, BP model can easily get trapped in local minima for the problems of pattern recognition and complex functions approximation [21], so that a local optimal solution is obtained. Second, the solutions are different for every train with the random initial weights.

Particle swarm optimization (PSO) developed by Kennedy and Eberhart in 1995 [22] has been widely used in engineering application for best solution search, and its superiority has attracted many researchers in forecasting [9–11]. However, in this chapter, we present a novel mutative scale chaotic sequence method to optimize the weights of BPNN considering the characteristic of ergodicity. Two kinds of chaotic mapping equations with tent map and logistic map have been used for generation of chaotic sequences. Besides, PSO is also developed in neural network optimization for concentration estimation by an E-nose for comparisons.

4.2 Materials and Methods

4.2.1 Electronic Nose

Our sensor array in E-nose system consists of four gas sensors from the TGS series including TGS2602, TGS2620, and TGS2201 with dual outputs (TGS2201A and TGS2201B). The characteristics of these sensors are listed in Table 4.1 which demonstrates the detected gases and applications. Also, we refer readers to the

Table 4.1 Characteristics of the sensors in use

Sensor type	Objectives	Applications
TGS2602	High sensitivity to VOCs, odorous gases, gaseous air contaminants	Air cleaners, ventilation control, air quality, VOC, and odor monitors
TGS2620	High sensitivity to alcohol and organic solvent vapors	Alcohol testers, organic vapor detectors/alarms, solvent detectors
TGS2201A	Sensitivity to diesel exhaust gas NO_x	Automobile ventilation control, air contaminants detection
TGS2201B	Sensitivity to gasoline exhaust gas CO, H_2, etc.	Automobile ventilation control, air contaminants detection

sensors' datasheets for more information on TGS sensors available in http://www.figaro.co.jp/en/product/index.php?mode=search&kbn=1. In addition, a module (SHT2230 of Sensirion in Switzerland) with two auxiliary sensors for the temperature (T) and humidity (H) is also used for compensation. The sensors were mounted on a custom-designed printed circuit board (PCB), along with associated electrical components. A 12-bit analog–digital (A/D) converter with type of TLC2543 is used as interface between the field-programmable gate array (FPGA) processor represented as the central processor unit (CPU) and the sensors. A synchronous dynamic access memory (SDRAM) connected with FPGA processor is used for data collection, storage, and processing. The E-nose system designed with FPGA processor and other peripheral circuits is connected to a personal computer (PC) via a Joint Test Action Group (JTAG) port which can be used to transfer data and debug programs. An input vector with six variables can be obtained in each observation; the multi-dimensional response dataset presents a nonlinear relation with the target gas concentration. Formaldehyde and benzene measurements were employed by an E-nose in the constant temperature and humidity chamber in which the temperature and humidity can be effectively controlled in terms of the target temperatures and humidity. Note that the precisions of the chamber for temperature and humidity are ±0.1 °C and ±5% RH. The self-designed electronic nose is shown in Fig. 4.1. The left picture in Fig. 1 denotes the impression of the product, and the right one in Fig. 4.1 denotes the internal PCB with integrated circuits including the main modules highlighted.

4.2.2 Data Acquisition

In terms of the indoor monitoring of the formaldehyde and benzene concentrations, formaldehyde and benzene were measured at the concentration range of 0–10 ppm, respectively, target temperatures of 15, 25, 30, and 35 °C, and target humidity of 40, 60 and 80% RH (relative humidity) in order to imitate the real environment indoor. For each measurement, the total measurement cycle time for one single

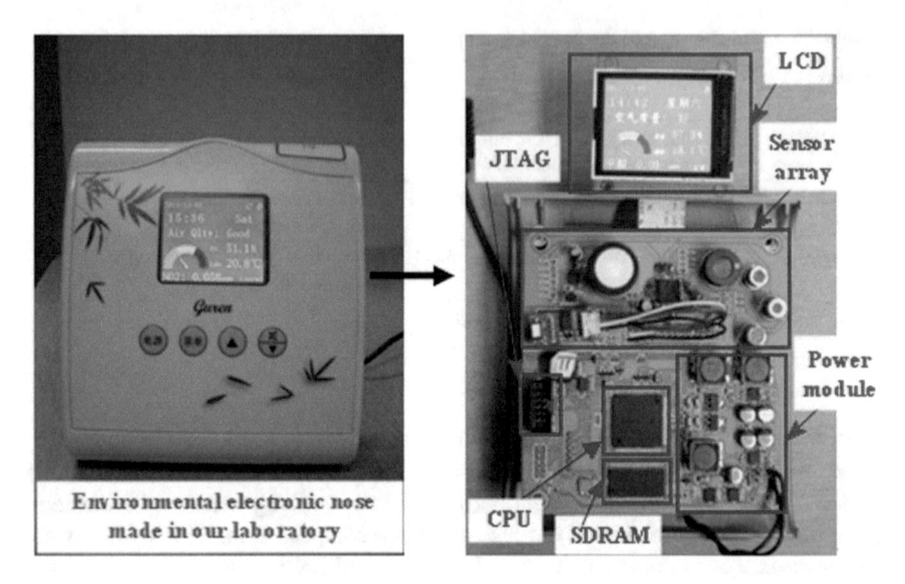

Fig. 4.1 Self-designed electronic nose for indoor air quality monitoring. The left picture is the integral electronic nose; the right picture is the internal PCB with integrated circuits with the main modules labeled

measurement was set to 20 min, i.e., 2 min for reference air (baseline), 8 min for gas sampling, and 10 min for cleaning of the chamber through injecting clean air before the next measurement begins. Totally, 116 observation samples were collected. For model building of formaldehyde monitoring, 71 training samples, 25 test samples, and 20 validation samples were divided into the whole sample set. Also, for model building of benzene monitoring, 40 training samples, 22 test samples, and 10 validation samples were divided into the whole sample set. The actual concentration of formaldehyde for each sample was obtained through the spectrophotometer analysis of the chemical sampling using the air sampler. The actual concentration of benzene is analyzed using gas chromatograph (GC). In detail, Table 4.2 presents the specific experimental samples of formaldehyde and benzene with different combinations of temperature (T) and relative humidity (RH). Twelve combinations {(15, 60), (15, 80), (20, 40), (20, 80), (25, 40), (25, 60), (25, 80), (30, 40), (30, 60), (30, 80), (35, 40), (35, 60)} in manner of (T, RH) were employed for covering the indoor conditions. From Table 4.2, we can find that various concentrations were employed for estimation model construction considering the various environments in dwellings. For each sample, a vector with six variables was extracted at the steady state response. Besides, a simple normalization method divided by 4095 was used for subsequent pattern analysis in voltage. Notice that 4095 is calculated as 212-1 with the principle of 12-bit A/D converter. The specific experimental platform is illustrated in Fig. 4.2 which has been used in our previous publication [23]. Five ports (ports_1–5) are used in the chamber. Port_1 is used for injection of contaminants, port_2 is used to clean the chamber through

Table 4.2 Concentration (ppm) conditions of all experimental samples in different combinations (T, RH) in which T denotes temperature and RH (%) denotes relative humidity

Conditions of formaldehyde samples												
(15, 60)	(15, 80)	(20, 40)	(20, 60)	(20, 80)	(25, 40)	(25, 60)	(25, 80)	(30, 40)	(30, 60)	(30, 80)	(35, 40)	(35, 60)
0.22	0.10	0.07	0.08	0.08	0.13	0.05	0.04	0.10	0.13	0.09	0.08	0.04
0.25	0.16	0.14	0.10	0.06	0.07	0.10	0.04	0.23	0.06	0.20	0.21	0.12
0.19	0.14	0.16	0.24	0.11	0.14	0.08	0.20	2.61	0.16	0.25	0.12	0.09
0.20	1.44	0.07	0.10	0.24	0.26	0.11	0.25	1.42	0.22	0.10	0.21	0.04
0.06	0.56	0.09	0.18	0.13	0.21	0.17	0.07		0.15	0.13	2.46	0.17
0.08		2.62	4.53	0.15	0.17	0.26	0.25		3.67	2.22	2.28	0.25
0.12		1.05	1.09	3.17	0.22	0.24	0.11		2.40	1.78		0.04
0.71				0.47	1.73	0.17	2.13					0.17
0.77					1.23	0.27	1.76					1.04
1.06						0.22						1.42
2.68						0.25						5.32
0.92						0.11						1.36
1.87						0.23						3.51
						0.25						
						0.12						
						1.22						
						2.38						
						1.75						
						3.00						
						1.28						
						2.99						
Conditions of benzene samples												
(15, 60)	(15, 80)	(20, 40)	(20, 60)	(20, 80)	(25, 40)	(25, 60)	(25, 80)	(30, 40)	(30, 60)	(30, 80)	(35, 40)	(35, 60)
0.17	0.17	0.17	0.17	0.17	0.17	0.17	0.17	0.17	0.17	0.17	0.17	0.17
0.28	0.28	0.28	0.28	0.28	0.28	0.28	0.28	0.28	0.28	0.28	0.28	0.28
0.49	0.49	0.49	0.49	0.49	0.49	0.49	0.49	0.49	0.49	0.49	0.49	0.49
0.91	0.91	0.91	0.91	0.91	0.91	0.91	0.91	0.91	0.91	0.91	0.91	0.91
0.71	0.71	0.71	0.71	0.71	0.71	0.71	0.71	0.71	0.71	0.71		0.71
0.11	0.06	0.09	0.08	0.20	0.19	0.15	0.15					

injection of fresh air, port_3 is set to control the relative humidity in the chamber by using a humidifier with a valve, port_4 is for data collection by connecting the electronic nose instrument to the PC with a JTAG, and port_5 is set to sampling by a gas sampler for sample's concentration analysis using spectrophotometer and GC.

4.2.3 Back-Propagation Neural Network

Back-propagation (BP) algorithm with a gradient descending strategy was proposed in [24]. Rumelhart further formulated the standard back-propagation (BP) algorithm for multilayer perceptrons. The architecture of a multilayer perceptron neural network is shown in Fig. 4.3 which presents a two hidden layered BP neural network.

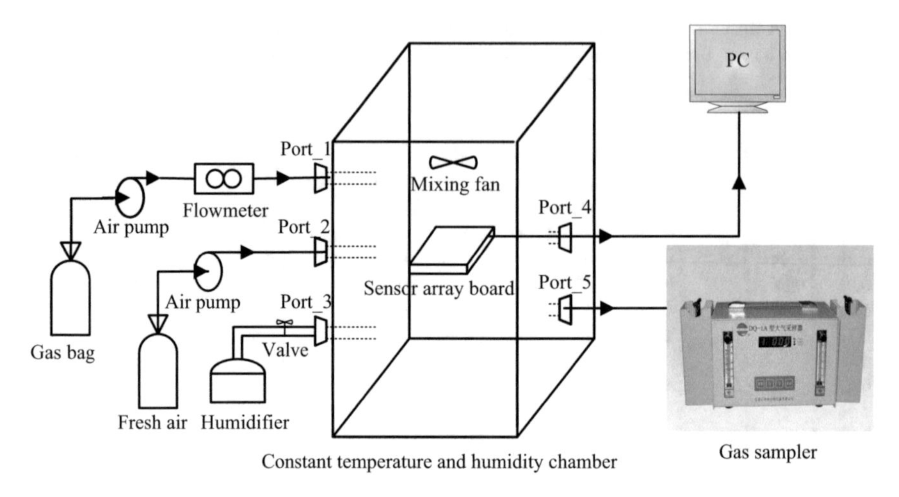

Fig. 4.2 Schematic of the experimental platform with our designed electronic nose system

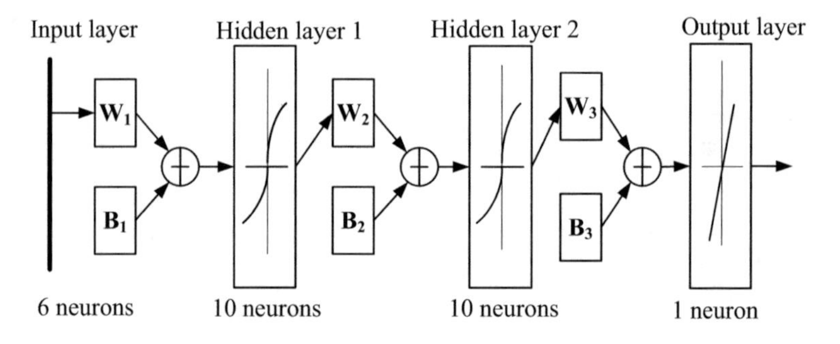

Fig. 4.3 Architecture of neural network with two hidden layers

BP neural network has been widely used to classify nonlinearly separable patterns in real application for its strong ability in recognition [25]. However, BP neural network still posses some inherent problems. First, BP model can easily get trapped in local minima for the problems of pattern recognition and fail to find the global optimal solution. Second, the initial weight matrix **W** and bias vectors **B** of back-propagation neural network are randomly produced for training so that different weights and biases would produce different trained neural networks. Thus, obtaining the global minimum of regression error by only using back-propagation neural network becomes little impractical. In our previous work, we have employed these problems using heuristic and bio-inspired methods [20].

4.2.4 Mutative Scale Chaotic Sequence Optimization

Chaos optimization is developed using chaotic variables. Three chaos map equations were studied in this chapter. The logistic map is shown by

$$z_k = \mu \cdot z_k \cdot (1 - z_k) \tag{4.1}$$

where z_k is the kth chaotic variable and k denotes the iteration number. Obviously, $z_k \in (0,1)$ under the conditions that the initial $z_0 \in (0,1)$ and the z_0 cannot be the digits of {0, 0.25, 0.75, 1}. Here, $\mu = 4$ can be a completely chaotic state.

The tent map [6] resembles the logistic map. It can also generate chaotic queue in (0, 1) in terms of the following form

$$z_k = \begin{cases} z_k/0.7, & z_k < 0.7 \\ 10/3 \cdot z_k \cdot (1 - z_k), & z_k \geq 0.7 \end{cases} \tag{4.2}$$

The Gauss map [6] which can also generate a chaotic queue in (0, 1) can be represented by

$$z_{k+1} = \begin{cases} 0, & z_k = 0 \\ 1/z_k - \lfloor 1/z_k \rfloor, & z_k \in (0,1) \end{cases} \tag{4.3}$$

where $\lfloor x \rfloor$ denotes the largest integer less than x.

The algorithm procedure of mutative scale chaotic sequence optimization neural network can be concluded as follows

Step 1: Initialization $g = 0$. Randomly generate one M-dimensional population $\mathbf{X}^0$ with N individuals within (0, 1), and determine the initial optimization boundary $[\boldsymbol{a}, \boldsymbol{b}]$ in the optimization space.

Step 2: Map the variable $\mathbf{X}_i^k$ into the optimization space, and obtain one new population $\mathbf{MX}_i^k$ using the following equation

$$MX_i^k = a_i^k + MX_i^k(b_i^k - a_i^k), \quad i = 1, \ldots, N;\ k = 1, \ldots, M \tag{4.4}$$

Step 3: Evaluate the new population **MX** using BPNN algorithm, and find the best individual **Xgbest**. The cost function has been defined as the maximum absolute relative error of the train and test samples shown by

$$f = \max\left\{1/n_1 \cdot \sum_{i=1}^{n_1} |ytr_i - ttr_i|/ttr_i,\ 1/n_2 \cdot \sum_{j=1}^{n_2} |yte_j - tte_j|/tte_j\right\} \times 100 \tag{4.5}$$

where n_1 and n_2 denote the number of train samples and test samples; **ytr** and **ttr** denote the predictive concentrations and actual concentrations of train samples; and **yte** and **tte** denote the predictive concentrations and actual concentrations of test samples. Note that a decoding of **MX** for the initial weights and bias of the neural network is necessary, because each individual is encoded as the weights and bias.

Step 4: Mutative scale of chaotic variable search.
If the best solution keeps invariant within T iterations, the shrink of the search boundary $[a, b]$ can be performed using the following strategy for specific search in a smaller space

$$\mathbf{a}^{g+1} = \mathbf{Xgbest} - \gamma^{g} \cdot (\mathbf{b}^{g} - \mathbf{a}^{g}) \quad (4.6)$$

$$\mathbf{b}^{g+1} = \mathbf{Xgbest} + \gamma^{g} \cdot (\mathbf{b}^{g} - \mathbf{a}^{g}) \quad (4.7)$$

$$\gamma^{g+1} = \beta_1 \cdot \gamma^{g} \quad (4.8)$$

where $\mathbf{a}^{g+1}$ and $\mathbf{b}^{g+1}$ denote the new search boundary, γ is the radius of search, and β_1 denotes decay coefficient less than 1.

Step 5: The constraints process of the boundary uses the following strategy

$$\text{If } a_i^k < -C^{g},\ a_i^k = -C^{g};\ \text{if } b_i^k > C^{g},\ b_i^k = C^{g} \quad (4.9)$$

$$C^{g+1} = \beta_2 \cdot C^{g} \quad (4.10)$$

where C denotes the maximum boundary and β_2 denotes the attenuation coefficient similar to simulated annealing.

Step 6: If the current solution satisfies the termination criteria or that the maximization iterations finished, stop; else go to step 7.

Step 7: Generate the new population $\mathbf{X}$ using the chaotic map equations, and go to step 2.

4.2.5 *Standard Particle Swarm Optimization (SPSO)*

In this standard PSO system, a number of particles cooperate to search for the best solutions by simulating the movement and flocking of birds [22]. These particles fly with a certain velocity and find the global best position after certain generations. At each generation t, the velocity is updated and the particle is moved to a new position. This new position is simply calculated as the sum of the previous position and the new velocity. The mathematical description of PSO is defined as follows.

Suppose the dimension for a searching space is D, the total number of particles is N, the position of the ith particle can be expressed as vector $\mathbf{X}_i = (x_{i1}, x_{i2}, \ldots, x_{iD})$, the best position of the ith particle searching until now is $\mathbf{P}_i = (p_{i1}, p_{i2}, \ldots, p_{iD})$, the best position of all the particles searching until now is $\mathbf{P}_g = (p_{g1}, p_{g2}, \ldots, p_{gD})$, the velocity of the ith particle is represented as $\mathbf{V}_i = (v_{i1}, v_{i2}, \ldots, v_{iD})$, then the standard PSO can be illustrated as

$$v_{id}(t+1) = w \cdot v_{id}(t) + c_1 \cdot r_{1i}^{p}(t) \cdot [p_{id}(t) - x_{id}(t)] + c_2 \cdot r_{2i}^{p}(t) \cdot [p_{gd}(t) - x_{id}(t)] \tag{4.11}$$

$$x_{id}(t+1) = x_{id}(t) + v_{id}(t+1), \quad 1 \leq i \leq N, \ 1 \leq d \leq D \tag{4.12}$$

where c_1, c_2 are the acceleration constants with positive values and w is called inertia factor;

$$r_{1i}^{p}(t) \leftarrow U(0,\ 1),\ r_{2i}^{p}(t) \leftarrow U(0,\ 1) \tag{4.13}$$

Noteworthy is that, placing a limit on the velocity $v_{\max}$ and adjusting the inertia weight w, the PSO can achieve better search performance.

$$v_p^i(t+1) = \begin{cases} v_{\max}^i, & v_p^i(t+1) > v_{\max}^i \\ -v_{\max}^i, & v_p^i(t+1) < -v_{\max}^i \\ v_p^i(k+1), & \text{otherwise} \end{cases} \tag{4.14}$$

In this chapter, the inertia weight w is updated through a cosine mechanism by

$$w(t) = w_{\text{start}} + w_{\text{end}} \cdot \cos(t/10) \tag{4.15}$$

where w_{start} is the initial value and w_{end} is the terminal value. In this work, w_{start} and w_{end} are set as 0.85 and 0.45, respectively.

4.2.6 Parameter Settings

Because the determination of the number of hidden layers and neurons in neural network cannot be fixed theoretically, in this chapter, a two hidden layered neural network whose structure is 6-10-10-1 was used for each gas monitoring in experience. Generally, the log-sigmoid function and pure linear function were used in the hidden layers and output layer, respectively. In terms of the network structure, matrix $\mathbf{W}_1$ (10×6), $\mathbf{W}_2$ (10×10), and $\mathbf{W}_3$ (1×10) represent the weights between the input layer and the hidden layer-1, the hidden layer-1 and the hidden layer-2, the hidden layer-2 and the output layer; vectors $\mathbf{B}_1$ (10×1), $\mathbf{B}_2$ (10×1) and $\mathbf{B}_3$ (1×1) represent the bias in hidden layer-1, hidden layer-2, and output layer. An individual $L = [l_1, l_2, \ldots, l_N]$ can be shown as follows

$$\overbrace{l_1,\ldots,l_{60}}^{\mathbf{W}_1},\overbrace{l_{61},\ldots,l_{70}}^{\mathbf{B}_1},\overbrace{l_{71},\ldots,l_{170}}^{\mathbf{W}_2},\overbrace{l_{171},\ldots,l_{180}}^{\mathbf{B}_2},\overbrace{l_{181},\ldots,l_{190}}^{\mathbf{W}_3},\ \overbrace{l_{191}}^{\mathbf{B}_3} \tag{4.16}$$

Therefore, the length N of one individual $\mathbf{L}$ can be calculated as $N = 6\times10 + 10 + 10 \times 10 + 10 + 10 \times 1+1 = 191$.

The training goal of neural network is dynamically set as 0.05–0.5 for formaldehyde and 0.005–0.05 for benzene. Considering the long running time of the whole algorithm, the size M of population is set as 50, the maximum iterations $G_1 = 100$ for formaldehyde, $G_2 = 10$ for benzene, and the permissible iterations $T = 10$ for stagnation. The initial boundaries of $\mathbf{a}$ and $\mathbf{b}$ are set as −20 and 20, and $C = 20$. Besides, the decay coefficient β_1 and attenuation coefficient β_2 are set as 0.98 and 0.95, respectively. The initial search radius γ is set as 0.2. The fitness function f in each optimization is set as the maximum relative error of the average relative training error and the average relative test error shown in Eq. 4.5. It is worth noting that the parameters settings are experienced and can be adjusted according to respective optimization problem.

4.2.7 On-Line Usage

The whole chaos and PSO-based neural network optimization and learning algorithms are implemented in PC for learning the best weights $\mathbf{W} = \{\mathbf{W}_1, \mathbf{W}_2, \mathbf{W}_3\}$ and $\mathbf{B} = \{\mathbf{B}_1, \mathbf{B}_2, \mathbf{B}_3\}$ of the neural network, and then $\mathbf{W}$ and $\mathbf{B}$ would be transferred to the E-nose system for prediction on line. In real-time application of the E-nose instrument, the forward computation process of neural network without back-propagation will be implemented in FPGA combined with learned neural network weights $\mathbf{W}$ and $\mathbf{B}$ for concentration estimation. The forward computation process in FPGA is illustrated as three steps:

Step 1: $\mathbf{y}_1 = 1/\left[1 + e^{-(\mathbf{W}_1 \cdot \mathbf{x} + \mathbf{B}_1)}\right]$
Step 2: $\mathbf{y}_2 = 1/\left[1 + e^{-(\mathbf{W}_2 \cdot \mathbf{y}_1 + \mathbf{B}_2)}\right]$
Step 3: $\mathbf{y}_3 = \mathbf{W}_3 \cdot \mathbf{y}_2 + \mathbf{B}_3$

where $\mathbf{x}$ is the real-time observation vector of the sensor array, $\mathbf{y}_1$ is the output vector of the first hidden layer with a log-sigmoid transfer function, $\mathbf{y}_2$ is the output vector of the second hidden layer with a log-sigmoid transfer function, and $\mathbf{y}_3$ is the final output of the neural network prediction with a pure linear function.

4.3 Results and Discussion

The estimation results for the formaldehyde and benzene experiments analyzed using the chaotic sequence optimization neural network and standard particle swarm optimization (PSO) neural network have been presented in this section. The model building is based on the training samples and test samples. The role of the test samples is used to control the overfitting risk of training samples and evaluate the fitness function of chaotic sequence and PSO optimization. We apply the maximum relative error of training and test samples as the fitness function to improve the robustness of the model. The validation samples are finally used to verify the efficiency of the model. Figure 4.4 illustrates the formaldehyde prediction results of the test samples (left) and validation samples (right) using four optimization methods combined with BPNN.

Similarly, Fig. 4.5 illustrates the benzene predictions of the test samples (left) and validation samples (right) using four optimization BPNN methods. From the trace of predictions and actual concentrations, we can find that the four methods can track the actual concentrations approximately. The predicted curve using chaos with logistic map can approach the actual curve better for formaldehyde concentration estimation. However, for benzene estimation, chaos method with Gaussian map presents a better prediction. For quantification of the prediction error, Table 4.3 presents the relative prediction error using four optimization methods. From this table, we can find that the chaotic sequence optimization with logistic map has the minimum prediction error 32.34% of formaldehyde validation samples and 26.03%

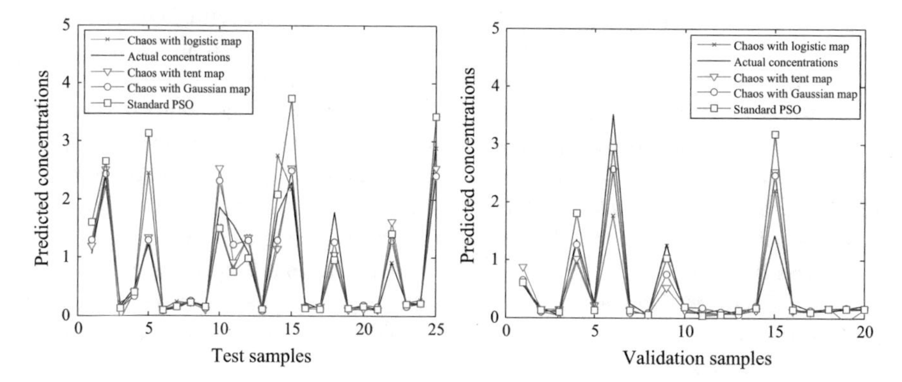

Fig. 4.4 Formaldehyde prediction results of the test samples (left) and validation samples (right) using four optimization methods

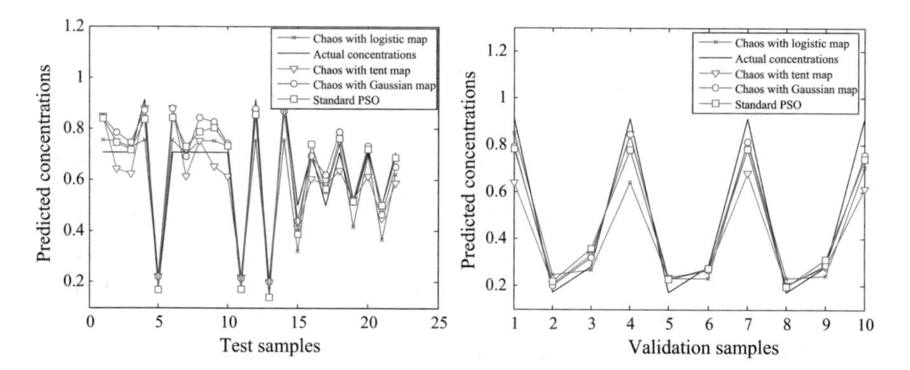

Fig. 4.5 Benzene prediction results of the test samples (left) and validation samples (right) using four optimization methods

Table 4.3 Relative prediction error using different optimization neural network methods

Estimation models	Relative prediction error (%)					
	Formaldehyde			Benzene		
	Train	Test	Validation	Train	Test	Validation
Single BP	46.55	40.38	58.37	27.18	16.53	29.71
Logistic map-BP	28.35	28.89	32.34	16.57	12.15	17.96
Tent map-BP	29.78	30.27	49.49	8.910	13.12	19.16
Gaussian map-BP	30.06	23.06	50.64	11.62	11.73	13.31
Standard PSO-BP	29.94	29.56	47.06	12.49	12.21	16.08

error decreased compared with the prediction error 58.37% using single neural network. And Gaussian map obtains the minimum prediction error 13.31% of benzene validation samples and 16.4% error decreased compared with the prediction error 29.71% using single neural network. Because the neural network is trained separately for each gas, therefore in real-time application of our electronic nose for formaldehyde and benzene estimation, the best neural network with minimum prediction error is selected for each gas monitoring.

From the analysis of estimation results, the optimization methods are effective for neural network optimization. From the comparisons of chaotic sequence methods and PSO on the results, we cannot definitely decide which one has better optimization ability because their prediction difference for each gas is not very obvious. However, from the characteristic of chaotic sequence, the ergodicity of chaos can help to find the global optimal in optimization after a number of generations. However, PSO has a fast convergence performance and get to a local optimal in a short time. So, researchers can decide which optimization can be used in terms of their need.

4.4 Summary

This chapter presents a novel chaos sequence optimization neural network method for concentration prediction of formaldehyde and benzene in dwellings by a portable electronic nose. For comparison, PSO is also used for optimization. Through the formaldehyde and benzene samples obtained in experiments combined with the self-designed electronic nose, we built the neural network prediction model combined with the chaos optimization and PSO, respectively. Comparing the prediction error of models based on chaotic sequence and PSO, we found that both methods are effective for weights optimization. The neural network has been improved to a large extent, and the concentration estimations for formaldehyde and benzene indoor have also been realized in our project.

References

1. A.P. Jones, Indoor air quality and health. Atmos. Environ. **33**, 4535–4564 (1999)
2. S.C. Lee, M. Chang, Indoor and outdoor air quality investigation at schools in Hong Kong. Chemosphere **41**, 109–113 (2000)
3. S. De Vito, E. Massera, M. Piga, L. Martinotto, G. Di Francia, On field calibration of an electronic nose for benzene estimation in an urban pollution monitoring scenario. Sens. Actuators B **129**, 750–757 (2008)
4. H. Huang, F. Haghighat, Modelling of volatile organic compounds emission from dry building materias. Build. Environ. **37**, 1349–1360 (2002)
5. H.G. Schuster, *Deterministic chaos: an introduction* (Physick-Verlag GmnH, Weinheim, Federal Republic of Germany, 1988)
6. B. Alatas, E. Akin, A. Bedri Ozer, Chaos embedded particle swarm optimization algorithms, Chaos. Solitons Fractals **40**, 1715–1734 (2009)
7. Q. Wu, A hybrid-forecasting model based on Gaussian support vector machine and chaotic particle swarm optimization. Expert Syst. Appl. **37**, 2388–2394 (2010)
8. Q. Wu, The hybrid forecasting model based on chaotic mapping, genetic algorithm and support vector machine. Expert Syst. Appl. **37**, 1776–1783 (2010)
9. X. Tang, L. Zhuang, C. Jiang, Prediction of silicon content in hot metal using support vector regression based on chaos particle swarm optimization. Expert Syst. Appl. **36**, 11853–11857 (2009)
10. C. Zhao, X. Sun, S. Sun, J. Ting, Fault diagnosis of sensor by chaos particle swarm optimization algorithm and support vector machine. Expert Syst. Appl. **38**, 9908–9912 (2011)
11. X. Tang, L. Zhuang, J. Cai, C. Li, Multi-fault classification based on support vector machine trained by chaos particle swarm optimization. Knowl.-Based Syst. **23**, 486–490 (2010)
12. X.Q. Zuo, Y.S. Fan, A chaos search immune algorithm with its application to neuro-fuzzy controller design. Chaos, Solitons Fractals **30**, 94–109 (2006)
13. Z. Guo, S. Wang, J. Zhuang, A novel immune evolutionary algorithm incorporating chaos optimization. Pattern Recogn. Lett. **27**, 2–8 (2006)
14. M.S. Simon, D. James, Z. Ali, Data analysis for electronic nose systems. Microchim. Acta **156**, 183–207 (2007)
15. J.E. Haugen, K. Kvaal, Electronic nose and artificial neural network. Meat Sci. **49**, S273–S286 (1998)
16. L. Carmel, N. Sever, D. Lancet, D. Harel, An eNose algorithm for identifying chemicals and determining their concentration. Sens. Actuators B **93**, 77–83 (2003)
17. E.G. Breijo, J. Atkinson, L.S. Sanchez, A comparison study of pattern recognition algorithms implemented on a microcontroller for use in a electronic tongue for monitoring drinking waters. Sens. Actuators A **172**, 570–582 (2011)
18. L.G. Sanchez, J. Soto, M.R. Martinez, A novel humid electronic nose combined with an electronic tongue for assessing deterioration of wine. Sens. Actuators A **171**, 152–158 (2011)
19. N. Paulsson, E. Larsson, F. Winquist, Extraction and selection of parameters for evaluation of breath alcohol measurement with an electronic nose. Sens. Actuators A **84**, 187–197 (2000)
20. L. Zhang, F. Tian, C. Kadri, G. Pei, H. Li, L. Pan, Gases concentration estimation using heuristics and bio-inspired optimization models for experimental chemical electronic nose. Sens. Actuators B **160**, 760–770 (2011)
21. M. Gori, A. Tesi, On the problem of local minima in back-propagation. IEEE Trans. Pattern Anal. Mach. Intell. **14**(1), 76–86 (1992)
22. J. Kennedy, R.C. Eberhart, Particle swarm optimization. Proc. IEEE Int. Conf. Neural Netw. **4**, 1942–1948 (1995)
23. L. Zhang, F. Tian, H. Nie, L. Dang, G. Li, Q. Ye, C. Kadri, Classification of multiple indoor air contaminants by an electronic nose and a hybrid support vector machine. Sens. Actuators B **174**, 114–125 (2012)

24. W. Wu, J. Wang, M.S. Cheng, Z.X. Li, Convergence analysis of online gradient method for BP neural networks. Neural Netw. **24**, 91–98 (2011)
25. J.R. Zhang, J. Zhang, A hybrid particle swarm Optimization back-propagation algorithm for feed-forward neural network training. Appl. Math. Comput. **185**, 1026–1037 (2007)

Chapter 5
Multilayer Perceptron-Based Concentration Estimation

Abstract Non-selective gas sensor array has different sensitivities to different chemicals because each gas sensor produces different voltage signals when exposed to analytes with different concentrations. Therefore, the characteristics of cross sensitivities and broad spectrum of non-selective chemical sensors stimulate the fast development of portable and low-cost electronic nose. Simultaneous concentration estimation of multiple kinds of chemicals is always a challenging task in electronic nose. Multilayer perceptron (MLP) neural network, as one of the most popular pattern recognition algorithms in electronic nose, will be studied further in this chapter. Two structures of single multiple-input multiple-output (SMIMO) and multi multiple-input single-output (MMISO)-based MLPs are presented for detection of six kinds of indoor air contaminants. Further, the network parameters optimization using eight computational intelligence optimization algorithms are presented. Experiments prove that the performance in accuracy and convergence of MMISO structure-based MLP are much better than SMIMO structure in concentration estimation for more general use of electronic nose.

Keywords Electronic nose · Concentration estimation · Multilayer perceptron
Computational intelligence optimization · Indoor air contaminants

5.1 Introduction

This chapter concerns applying E-nose to detect six kinds of chemicals which have been studied widely for their potential harms to people's health [1, 2]: formaldehyde, benzene, toluene, and ammonia exhausted from new furniture, oil paint, building materials of residuals, and carbon monoxide and nitrogen dioxide produced by the smoking of cigarettes, wood burning stoves, and car exhaust.

Artificial neural networks (ANNs), especially multilayer perceptrons (MLPs) based on back-propagation (BP) learning algorithm, are one of the most popular pattern recognition (qualitative) and concentration estimation (quantitative) models in E-noses for environmental monitor applications, as remarked in [3]. The

L. Zhang et al., *Electronic Nose: Algorithmic Challenges*,
https://doi.org/10.1007/978-981-13-2167-2_5

quantification models based on MLPs have been specially proposed for concentration estimation of pollutant gases [4–9]. In our previous research, we have also proposed the standardization model for sensor array's signal difference caused by new sensor replacement [10] and sensor drift prediction model [11] based on ANN.

There are two main facets of MLP to be discussed in E-nose for concentration estimation of multiple kinds of gases. The first thing is the parameters learning of multilayer perceptron. MLP used in E-nose is generally trained by BP learning algorithm which has advantages of fast convergence and mean square error minimum. The second thing is the structure of multilayer perceptron. Three-layered MLP (input layer, one hidden layer, and output layer) is a satisfactory structure to solve a common regression problem in general. Therefore, the performance study of MLP in this chapter is developed fully on the basis of these two facets. Particularly, two structures of single multiple-input multiple-output (SMIMO) and multi multiple-input single-output (MMISO)-based MLP are studied for quantification of multiple kinds of indoor air contaminants.

Considering the disadvantage of easily trapped into local optimum of BP algorithm learned with initially random weights, we introduce eight computational intelligence optimization algorithms including standard genetic algorithm (SGA) [12], standard particle swarm optimization (SPSO) [13], IWA-PSO [14], ARPSO [15], DRPSO [16], PSOBC [17], and PSOAGS [5] for parameters optimization of MLP. In addition, we also introduce three linear regression methods including multivariate linear regression (MLR), partial least square (PLS) [18, 19], and principal component regression (PCR) [20, 21] which have been reported in E-nose system for concentration estimations and comparisons. The SMIMO-based MLP with parameters optimization has been studied in E-nose in our previous research [5]. This chapter further employs the performance of MMISO-based MLP with parameters optimization in solution of a more complex problem and presents the comparison with SMIMO structure for more general use of a portable E-nose.

5.2 E-Nose Systems and Data Acquisition

5.2.1 Low-Cost Electronic Nose System

The E-nose system has been introduced in [5]. Consider the characteristics of cross-sensitivity, broad spectrum response, and low cost of metal oxide semiconductor gas sensors; our sensor array in E-nose system consists of only four metal oxide semiconductor (MOS) gas sensors manufactured in Figaro Inc. including TGS2602, TGS2620, TGS2201A, and TGS2201B. The heating voltage of TGS2620 and TGS2602 is 4 V, and the heating voltage of TGS2201A/B is 7 V. The supplied power voltage of system is DC12 V. In addition, considering that MOS gas sensors are sensitive to environmental temperature and humidity, a module (STD2230-I^2C of Sensirion in Switzerland) with two auxiliary sensors for temperature (T) and relative humidity (RH) measurement is also used in sensor array. A 12-bit analog–digital converter is used as the interface between the field-programmable gate array (FPGA) processor and the sensor array. FPGA is used for data collection, storage, and processing in E-nose. The E-nose system is then connected to a personal computer (PC) via a Joint Test Action Group (JTAG) port which can be used for transferring data and debugging programs. The E-nose system and the experimental platform developed in our laboratory are described in Chap. 4.

5.2.2 Experimental Setup

The gaseous experiments of E-nose in this chapter were employed in the constant temperature and humidity chamber (with type of LRH-150S), which can automatically adjust the temperature and humidity by setting up the target temperature and humidity in advance. The target gases were collected in a gas bag. The gas was injected into the chamber through a flowmeter. A fan is fixed in the chamber for purging, and the gas can diffuse evenly. Totally, 10 min are consumed in each experiment and one sample is obtained. The specific experimental procedures including four stages can be generally illustrated as follows.

(1) Gas preparing and collection. Collect each target gas in a bag, and dilute each target gas using pure nitrogen (N_2).
(2) Data collection (major part). In this stage, there are several steps shown as follows.

Step 1: Set the initial temperature and humidity of the chamber. For simulation of the indoor environment, 15, 20, 25, 30, and 35 °C are considered as target temperature, and 40, 60, and 80% are considered as target humidity. The specific combinations of temperature and humidity considered in the experiments are {(15, 60), (15, 80), (20, 40), (20, 60), (20, 80), (25, 40), (25, 40), (25, 60), (25, 80), (30, 40), (30, 60), (30, 80), (35, 60)}.
Step 2: Turn on the electronic nose system until the temperature and humidity in the chamber reach the initial setting by authors, and then collect sensor array's baseline for 2 min. Note that the baseline denotes the initial state of sensor when exposed to clean air.
Step 3: Inject target gas by using a flowmeter with time control. Accordingly, turn on the gas sampler and it will keep 10 min sampling for true concentration analysis in this experiment. Then, the sensors will have quick response to target gas and until the sensors reach steady state response after about 8 min. Therefore, one experiment of sample collection would sustain 10 min. The typical sensor response curves when exposed to each chemical of some concentration are illustrated in Fig. 5.1. The concentrations corresponding to the data shown in Fig. 5.1a–f are 0.18, 0.28, 0.14, 6.0, 0.5, and 1.62 ppm, respectively. Note that the chapter focuses on the study of concentration estimation model, while the filtering and feature selection are not studied in

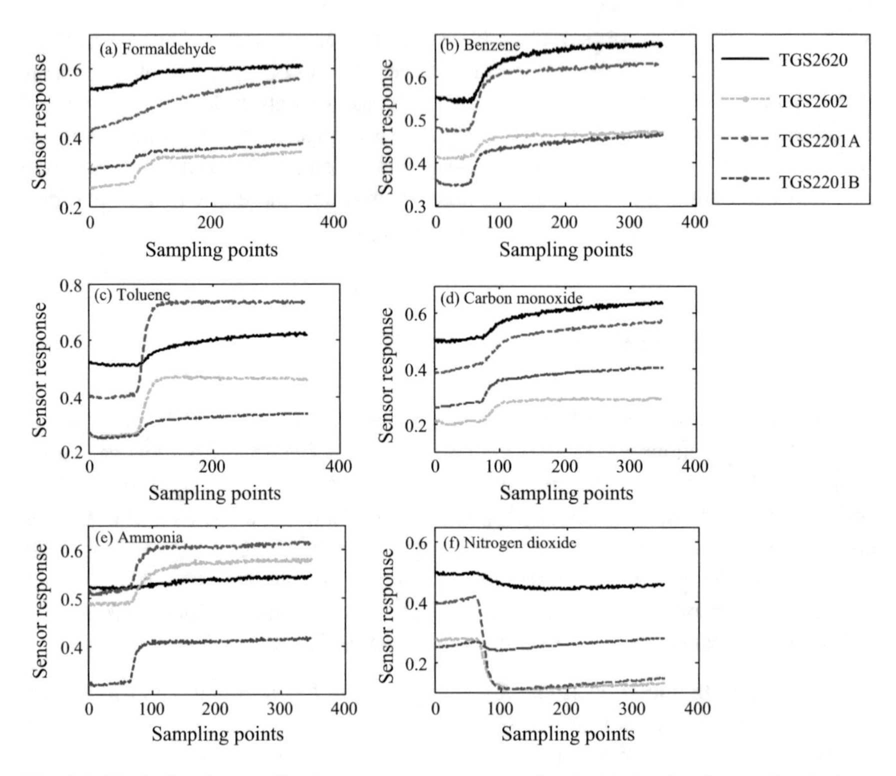

Fig. 5.1 Typical and normalized sensor response curves for six kinds of pollutant chemicals

deep. Therefore, the sensor response has been normalized within [0, 1] by dividing the maximum value (i.e., $x = x/x_{max}$) for easier analysis of neural network. The steady state response in Fig. 5.1 is extracted to represent the chemical feature for pattern analysis in a more effective way in actual application.

(3) Air exhaust and chamber cleaning.
After one experiment of sample collection, air exhaust by an air pump is necessary for chamber cleaning to recover the sensor to the initial state as soon as possible.
(4) Data transferring to PC.
Sensor response data in this experiment is obtained at this time through a JTAG connected between the electronic nose and PC. Then, the collected data can be transferred to the PC in a convenient way for data analysis and neural network learning in computer.
Finally, the learned results (i.e., hyper-parameters of neural networks) in computer would be transferred to the E-nose system for on-line detection of the target contaminants.

5.2.3 *Description of Dataset*

Six kinds of indoor contaminants including formaldehyde (CH_2O), benzene (C_6H_6), toluene (C_7H_8), carbon monoxide (CO), ammonia (NH_3), and nitrogen dioxide (NO_2) have been studied in this chapter. The electronic nose experiments of each analyte are employed in terms of the illustrated experimental procedures. The specific information (i.e., target temperature, target humidity, and concentration) of each experimental sample for each chemical is presented in Table 5.1. For instance, (15, 60) denotes the target temperature 15 °C and relative humidity 60%. In addition, the data statistics including the number of total samples, training samples and testing samples, the lowest concentration (ppm) and the highest concentration (ppm) of each gas are presented in Table 5.2. Note that the approximated 2/3 training samples are randomly selected from the total samples of each gas for MLP learning.

5.3 MLP-Based Quantization Models

Multilayer perceptron neural network is used for learning on a large dataset between sensor array data matrix **P** and the target concentration **Target** using back-propagation algorithm. The learned hyper-parameters are memorized in weights **W** and bias **B**. Then, the learned parameters of MLP are used for concentration estimation of unknown gas samples with a portable electronic nose. Two structures (i.e., SMIMO and MMISO) for concentration estimation of six kinds of indoor contaminants studied in this section are shown in Fig. 5.2.

5.3.1 *Single Multi-Input Multi-Output (SMIMO) Model*

SMIMO-based MLP model for concentration estimation of three kinds of chemicals has been studied in the previous research [5]. SMIMO is a straightforward method for prediction of multiple kinds of gases. Assume that the size of sensor array is m (the number of sensors) and the number of chemicals in detection is q; then, the number of input neurons and output neurons in SMIMO is also m and q, respectively. Namely, each output node represents a kind of gas. The structure of SMIMO-based MLP with one hidden layer is presented in Fig. 5.3a.

The input training matrix **P** and target concentration matrix **Target** of SMIMO for the training samples of all chemicals are shown in Eq. 5.1.

Table 5.1 Electronic data of indoor contaminants experiments in 12 conditions (temperature, relative humidity)

Formaldehyde data (CH_2O) (concentration unit: ppm)											
(15,60)	(15,80)	(20,40)	(20,60)	(20,80)	(25,40)	(25,60)	(25,80)	(30,40)	(30,60)	(30,80)	(35,60)
0.04	0.14	0.07	0.10	0.08	0.13	0.24	0.06	0.23	0.13	0.09	0.04
0.05	0.16	0.07	0.07	0.06	0.45	0.10	0.25	0.39	0.15	0.02	0.09
0.08	0.72	0.16	0.15	0.11	0.31	0.26	1.04	1.37	0.22	0.25	0.81
0.12	0.10	1.32	0.09	0.28	0.52	2.11	0.02	0.58	2.06	0.09	0.16
0.64	0.34	0.60	0.23	1.10	0.22	0.37	0.11	0.52	0.08	0.13	0.58
0.21	1.22	0.61	0.17	0.13	0.49	0.05	0.04	0.02	0.23	0.43	0.04
0.25	0.13	0.16	0.16	0.15	0.07	0.01	0.27	0.09	0.27	0.76	0.12
0.06	0.26	2.62	0.20	0.25	0.30	0.11	0.33	0.12	0.29	1.15	0.45
0.05	0.52	0.45	0.21	0.35	0.26	0.17	0.79	0.56	0.31	2.42	0.39
0.18	0.69	0.05	0.22	0.59	0.23	0.24	1.01	0.68	0.56	2.01	0.61
0.22	2.29	0.08	0.24	1.16	0.04	0.17	1.29	0.92	1.01	2.30	1.62
0.12	2.45	1.16	0.24	1.88	1.01	0.27	1.93	1.31	1.06		0.22
0.19	0.12	3.13	0.60	1.17	1.09	0.12	2.17	2.37	1.65		0.31
0.20	1.06	0.48	0.77	1.83	2.62	0.17	0.26	0.01	1.84		0.32
0.52	1.44	0.60	1.23		0.06	0.24	1.01	0.09	0.08		0.39
…	…	…	…		…	…	…	…	…		…
Benzene data (C_6H_6) (concentration unit: ppm)											
(15,60)	(15,80)	(20,40)	(20,60)	(20,80)	(25,40)	(25,60)	(25,80)	(30,40)	(30,60)	(30,80)	(35,60)
0.17	0.17	0.17	0.17	0.17	0.17	0.17	0.17	0.17	0.17	0.17	0.17
0.28	0.28	0.28	0.28	0.28	0.28	0.28	0.28	0.28	0.28	0.28	0.28
0.49	0.49	0.49	0.49	0.49	0.49	0.49	0.49	0.49	0.49	0.49	0.49
0.91	0.91	0.91	0.91	0.91	0.91	0.91	0.91	0.91	0.91	0.91	0.91
0.71	0.71	0.71	0.71	0.71	0.71	0.71	0.71	0.71	0.71	0.71	0.71
0.11	0.06	0.09	0.08	0.20	0.19	0.15	0.15	0.19	0.15	0.20	0.14
0.18	0.20	0.07	0.15	0.06	0.10	0.13	0.06	0.18	0.18	0.13	
0.25	0.21	0.25	0.21	0.14	0.21	0.14	0.14	0.08	0.10	0.24	
0.18	0.11	0.26	0.36	0.16	0.18	0.19	0.16	0.24	0.19	0.22	

(continued)

Table 5.1 (continued)

0.24 0.32 0.11	0.30 0.22 0.26	0.18 0.06 0.11	0.42 0.43	0.21 0.21	0.33 0.16	0.20 0.10 0.17	0.16 0.20 0.21	0.25 0.18 0.24	0.30 0.41		
Toluene data (C_7H_8) (concentration unit: ppm)											
(15,60)	(15,80)	(20,40)	(20,60)	(20,80)	(25,40)	(25,60)	(25,80)	(30,40)	(30,60)	(30,80)	(35,60)
0.05 0.08 0.14 0.06	0.05 0.06 0.14 0.08	0.05 0.08 0.14 0.06	0.05 0.06 0.14 0.08	0.06 0.08 0.14 0.05	0.05 0.06 0.14 0.08	0.05 0.06 0.14 0.08	0.05 0.06 0.14 0.08	0.05 0.06 0.14 0.08	0.05 0.06 0.14 0.08	0.05 0.06 0.14 0.08	0.05 0.06 0.14 0.08
Carbon monoxide data (CO) (concentration unit: ppm)											
(15,60)	(15,80)	(20,40)	(20,60)	(20,80)	(25,40)	(25,60)	(25,80)	(30,40)	(30,60)	(30,80)	(35,60)
6 11 43 23	4 23 43 12	6 12 41 22 13	5 22 43 11 9	5 22 44 12 20	6 24 46 14 10	4 8 10 21 12	5 23 45 33 48	5 8 23 37 6	14 29 49 55 25	4 16 48 13 16	5 13 20 29 20
Ammonia data (NH_3) (concentration unit: ppm)											
(15,60)	(15,80)	(20,40)	(20,60)	(20,80)	(25,40)	(25,60)	(25,80)	(30,40)	(30,60)	(30,80)	(35,60)
0.10 0.50 0.25	0.28	0.34 1.72 0.80	0.80 0.79	0.98 0.44	0.09 0.53 0.12	0.33 0.79 0.55	0.27 0.73	0.66 0.09	0.79 0.92 1.18	0.20 0.36	0.28 2.15 0.27
Nitrogen dioxide data (NO_2) (concentration unit: ppm)											
(15,60)	(15,80)	(20,40)	(20,60)	(20,80)	(25,40)	(25,60)	(25,80)	(30,40)	(30,60)	(30,80)	(35,60)
0.09 0.20 1.62 0.66	–	0.03 0.92 0.77	0.16 0.84 0.28	0.10 0.54 0.18	0.12 0.31 0.20	0.15 0.22 0.70 0.02	0.03 0.05 0.87 1.59 0.07 0.17	–	0.21 0.61 1.36	–	–

Table 5.2 Data statistics

Gases	*N*_total	*N*_training	*N*_testing	Lowest (ppm)	Highest (ppm)
CH_2O	126	84	42	0.04	5.32
C_6H_6	72	50	22	0.17	0.91
C_7H_8	66	45	21	0.05	0.14
CO	58	42	16	5.00	49.0
NH_3	29	20	9	0.09	2.15
NO_2	30	20	10	0.03	1.62

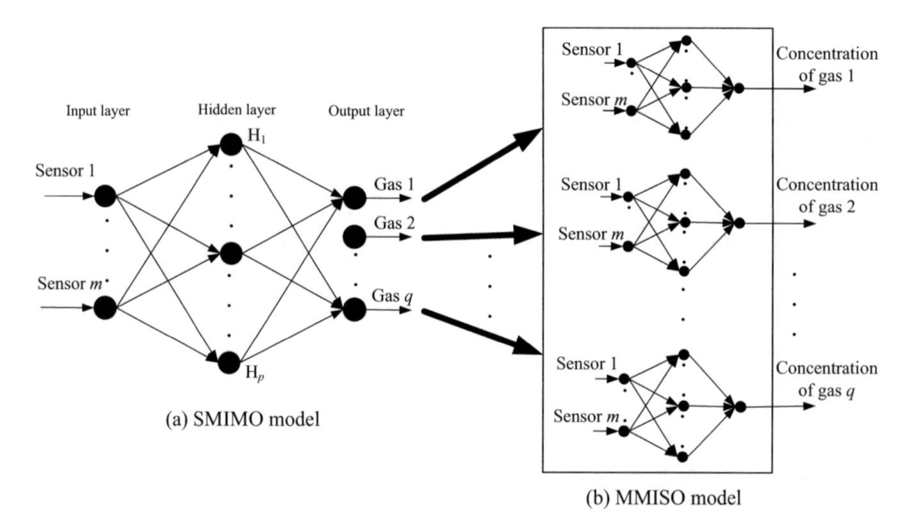

Fig. 5.2 Structures of SMIMO (**a**) and MMISO (**b**) based MLP concentration estimation models

$$\mathbf{P} = \left[\begin{array}{cccc|cccc|c|ccc} \multicolumn{4}{c}{\overbrace{\qquad\qquad}^{\text{gas\_1}}} & \multicolumn{4}{c}{\overbrace{\qquad\qquad}^{\text{gas\_2}}} & & \multicolumn{3}{c}{\overbrace{\qquad\qquad}^{\text{gas\_}q}} \\ S_{1,1} & S_{1,2} & \cdots & S_{1,n_1} & S_{1,n_1+1} & S_{1,n_1+2} & \cdots & S_{1,n_1+n_2} & \cdots & S_{1,n_1+\ldots+n_{q-1}+1} & \cdots & S_{1,n_1+\ldots+n_q} \\ S_{2,1} & S_{2,2} & \cdots & S_{2,n_1} & S_{2,n_1+1} & S_{2,n_1+2} & \cdots & S_{2,n_1+n_2} & \cdots & S_{2,n_1+\ldots+n_{q-1}+1} & \cdots & S_{2,n_1+\ldots+n_q} \\ \vdots & \vdots & \vdots & \vdots & \vdots & \vdots & \vdots & \vdots & \vdots & \vdots & \vdots & \vdots \\ S_{m,1} & S_{m,2} & \cdots & S_{m,n_1} & S_{m,n_1+1} & S_{m,n_1+2} & \cdots & S_{m,n_1+n_2} & \cdots & S_{m,n_1+\ldots+n_{q-1}+1} & \cdots & S_{m,n_1+\ldots+n_q} \end{array}\right]$$

$$\overset{\text{MLP learning}}{\Rightarrow} \mathbf{Target} = \left[\begin{array}{cccc|cccc|c|ccc} \multicolumn{4}{c}{\overbrace{\qquad\qquad}^{\text{gas\_1}}} & \multicolumn{4}{c}{\overbrace{\qquad\qquad}^{\text{gas\_2}}} & & \multicolumn{3}{c}{\overbrace{\qquad\qquad}^{\text{gas\_}q}} \\ t_{1,1} & t_{1,2} & \cdots & t_{1,n_1} & 0 & 0 & \cdots & 0 & \cdots & 0 & \cdots & 0 \\ 0 & 0 & \cdots & 0 & t_{2,n_1+1} & t_{2,n_1+2} & \cdots & t_{2,n_1+n_2} & \cdots & 0 & \cdots & 0 \\ \vdots & \vdots & \vdots & \vdots & \vdots & \vdots & \vdots & \vdots & \vdots & \vdots & \vdots & \vdots \\ 0 & 0 & \cdots & 0 & 0 & 0 & \cdots & 0 & \cdots & t_{q,n_1+\ldots+n_{q-1}+1} & \cdots & t_{q,n_1+\ldots+n_q} \end{array}\right] \tag{5.1}$$

where each column in **P** represents one feature vector corresponding to one observation sample, $S_{i,j}$ ($i = 1,\ldots, m$, $j = 1,\ldots, n_1 + n_2 + \cdots + n_q$) is the extracted

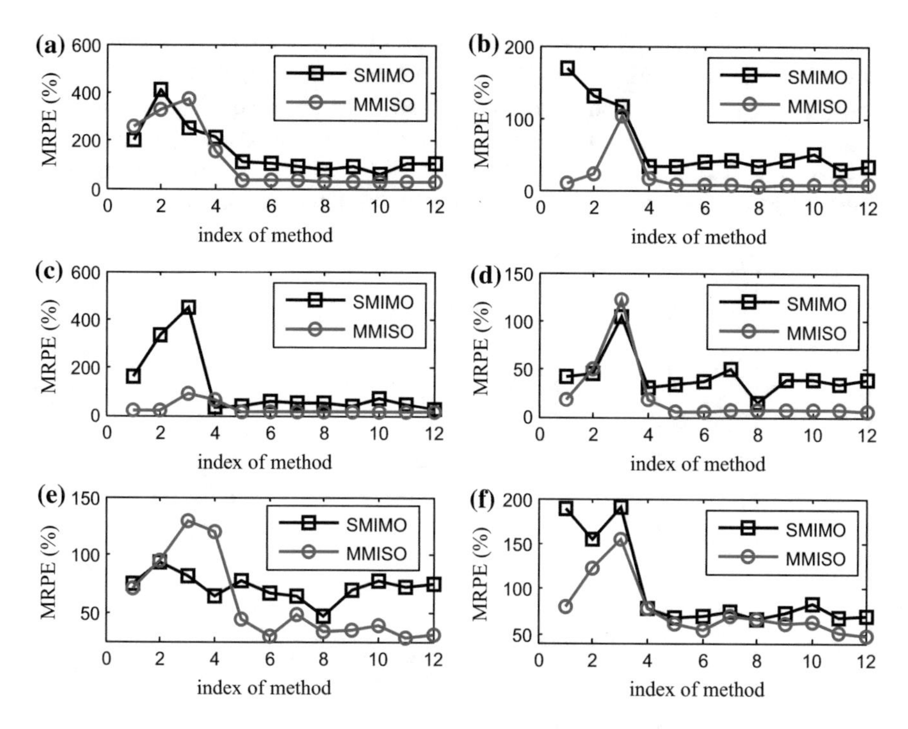

Fig. 5.3 Estimation performance comparisons of SMIMO and MMISO structures using 12 methods. Index of method: method 1: MLR; method 2: PLS; method 3: PCR; method 4: single BP; method 5: SGA-BP; method 6: SPSO-BP; method 7: IWAPSO-BP; method 8: APSO-BP; method 9: DRPSO-BP; method 10: ARPSO-BP; method 11: PSOBC-BP; method 12: PSOAGS-BP. Prediction of chemicals: **a** formaldehyde, **b** benzene, **c** toluene, **d** carbon monoxide, **e** ammonia, and **f** nitrogen dioxide

feature at the steady state sensor response of each sample, and $t_{1,\ k}$ ($k = 1,\ldots, n_1$), $t_{2,k}$ ($k = n_1 + 1,\ldots, n_1 + n_2$),..., $t_{q,k}$ ($k = n_1 +\cdots+n_{q\text{-}1} + 1,\ldots, n_1 +\cdots+ n_q$) are the target concentrations of gas_1, gas_2,..., gas_q, respectively.

Due to that the gas_1, gas_2,..., gas_q are single gas in experiments, the target concentration vector of gas_i in gas_j ($i = 1,\ldots, q; j = 1,\ldots, q; i \neq j$) should be $\vec{\mathbf{0}}$ in the target concentration matrix **Target**.

Through the structure of SMIMO model, the neural network parameters learning are not flexible and difficult to be adjusted neatly with different kinds of gases due to that the prediction has been integrated into one single neural network.

5.3.2 Multiple Multi-Input Single-Output (MMISO) Model

For further study, MMISO-based MLP model is an improvement of SMIMO model. In MMISO model, there are q single MLP neural networks for concentration estimation of q kinds of gases. The structure of MMISO with one hidden layer individually is presented in Fig. 5.2b. For each individual MLP (MLP_1, MLP_2,..., MLP_q), m input neurons and one output neuron are contained. The training data $\mathbf{P}_{gas\_1}$, $\mathbf{P}_{gas\_2}$,..., $\mathbf{P}_{gas\_q}$ and the corresponding target concentration vectors $\mathbf{Target}_{gas\_1}$, $\mathbf{Target}_{gas\_2}$,..., $\mathbf{Target}_{gas\_q}$ of MLP_1, MLP_2,..., MLP_q for quantification of gas_1, gas_2,..., gas_q, respectively, are illustrated in Eq. 5.2.

$$\begin{cases} \mathbf{P}_{gas\_1} = \begin{bmatrix} S_{1,1}^{(1)} & S_{1,2}^{(1)} & \cdots & S_{1,n_1}^{(1)} \\ S_{2,1}^{(1)} & S_{2,2}^{(1)} & \cdots & S_{2,n_1}^{(1)} \\ \vdots & \vdots & \vdots & \vdots \\ S_{m,1}^{(1)} & S_{m,2}^{(1)} & \cdots & S_{m,n_1}^{(1)} \end{bmatrix} \overset{MLP_1 \text{learning}}{\Rightarrow} \mathbf{Target}_{gas\_1} = \begin{bmatrix} t_{1,1}^{(1)} & t_{1,2}^{(1)} & \cdots & t_{1,n_1}^{(1)} \end{bmatrix} \\ \mathbf{P}_{gas\_2} = \begin{bmatrix} S_{1,1}^{(2)} & S_{1,2}^{(2)} & \cdots & S_{1,n_2}^{(2)} \\ S_{2,1}^{(2)} & S_{2,2}^{(2)} & \cdots & S_{2,n_2}^{(2)} \\ \vdots & \vdots & \vdots & \vdots \\ S_{m,1}^{(2)} & S_{m,2}^{(2)} & \cdots & S_{m,n_2}^{(2)} \end{bmatrix} \overset{MLP_2 \text{learning}}{\Rightarrow} \mathbf{Target}_{gas\_2} = \begin{bmatrix} t_{1,1}^{(2)} & t_{1,2}^{(2)} & \cdots & t_{1,n_2}^{(2)} \end{bmatrix} \\ \vdots \qquad\qquad\qquad\qquad\qquad\qquad\qquad\qquad \vdots \\ \mathbf{P}_{gas\_q} = \begin{bmatrix} S_{1,1}^{(q)} & S_{1,2}^{(q)} & \cdots & S_{1,n_q}^{(q)} \\ S_{2,1}^{(q)} & S_{2,2}^{(q)} & \cdots & S_{2,n_q}^{(q)} \\ \vdots & \vdots & \vdots & \vdots \\ S_{m,1}^{(q)} & S_{m,2}^{(q)} & \cdots & S_{m,n_q}^{(q)} \end{bmatrix} \overset{MLP_q \text{learning}}{\Rightarrow} \mathbf{Target}_{gas\_q} = \begin{bmatrix} t_{1,1}^{(q)} & t_{1,2}^{(q)} & \cdots & t_{1,n_q}^{(q)} \end{bmatrix} \end{cases} \tag{5.2}$$

where $S_{i,j}^{(k)}$, $i = 1,\ldots,m;\ j = 1,\ldots,n_k;\ k = 1,\ldots,q$ in $\mathbf{P}_{gas\_k}$ represents an extracted steady state feature in each sensor response of gas_k, each column in $\mathbf{P}_{gas\_k}$, k = 1,..., q represents a feature vector of one observation sample, and $t_{i,j}^{(k)}, i = 1,\ldots,m;\ j = 1,\ldots,n_k;\ k = 1,\ldots,q$ is the corresponding target concentration of each column vector in $\mathbf{P}_{gas\_k}$.

Note that the MLP_1, MLP_2,..., MLP_q of each gas in MMISO model are learned individually and independently. The convergence performance (i.e., training goal, MSE) of MLP by BP training is also related with the training data and the target concentrations, while the target concentrations of each gas are also not in the same order of magnitude (i.e., the concentration ranges of CH_2O, C_6H_6, C_7H_8, CO, NH_3, and NO_2 are within 0–6, 0.1–1, 0–0.2, 5–49, 0–3, and 0–2 ppm). Therefore, the MMISO model can be adjusted conveniently in terms of the characteristic of chemicals, avoid the disadvantages of SMIMO model, and improve the performance of convergence and prediction significantly.

5.3.3 Model Optimization

BP neural network learning algorithm in this chapter is the Levenberg–Marquardt (LM) algorithm [22] which takes the rule of mean square error minimum. LM is a fast algorithm with strong convergence ability and can get higher regression accuracy in learning of BP neural networks. LM algorithm is in fact a combination of Newton method and gradient descent. Besides, there is a special training function "trainlm.m" in MATLAB toolbox that we are using for convenient LM learning of the BP neural network. LM has been recognized as an effective learning algorithm in regression of this chapter, but not that good in pattern recognition. However, the adjustments of hyper-parameters (i.e., weights and bias) are generally prone to premature convergence and trap into local optimum. Therefore, eight population-based intelligent optimization algorithms, i.e., genetic algorithm (GA), standard particle swarm optimization (SPSO), inertia weight approach PSO (IWA-PSO), adaptive PSO (APSO), attractive and repulsive PSO (ARPSO), PSO based on diffusion and repulsion (DRPSO), PSO based on bacterial chemotaxis (PSOBC), and PSO based on adaptive genetic strategy (PSOAGS) have been studied generally for hyper-parameters optimization in this chapter. Note that the computational intelligence optimizations can also not guarantee the global optimum theoretically. However, it can effectively improve the performance of convergence of BP neural network in engineering applications, promote the progress of E-nose accuracy in quantitative concentration estimation greatly, and satisfy the actual necessity.

Assume a structure *m*-*h*-*o* of MLP with three layers (input layer, one hidden layer, and output layer), then, the weights matrix connected between the input layer and the hidden layer is $\mathbf{W}_1$ $(h \times p)$, the bias vector of the hidden layer is $\mathbf{B}_1$ $(h \times 1)$, the weights matrix connected between the hidden layer and the output layer is $\mathbf{W}_2$ $(o \times h)$, and the bias vector of the output layer is $\mathbf{B}_2$ $(o \times 1)$. The elements in weights and bias are viewed as genes in optimization, and the weights and bias can be decoded as a vector which is an individual in the population as follows

$$\underbrace{x_1, x_2, \ldots, x_{m\times h}}_{\mathrm{W}_1}, \underbrace{x_{m\times h+1}, \ldots, x_{m\times h+h}}_{\mathrm{B}_1} \underbrace{x_{m\times h+h+1}, \ldots, x_{m\times h+h+h\times o}}_{\mathrm{W}_2}, \underbrace{x_{m\times h+h+h\times o+1}, \ldots, x_L}_{\mathrm{B}_2} \tag{5.3}$$

where *m*, *h*, and *o* are the number of input neurons, hidden neurons, and output neurons, respectively. Therefore, the length L of an individual is $L = m \times h + h + h \times o \times o$.

The cost function in the form of mean relative prediction error (MRPE, %) in optimization is illustrated as

$$CostFun = \max\left\{\frac{1}{M_1}\sum_{i=1}^{M_1}\left|\frac{Y_i^{tr} - T_i^{tr}}{T_i^{tr}}\right|, \frac{1}{M_2}\sum_{j=1}^{M_2}\left|\frac{Y_j^{te} - T_j^{te}}{T_j^{te}}\right|\right\} \times 100 \tag{5.4}$$

where M_1 denotes the number of training samples, M_2 denotes the number of testing samples, $\mathbf{Y}^{tr}$ and $\mathbf{T}^{tr}$ represent the estimated and target concentrations of the training samples, and $\mathbf{Y}^{te}$ and $\mathbf{T}^{te}$ represent the estimated and target concentrations for the testing samples. The smaller value of *CostFun* corresponds to a lower MRPE and represents a better estimation performance.

Besides, the mean square error of prediction (MSEP, %) is also used for the evaluation of the concentration estimation model. The MSEP in percentage can be represented as

$$\text{MSEP} = \frac{1}{M_2}\sum_{i=1}^{M_2}(Y_i^{te} - T_i^{te})^2 \times 100 \tag{5.5}$$

Algorithm 5.1 MMISO-MLP

Input: Electronic nose data: $\mathbf{P}_{gas\_1}, \mathbf{P}_{gas\_2},\ldots,\mathbf{P}_{gas\_q}$; $\mathbf{Target}_{gas\_1}, \mathbf{Target}_{gas\_2},\ldots,\mathbf{Target}_{gas\_q}$;

MLP: the number *m*, *h* and *o* of input neurons, hidden neurons and output neurons, and MSE (training goal);

Optimization: the length *L* of each individual, the population size *M*, and the maximum generations *G*;

(1) **switch** index (1,…,*q*)of gas

(2) **case** 1, **do**

determine the input matrix **P** and output **T** of MLP as $\mathbf{P}_{gas\_1}$ and $\mathbf{Target}_{gas\_1}$;

case 2, **do**

determine the input matrix **P** and output **T** of MLP as $\mathbf{P}_{gas\_2}$ and $\mathbf{Target}_{gas\_2}$;

⋮

case *q*, **do**

determine the input matrix **P** and output **T** of MLP as $\mathbf{P}_{gas\_q}$ and $\mathbf{Target}_{gas\_q}$;

(3) Randomly initialize the population and set *Generation*=0;

(4) **while** *Generation* <*G* **do**

(5) *Generation*=*Generation*+1;

(6) **for** *i*=1 to *M* **do**

(7) Encode the individual as initial weights $\mathbf{W}_1$, $\mathbf{W}_2$ and bias $\mathbf{B}_1$, $\mathbf{B}_2$ according to Eq. 5.3;

(8) Evaluates the individual through MLP training and calculate the cost function Eq. 5.4;

(9) **end for**

(10) Save the best individual from the population;

(11) Update the population using one of the eight computational intelligence search algorithms.

(12) **end while**;

(13) Select the best individual from the total *G* generations

(14) Encode the individual as initial weights $\mathbf{W}_1$, $\mathbf{W}_2$ and bias $\mathbf{B}_1$, $\mathbf{B}_2$ according to Eq. 5.3;

(15) **end switch**

This chapter studies the concentration estimation of six kinds of chemicals by portable electronic noses. Therefore, in SMIMO structure-based MLP model, the network structure of SMIMO is set as 6-25-6 (6 input neurons, 25 hidden neurons, and 6 output neurons). Six neural networks with the structure 6-10-1 are contained in MMISO.

For each computational intelligence optimization, the size M of population is set as 50 and the maximum generation G is set as 30 in experiments for equal comparison. The pseudo-code implementation of the MMISO-based concentration estimation model with different computational intelligence optimization methods is illustrated in **Algorithm 5.1**.

5.4 Results and Discussion

In this chapter, 12 methods including 3 linear methods (MLR, PLS, and PCR) and 9 nonlinear methods (single BP, SGA-BP, SPSO-BP, IWAPSO-BP, APSO-BP, DRPSO-BP, ARPSO-BP, PSOBC-BP, and PSOAGS-BP) often reported in E-nose are studied on concentration estimation of chemicals by an E-nose based on MMISO structures. The results based on SMIMO structure can be referred to as the paper [23].

The estimation results (MRPE) of electronic nose are presented in Table 5.3. The first two methods with the lowest average MRPE of 23.13 and 23.54% are obtained using PSOAGS-BP and PSOBC-BP based on MMISO model. The lowest MRPEs for prediction of formaldehyde, benzene, toluene, carbon monoxide, ammonia, and nitrogen dioxide are 30.55, 6.587, 16.15, 6.115, 31.41, and 47.95%, respectively.

Table 5.3 Mean relative prediction errors (MRPEs) of indoor contaminants using MMISO-based structure experiments

Methods based on MMISO structure	MRPE (%)						
	CH_2O	C_6H_6	C_7H_8	CO	NH_3	NO_2	Average
MLR	258.8	10.07	20.82	18.56	70.99	80.67	76.65
PLS	330.1	21.93	22.31	50.03	94.84	122.7	106.9
PCR	372.7	104.1	95.69	122.9	129.1	154.9	163.2
Single BP	158.7	15.51	65.56	17.67	120.5	77.61	75.93
SGA-BP	34.25	7.270	18.71	6.264	44.06	60.98	28.59
SPSO-BP	32.36	6.935	18.71	5.103	29.43	54.70	24.54
IWAPSO-BP	32.67	7.323	17.63	7.651	47.79	69.69	30.46
APSO-BP	31.66	5.679	18.01	6.539	34.12	65.59	26.93
DRPSO-BP	31.38	7.196	17.95	6.677	35.40	60.71	26.55
ARPSO-BP	31.56	6.879	17.16	7.377	38.32	63.41	27.45
PSOBC-BP	31.18	6.951	17.83	6.426	28.11	50.79	**23.54**
PSOAGS-BP	30.55	6.587	16.15	6.115	31.41	47.95	**23.13**

The boldface type denotes the best performance.

Table 5.4 presents the concentration estimation analysis using the evaluation of MSEP in percentage based on MMISO structure. The MSEP of PSOAGS-BP method for formaldehyde, benzene, toluene, carbon monoxide, ammonia, and nitrogen dioxide are 7.10, 0.26, 0.033, 3.09, 4.30, and 5.58%. The lowest average MSEPs are 3.33 and 3.45% obtained by PSOBC-BP and PSOAGS-BP, which have significant improvement than SMIMO-based structure. The results from Tables 5.3 and 5.4 demonstrate that the proposed MMISO-based structure is very effective for concentration estimation of multiple kinds of indoor air contaminants.

For directly perceived through the senses, we have presented the performance comparisons of SMIMO and MMISO models using 12 estimation methods in Fig. 5.3. Figure 5.3a–f illustrates the estimation MRPEs of formaldehyde, benzene, toluene, carbon monoxide, ammonia, and nitrogen dioxide, respectively, using 12 methods (index of method in the horizontal axis of each subfigure: method 1: MLR; method 2: PLS; method 3: PCR; method 4: single BP; method 5: SGA-BP; method 6: SPSO-BP; method 7: IWAPSO-BP; method 8: APSO-BP; method 9: DRPSO-BP; method 10: ARPSO-BP; method 11: PSOBC-BP; method 12: PSOAGS-BP). It is clear that the MRPE of estimation for each gas of MMISO is much lower than SMIMO structure-based MLP model using nonlinear estimation methods. This demonstrates that the MMISO is much better than SMIMO structure in detection of multiple kinds of gases by portable E-noses due to that cross errors have been occurred in SMIMO during the learning process due to the correlation among sensor signals.

It is worth noting that though the MRPEs of estimation seem to be still large in measurement by using the proposed method, while the results of this chapter are actual experimental test, the seemingly large error results from the electronic nose system and experimental data. Currently, the electronic nose system is developed

Table 5.4 Mean square errors of prediction (MSEPs) for indoor contaminants concentration estimation using MMISO-based structure experiments

Methods based on MMISO structure	MSEP (%)						
	CH_2O	C_6H_6	C_7H_8	CO	NH_3	NO_2	Average
MLR	825.6	0.430	0.430	26.56	15.90	20.69	148.3
PLS	1350	1.920	0.470	183.9	28.46	46.94	268.6
PCR	1716	46.97	0.870	1095	52.48	74.79	497.7
Single BP	120.7	0.950	0.475	24.00	85.31	68.92	50.06
SGA-BP	10.61	0.390	0.081	3.39	7.62	10.03	5.353
SPSO-BP	11.23	0.330	0.052	3.450	16.31	4.46	5.972
IWAPSO-BP	11.26	0.320	0.041	4.320	11.27	13.18	6.732
APSO-BP	9.750	0.270	0.044	2.180	4.660	9.35	4.376
DRPSO-BP	10.05	0.410	0.042	3.020	14.38	8.78	6.114
ARPSO-BP	19.15	0.300	0.034	2.080	6.170	8.030	5.961
PSOBC-BP	10.59	0.340	0.033	2.130	4.890	1.980	**3.327**
PSOAGS-BP	7.710	0.260	0.033	3.090	4.030	5.580	**3.451**

The boldface type denotes the best performance.

based on a metal oxide semiconductor (MOS) gas sensor array which has characteristic of low cost and broad spectrum but low accuracy. The cross-sensitivity of MOS gas sensors combined with pattern recognition can make electronic nose detect multiple kinds of contaminants. Generally, the electronic nose with one electrochemical gas sensor developed in our laboratory has a high accuracy in measurement of formaldehyde concentration. The MRPE is around 20% for electrochemical sensor, but it can only be used for formaldehyde test. The defect of E-nose based on electrochemical sensor is the single gas detection and higher cost. Besides, the experimental data is also an aspect in the estimation performance. In this chapter, we have used all the data without any possible outlier or noise removal for analysis, considering that the robustness (generality of prediction) is also important in detection. A good accuracy in computer cannot guarantee a successful prediction in actual application due to the complex environmental conditions. Instead, the existence of noise in the learning dataset can enhance the ability of noise counteraction in an E-nose. And also, the disadvantage of easily trapped into local minimum for BP neural network can be improved by the intelligent optimizations which have been significant in electronic nose detection from the presented results, though the computational intelligence optimizations can also not guarantee the seriously global optimum theoretically. Therefore, we can say that the estimation performance in this chapter can be accepted with an average MRPE of 23% and an average MSEP of 3%.

On the other hand, the time consumptions of SMIMO and MMISO for running all the estimation algorithms for each gas are also presented in Table 5.5 for comparing their efficiency in application. Due to that, SMIMO realizes the estimation of each gas at the same time and the SMIMO model endures a larger burden for MLP training because of larger dataset which consists of all training data, and the total time for SMIMO is calculated as approximately 27 h. Due to that, the estimation of each gas is independent based on the MMISO structure; 2.94, 0.54, 0.32, 3.22, 1.75, and 1.15 h are consumed for parameters learning of formaldehyde, benzene, toluene, carbon monoxide, ammonia, and nitrogen dioxide, respectively. Totally, 9.9 h are consumed for all chemicals based on MMISO. Therefore, it is clear that the efficiency of MMISO is also much higher than SMIMO. The estimation performance and efficiency fully prove that MMISO model is a better selection for concentration estimation of multiple kinds of gases in electronic nose development.

Table 5.5 Time consumption for running all algorithms experiments

Model	Time consumption for running all algorithms in computer (unit: hour)						
	CH_2O	C_6H_6	C_7H_8	CO	NH_3	NO_2	Total time
SMIMO	–	–	–	–	–	–	27.0
MMISO	2.94	0.54	0.32	3.22	1.75	1.15	9.90

Note that the running time presented in Table 5.5 is only for algorithm learning in computer, but not in electronic nose system, considering the higher cost of CPU and the real-time characteristic in an E-nose. The learned results (i.e., the weights of neural network) will be transferred to the E-nose system and perform the feed-forward computation of multilayer perceptron instead. The computation velocity is in the level of microsecond (μs), and therefore, it can still promise the real-time characteristic of E-nose detection.

5.5 Summary

This chapter presents a performance study of SMIMO- and MMISO-based MLP with parameters optimization for quantifying the concentration of six kinds of indoor air contaminants by a portable E-nose. There are three major findings in experiments that concentration estimation of multiple kinds of chemicals by an E-nose with non-selective chemical sensors is a completely nonlinear problem when compared with those three linear methods, computational intelligence optimizations are very effective in improving the parameters learning of MLP based on BP algorithm, and MMISO structure-based MLP has better superiorities for quantifying multiple kinds of chemicals than SMIMO in terms of the estimation accuracy and the time consumption of algorithm convergence.

References

1. K. Sakai, D. Norbäck, Y. Mi, E. Shibata, M. Kamijima, T. Yamada, Y. Takeuchi, A comparison of indoor air pollutants in Japan and Sweden: formaldehyde, nitrogen, dioxide, and chlorinated volatile organic compounds. Environ. Res. **94**, 75–85 (2004)
2. S.C. Lee, M. Chang, Indoor and outdoor air quality investigation at schools in Hong Kong. Chemosphere **41**, 109–113 (2000)
3. M. Pardo, G. Sberveglieri, Remarks on the use of multilayer perceptrons for the analysis of chemical sensor array data. IEEE. Sens. J. **4**(3), 355–363 (2004)
4. M. Pardo, G. Faglia, G. Sberveglieri, M. Corte, F. Masulli, M. Riani, A time delay neural network for estimation of gas concentrations in a mixture. Sens. Actuators, B Chem. **65**, 267–269 (2000)
5. L. Zhang, F. Tian, C. Kadri, G. Pei, H. Li, L. Pan, Gases concentration estimation using heuristics and bio-inspired optimization models for experimental chemical electronic nose. Sens. Actuators, B Chem. **160**, 760–770 (2011)
6. L. Zhang, F. Tian, S. Liu, H. Li, L. Pan, C. Kadri, Applications of adaptive Kalman filter coupled with multilayer perceptron for quantification purposes in electronic nose. J. Compu. Inf. Syst **8**(1), 275–282 (2012)
7. L. Zhang, F. Tian, S. Liu, J. Guo, B. Hu, Q. Ye, L. Dang, X. Peng, C. Kadri, J. Feng, Chaos based neural network optimization for concentration estimation of indoor air contaminants by an electronic nose. Sens. Actuators, A Phys. **189**, 161–167 (2013)

8. D. Gao, M. Chen, Y. Ji, Simultaneous estimation of classes and concentrations of odours by an electronic nose using combinative and modular multilayer perceptrons. Sens. Actuators, B Chem. **107**, 773–781 (2005)
9. D. Gao, Z. Yang, C. Cai, F. Liu, Performance evaluation of multilayer perceptrons for discriminating and quantifying multiple kinds of odors with an electronic nose. Neural Netw. **33**, 204–215 (2012)
10. L. Zhang, F. Tian, X. Peng, L. Dang, G. Li, S. Liu, C. Kadri, Standardization of metal oxide sensor array using artificial neural networks through experimental design. Sens. Actuators, B Chem. **177**, 947–955 (2013)
11. L. Zhang, F. Tian, S. Liu, L. Dang, X. Peng, X. Yin, Chaotic time series prediction of E-nose sensor drift in embedded phase space. Sens. Actuators, B Chem. **182**, 71–79 (2013)
12. V. Maniezzo, Genetic evolution of the topology and weight distribution of neural networks. IEEE. Trans. Neural Netw. **5**(1), 39–53 (2002)
13. J. Kennedy, R.C. Eberhart, *Particle Swarm Optimization*, in Proceedings of IEEE International Conference on Neural Networks, vol. 4 (1995), pp. 1942–1948
14. R.C. Eberhart, Y. Shi, Comparing inertia weights and constriction factors in particle swarm optimization. IEEE. Evol. Comput. **1**, 84–88 (2002)
15. J. Riget, J. S. Vesterstrom, *A Diversity-Guided Particle Swarm Optimizer-The ARPSO*. Technical Report in Department of Computer Science (2002)
16. W. Jiang, Y. Zhang, *A Particle Swarm Optimization Algorithm Based on Diffusion-Repulsion and Application to Portfolio Selection*, IEEE International Symposium on Information Science and Engineering (2008)
17. B. Niu, Y. Zhu, An improved particle swarm optimization based on bacterial chemotaxis. IEEE. Intell. Control. Autom. **1**, 3193–3197 (2006)
18. S. Wold, M. Sjöström, L. Eriksson, PLS-regression: a basic tool of chemometrics. Chemometr. Intell. Lab. Syst. **58**, 109–130 (2001)
19. J.H. Sohn, M. Atzeni, L. Zeller, G. Pioggia, Characterisation of humidity dependence of a metal oxide semiconductor sensor array using partial least squares. Sens. Actuators, B Chem. **131**, 230–235 (2008)
20. T. Nes, H. Martens, Principal component regression in NIR analysis: viewpoints, background details and selection of components. J. Chemom. **2**, 155–167 (1988)
21. J. Getino, M.C. Horrillo, J. Gutiérrez, L. Arés, J.I. Robla, C. García, I. Sayago, Analysis of VOCs with a tin oxide sensor array. Sens. Actuators, B Chem. **43**, 200–205 (1997)
22. T. Hagan, H. Demuth, M. Beale, *Neural. Netw. Des.* (PWS Publishing, Boston, MA, 1996)
23. L. Zhang, F. Tian, Performance study of multilayer perceptrons in low-cost electronic nose. IEEE. Trans. Instru. Meas. **63**, 1670–1679 (2014)

Chapter 6
Discriminative Support Vector Machine-Based Odor Classification

Abstract This chapter presents a laboratory study of multi-class classification problem for multiple indoor air contaminants. The effectiveness of the proposed HSVM model has been rigorously evaluated. In addition, we have also compared with existing methods including Euclidean distance to centroids (EDC), simplified fuzzy ARTMAP network (SFAM), multilayer perceptron neural network (MLP) based on back-propagation, individual FLDA, and single SVM. Experimental results demonstrate that the HSVM model outperforms other classifiers in general. Also, HSVM classifier preliminarily shows its superiority in solution to discrimination in various electronic nose applications.

Keywords Electronic nose · Classification · Multi-class problem
Hybrid support vector machine · Fisher linear discrimination analysis

6.1 Introduction

Patterns from known odorants are employed to construct a database and train a pattern recognition model using some learning rules, such that unknown odorants can be classified and discriminated [1]. In pattern analysis, one or more features in steady state responses were selected and a vector is formulated as the pattern of each observation. In discrimination, a classification model is first developed on the training patterns; then, the performance of the model is evaluated by means of the independent testing samples; the final classification accuracy can be calculated by comparing their predicted categories with their own true categories. So far, many pattern recognition models based on intuitive, linear, and nonlinear supervised techniques have been explored in E-nose data. In this chapter, we have systematically studied different linear and nonlinear tools and try to find an optimal model for gases classification. Among a large number of classification models, we select five representative methods for comparisons. They are Euclidean distance to centroids (EDC) [2], fuzzy ARTMAP network [3, 4], multilayer perceptron neural

L. Zhang et al., *Electronic Nose: Algorithmic Challenges*,
https://doi.org/10.1007/978-981-13-2167-2_6

network (MLP) [5–7], Fisher linear discrimination analysis (FLDA) [8], and support vector machine (SVM) [9, 10].

EDC, which assigns samples to the class with the minimum distance, is a very intuitive classification method. For each class and each variable, the centroid is calculated over all samples in that class. It is assumed that the distribution of samples around the centroid is symmetrical in the original variable space for each class. However, it cannot make use of the full discriminatory power available in all the variables so that this method actually obtains worse classification. Artificial neural networks (ANNs), especially fuzzy ARTMAP and MLP based on back-propagation learning rule, have been recognized to be successful in pattern recognition system (PARC). Fuzzy ARTMAP is a constructive neural network model developed upon adaptive resonance theory and fuzzy set theory [11, 12]. It allows knowledge to be added during training if necessary so that it has also been used for pattern recognition. Back-propagation multilayer perceptron neural network, which is a nonlinear, nonparametric, and supervised method, performed well in a variety of application [13, 14]. When it comes to the drawbacks of MLP, back-propagation algorithm has a limited capability to compensate for undesirable characteristics of the sensor system (e.g., temperature, humidity variations, and drift) and it is trained "off-line" and unable to adapt autonomously to the changing environment. Consequently, recalibration is still necessary for different periods. Although ARTMAP can realize "on-line" training through testing the new measurements, the problem is that it does not know the specific component or category in each new measurement. And also, the robustness and real-time characteristic of ARTMAP will be lost when compared with MLP in real applications. LDA, as a supervised method, has been used for feature extraction and variable selection [15] in a dataset like the unsupervised principal component analysis (PCA). Both of them extract features by transforming the original parameter vectors into a new feature space through a linear projection. Besides, LDA has also been used for discrimination. However, when the actual problem becomes completely nonlinear (e.g., the sensor array system), it will become unqualified. SVM, which was first introduced by Vapnik, is a relatively new machine learning technique [16, 17]. It has been proven advantageous in handling classification tasks with excellent generalization performance and robustness. For improvement of SVM, LDA as feature extraction method has been combined with SVM for fault diagnosis and hepatitis disease diagnosis [18]. Unfortunately, the sensor array produces a response vector for each observation, but not a matrix or dataset in real-time E-nose monitoring. In other words, a certain sampling time for a dataset collection should be needed for easy analysis by LDA or PCA which would make an on-line/real-time use of an E-nose impossible. Since the feature extraction by LDA or PCA cannot operate in real-time processing, the hybrid classification model would also become meaningless.

Particularly, most classification models can successfully solve a simple two-class problem. In this chapter, we present a laboratory study of a multi-class problem for classification of six contaminants using a hybrid discrimination model based on Fisher linear discrimination analysis (FLDA) and support vector machine (SVM) for monitoring and realizing a real-time gases category decision in people's dwellings by an E-nose. The role of FLDA is equivalent to a pre-classification by transforming the original data into a new feature space with more linearly independent variables correlated with each classifier, and more prior information about each class in the new feature space would be obtained. Thus, it makes SVM easier for final discrimination in the new feature space. For clarity, the hybrid model of FLDA and SVM in this chapter is called HSVM. The comparison results with EDC, simplified fuzzy ARTMAP (SFAM), MLP, individual FLDA, and single SVM demonstrate the potential ability of HSVM in E-nose.

6.2 Classification Methodologies

6.2.1 Euclidean Distance to Centroids (EDC)

EDC is a very intuitive classification method through assigning the nearest samples with the centroid to the corresponding class [2]. For each class, the mean (centroid) is calculated over all samples in that class. The Euclidean distance between sample i and the class k centroid is calculated as

$$d_{ik} = \sqrt{(\boldsymbol{x}_i - \bar{\boldsymbol{x}}_k) \cdot (\boldsymbol{x}_i - \bar{\boldsymbol{x}}_k)^{\mathrm{T}}} \tag{6.1}$$

where $\boldsymbol{x}_i$ is the i-th sample described by a row vector with n variables (n denotes the number of sensors), $\bar{\boldsymbol{x}}_k$ is the centroid of class k, and ^{T} denotes the transpose of a vector. The sample with the minimum distance d will be assigned to a specific class.

6.2.2 Simplified Fuzzy ARTMAP Network (SFAM)

ARTMAP consists of two modules (fuzzy ART and inter ART) that create stable recognition categories in response to the input patterns. Fuzzy ART receives a stream of input features representing the pattern map to the output classes in the category layer. Fuzzy ART module has three layers: F_0, F_1, and F_2. Inter ART module works by increasing the small vigilance parameter ε of fuzzy ART for updating the prediction error in the output category layer. We refer interested readers to [19] for the basic mathematical descriptions of SFAM. For parameter settings, the related parameters in [19] such as vigilance $\rho = 0.9$, $\alpha = 0.2$, learning rate $\beta = 1$, and $\varepsilon = 0.001$ are used in this chapter; the number of maximum categories and training times is set to 100, respectively.

6.2.3 Multilayer Perceptron Neural Network (MLP)

A typical multilayer perceptron consists of an input layer, one hidden layer, and one output layer. The input and output elements denote the observations composed of six variables (sensor) and the known category (labels) of each observation. Detailed description of MLP is out of the scope of this present study; for that, we refer the readers to [20]. In this chapter, we use three-bit binary codes to represent the identities of six categories. The identities of formaldehyde, benzene, toluene, carbon monoxide, ammonia, and nitrogen dioxide were labeled as $(0, 0, 1)^T$, $(0, 1, 0)^T$, $(0, 1, 1)^T$, $(1, 0, 0)^T$, $(1, 0, 1)^T$, and $(1, 1, 0)^T$, respectively. The number of nodes in input layer, hidden layer, and output layer was set as 6, 35, and 3, respectively. The output value p for each node should be adjusted as if $p \geq 0.5$, $p = 1$; else $p = 0$. The activation functions of the hidden layer and output layer we have used in classification are "logsig" and "purelin". The training goal and training times are set to 0.05 and 1000, respectively.

6.2.4 Fisher Linear Discriminant Analysis (FLDA)

Fisher linear discrimination analysis easily handles the case where the within-class frequencies are unequal, and their performances have been examined on randomly generated test data. This method maximizes the ratio of between-class variance to the within-class variance in any particular dataset, thereby guaranteeing maximum separability and also producing a linear decision boundary between two classes. A brief mathematical description for a two-class problem is shown as follows

Assume we have a set of n-dimensional dataset $\mathbf{X} = \{\mathbf{X}_1, \mathbf{X}_2\}$, where $\mathbf{X}_1$ belongs to class 1 which contains N_1 column vectors and $\mathbf{X}_2$ belongs to class 2 which contains N_2 column vectors. The centroid of each class is calculated by

$$\mu_i = 1/N_i \cdot \sum X_i, \quad i = 1, 2 \tag{6.2}$$

The within-scatter matrix of class i is shown by

$$S_i = \sum_{j=1}^{N_i} (X_{i,j} - \mu_i)(X_{i,j} - \mu_i)^T, \quad i = 1, 2 \tag{6.3}$$

Then, the within-class scatter matrix S_w and the between-class scatter matrix S_b can be calculated by

$$S_w = \sum_{i=1}^{2} S_i \tag{6.4}$$

$$S_b = \sum_{i=1}^{2} N_i \cdot (\mu_i - \bar{X}) \cdot (\mu_i - \bar{X})^{\mathrm{T}} \tag{6.5}$$

where $\bar{X}$ denotes the centroid of the total dataset $\boldsymbol{X}$.

Finally, the Fisher criterion in terms of S_{w} and S_{b} is expressed as

$$J(W) = W^{\mathrm{T}} S_{\mathrm{b}} W / W^{\mathrm{T}} S_{\mathrm{w}} W \tag{6.6}$$

where $\boldsymbol{W}$ is the transformation matrix which can be calculated by solving the eigenvalue problem

$$W^* = \arg\max\{J(W)\} = S_{\mathrm{w}}^{-1} \cdot (\mu_1 - \mu_2) \tag{6.7}$$

6.2.5 Support Vector Machine (SVM)

Support vector machines perform structural risk minimization in the framework of regularization theory. For linearly inseparable cases, SVM applies a nonlinear kernel function to transform the input space to a higher-dimensional feature space so that the classes may be linearly separable prior to calculate the separating hyperplane. This kernel function can be polynomial, Gaussian radial basis function (RBF) or sigmoid function. In this work, a linearly inseparable case is considered, and only Gaussian RBF kernel function was attempted for the classification due to its good generalization and without the guidance from those prior experiences. Therefore, this problem aims at solving a quadratic optimization in a higher-dimensional feature space. The Lagrangian function is shown by

$$L_{\mathrm{LSSVM}}(\alpha) = \sum_{i=1}^{N} \alpha_i - 1/2 \cdot \sum_{i,j=1}^{N} \alpha_i \alpha_j y_i y_j \phi(x_i)^{\mathrm{T}} \phi(x_j) \tag{6.8}$$

which needs to be minimized under the constraints: $\alpha_i > 0$ and $\sum_{i=1}^{N} \alpha_i y_i = 0$.

By introducing a kernel function

$$K(x_i, x_j) = \phi(x_i)^{\mathrm{T}} \phi(x_j) \tag{6.9}$$

the Lagrangian function can be rewritten by

$$L_{\mathrm{LSSVM}}(\alpha) = \sum_{i=1}^{N} \alpha_i - 1/2 \cdot \sum_{i,j=1}^{N} \alpha_i \alpha_j y_i y_j K(x_i, x_j) \tag{6.10}$$

The Gaussian RBF kernel function can be represented as

$$K(x_i, x_j) = \exp(-\|x_i - x_j\|^2/\sigma^2) \tag{6.11}$$

where σ^2 is the kernel parameter which determines the bandwidth of RBF. The decision function can be expressed as

$$f(x) = \mathrm{sgn}\left(\sum_{i=1}^{N} \alpha_i K(x_i, x) + b\right) \tag{6.12}$$

where α and b are the optimal decision parameters.

6.3 Experiments

6.3.1 Experimental Setup

The details of the E-nose module have been illustrated in [21]. Briefly, four metal oxide semiconductor gas sensors (TGS2602, TGS2620, TGS2201A, and B from Figaro company) and an extra module (HTD2230-I^{2}C) with two auxiliary sensors for temperature and humidity compensations were used in our E-nose. The sensors were mounted on a custom designed printed circuit board (PCB), along with associated electrical components. A 12-bit analog–digital converter (A/D) is used as interface between the field-programmable gate array (FPGA) processor and the sensors. The system can be connected to the PC via a Joint Test Action Group (JTAG) port. The sensor array will produce a group of odorant pattern with six variables including temperature, humidity, TGS2620, TGS2602, TGS2201A, and TGS2201B in each observation. The reasons for selection of these four gas sensors can be concluded as two aspects. First, they have a good sensitivity to indoor air contaminants. The sensitivity can be indicated as the ratio of sensor resistance (Rs) at various concentrations and sensor resistance (Ro) in fresh air or 300 ppm of ethanol. The corresponding parameters including the basic circuits, heater voltage, heater current, and standard test curves for each sensor of this work have been provided with the datasheet (.pdf) in the supplementary data. The species monitored by the sensor array contain carbon monoxide, nitric oxide, nitrogen dioxide, ammonia, toluene, ethanol, hydrogen, methane, hydride, and VOCs. Second, they have a long-term stability and good reproducibility. Also, we refer readers to the sensors' datasheets available in http://www.figaro.co.jp/en/product/index.php?mode=search&kbn=1 for more information on the other TGS sensors.

6.3.2 Dataset

Six familiar chemical contaminants indoor including formaldehyde (HCHO), benzene (C_6H_6), toluene (C_7H_8), carbon monoxide (CO), ammonia (NH_3), and nitrogen dioxide (NO_2) are investigated in this work. The experiments were employed by an E-nose in the constant temperature and humidity chamber whose type is LRH-150S. The accuracy for temperature and humidity of the chamber is ±0.5 °C and ±5%. In gases preparation, HCHO, C_6H_6, and C_7H_8 are liquor, and CO, NH_3, and NO_2 are standard gas. In each gas measurements, a gas bag collected with target gas and nitrogen (N_2) was prepared for injection into the chamber. Note that N_2 is used to dilute the gas concentration in the gas bag, and we get various concentrations by setting different injection time (injection speed to 5 l/min). The true concentrations for HCHO and NH_3 were measured using spectrophotometer, C_6H_6 and C_7H_8 were employed using gas chromatography, and CO and NO_2 were obtained using the reference instruments whose measurement accuracy is within ±3%. For each experiment, 12 min (e.g., 2 min for baseline and 10 min for response) was consumed and extra 15 min was also needed for cleaning the chamber by injecting pure air. Totally, 260, 164, 66, 58, 29, and 30 samples with target temperature, humidity, and various concentrations were collected for HCHO, C_6H_6, C_7H_8, CO, NH_3, and NO_2, respectively. These samples were measured with different combinations of the target temperatures 15, 25, 30, 35 °C and relative humidity (RH) of 40, 60, 80% which can approximately simulate the indoor temperature and humidity for improving the classifier robustness of the E-nose. Twelve combinations {(15, 60), (15, 80), (20, 40), (20, 80), (25, 40), (25, 60), (25, 80), (30, 40), (30, 60), (30, 80), (35, 60)} in manner of (T, RH), in which T denotes temperature and RH (%) denotes relative humidity, were employed for covering the indoor conditions. Note that the sensor responses have been normalized, and the measured concentrations of each selected sample for HCHO, C_6H_6, C_7H_8, CO, NH_3, and NO_2 were 0.18, 0.28, 0.14, 6.0, 0.50, and 1.62 ppm, respectively. The normalization is that sensor responses were directly divided by 4095. It is worth noting that the digit of 4095 (that is, $2^{12} - 1$) is the maximum value of the 12-bit A/D output for each sensor. In feature extraction, one value at the steady state response for each sensor was selected as the corresponding feature in such a simple way. Therefore, a vector with six variables including temperature and relative humidity is extracted as the feature vector of that sample for subsequent pattern recognition.

6.3.3 Multi-class Discrimination

The commonly used two methods for solving multi-class problems are "one-against-all" and "one-against-one" [22]. In this work, the "one-against-one" strategy (OAO) is used in HSVM to build the $k = 6$ classes classifier for the recommendation that it would be a better choice for $k \leq 10$ [10]. Thus, this strategy builds $k(k - 1)/2 = 15$ sub-classifiers (FLDA classifier or SVM classifier) trained using input patterns of two classes. Consequently, a complex multi-class problem can be untied through solutions of multiple two-class classifiers with a voting scheme in decision that if the indicator function of each sub-classifier says that x belongs to class i, then the vote for class i is increased by one; otherwise, the vote for class j is increased by one.

In terms of OAO strategy, 15 FLDA and 15 SVM classifiers in the HSVM model should be designed separately in a six-category classification problem. Here, FLDA transforms the original data into a new feature space composed of more linearly independent variables which can be recognized as the new variables with more characteristic of linear separability and prior information of each sub-classifier which are easier for SVM classification. The implementation process of HSVM in E-nose data can be illustrated as follows.

Assume that the number of HCHO, C_6H_6, C_7H_8, CO, NH_3, and NO_2 training samples is n_1, n_2, n_3, n_4, n_5, and n_6, respectively. Thus, the original training input data matrix $\mathbf{X_{original}}$ can be constructed in order by

$$\mathbf{X_{original}} = \{\mathbf{X}_1, \mathbf{X}_2, \mathbf{X}_3, \mathbf{X}_4, \mathbf{X}_5, \mathbf{X}_6\} \tag{6.13}$$

where the i-th matrix $\mathbf{X}_i$ is $6 \times n_i$, $i = 1,\ldots,6$; thus, $\mathbf{X_{original}}$ is a matrix of $6 \times \sum_{i=1}^{6} n_i$, and each column denotes one observation vector.

The training goal (category label) is constructed in order as

$$\text{label} = \{\overbrace{1,\ldots,1}^{n_1}, \overbrace{2,\ldots,2}^{n_2}, \overbrace{3,\ldots,3}^{n_3}, \overbrace{4,\ldots,4}^{n_4}, \overbrace{5,\ldots,5}^{n_5}, \overbrace{6,\ldots,6}^{n_6}\} \tag{6.14}$$

Assume that the total transformation matrix of the 15 FLDA classifiers is expressed by

$$\mathbf{W} = \{\boldsymbol{w}_1, \boldsymbol{w}_2, \ldots, \boldsymbol{w}_{15}\} \tag{6.15}$$

where $\boldsymbol{w}_j$ ($j = 1,\ldots, 15$) is a column vector of 6×1 representing the transformation of each sub-classifier between two classes which can be directly used for classification. Therefore, $\mathbf{W}$ is a matrix with a size of 6×15. Then, the input data $\mathbf{X_{HSVM}}$ of HSVM can be reconstructed by projection

$$\mathbf{X_{HSVM}} = \mathbf{W}^{\mathrm{T}}\mathbf{X_{original}} \tag{6.16}$$

Similarly, the original testing input data matrix should also be reconstructed in terms of the principle of training input data. Consequently, the training and test input pattern of SVM has now been correlated with the initial FLDA classification, and more linearly independent variables correlated with sub-classifiers were produced through a linear projection without shifting the number of original patterns.

The new patterns preprocessed by FLDA will then be used for SVM classification with structural risk minimization.

6.3.4 *Data Analysis*

The performance of E-nose data classification was assessed in terms of the classification accuracy of test samples. The classification accuracy is defined as a percentage of correct classifications in all test samples. Also, the average accuracy of training and test samples were calculated for insight of the whole data. To validate the robustness and generalization of all classifiers considered in this work, three proportions 30–70, 50–50, and 80–20% of training and test samples were analyzed, respectively. For selection of training set in terms of some proportion, a Kennard–Stone sequential (KSS) algorithm [23] based on the multivariate Euclidean distance was used, and the remaining samples were recognized as test samples. The distribution with three proportions of training set and test set for each class is represented in Table 6.1.

Note that, all the classification models were only performed on the training sets; then, the trained parameters were applied to the testing sets. All algorithms for multi-class discrimination were implemented in MATLAB 2009a, operating on a laboratory computer equipped with Inter i3 CPU 530, 2.93 GHz processors and 2 GB of RAM.

6.4 Results and Discussion

6.4.1 *Experimental Results*

To evaluate the effectiveness of the hybrid model HSVM, the E-nose data were analyzed by using all the classifiers considered in our project. We first presented the principal component analysis (PCA) results of the original training sets. Figure 6.1 illustrates three 2-D scatter sub-plots (PC-1 vs. PC-2, PC-1 vs. PC-3, and PC-2 vs. PC-3) and a 3-D scatter sub-plot of the first three principal components by running

Table 6.1 Distribution of training set and testing set

Training–testing proportion (%)	Number of samples in the subset											
	Training set						Testing set					
	HCHO	C_6H_6	C_7H_8	CO	NH_3	NO_2	HCHO	C_6H_6	C_7H_8	CO	NH_3	NO_2
30–70	78	49	20	18	9	9	182	115	46	40	20	21
50–50	130	82	33	29	15	15	130	82	33	29	14	15
80–20	208	131	53	46	23	24	52	33	13	12	6	6

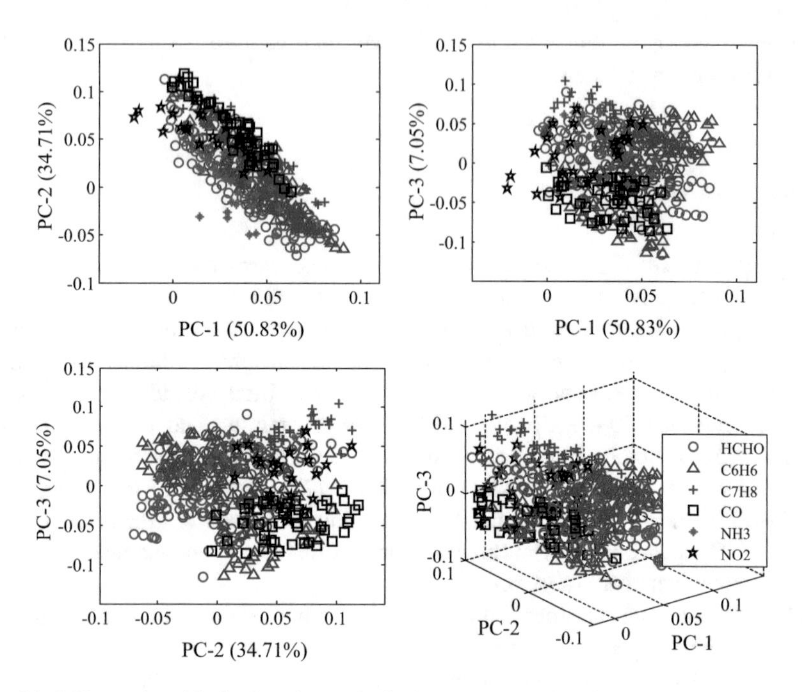

Fig. 6.1 PCA results with the first three principal components of 80% training set

PCA on the 80% training set. From the 2-D and 3-D PCA plots, we can get that the first three PCs can totally account for 92.59% information of the training data. Obviously, the multi-class problem in this work belongs to a completely linear-inseparable case because of the serious overlaps among all classes. Especially, the patterns of HCHO and C_6H_6 as indoor air contaminants are completely inseparable with other gas patterns from the PCA results. It is noteworthy that PCA is an unsupervised method which transforms the original data into the space of the principal components through a linear projection [24]. The analysis of the PCA results confirms the necessity of nonlinear classifiers employment due to that they can make the linearly inseparable problem separable in a high-dimensional space through a nonlinear transform.

Table 6.2 presents the discrimination results of 50% of testing samples using the HSVM classification model developed on the remaining 50% of training samples. The digits with bold type in diagonal line denote the number of correctly classified samples, while others denote the number of misclassified samples.

For quantification of classification accuracy and present the comparisons with other classification models, Table 6.3 shows the classification accuracy including the train set and test set with a proportion of 50%. The average accuracy of the six classes for each model and the classification accuracy for the total train set and test set separately are also given.

Table 6.2 Multi-classification results of *N* testing samples which occupy 50% of the total samples using HSVM classification method

Class	*N*	Classified as					
		HCHO	C_6H_6	C_7H_8	CO	NH_3	NO_2
HCHO	130	**116**	5	6	1	2	0
C_6H_6	82	9	**66**	5	1	1	0
C_7H_8	33	0	0	**33**	0	0	0
CO	29	0	0	0	**29**	0	0
NH_3	14	0	0	1	0	**13**	0
NO_2	15	2	5	0	0	0	**8**

The boldface type denotes the best performance.

From Table 6.3, we can clearly find that with the increasing number of training samples, the classification accuracy increases also. For each model, the discrimination of NO_2 was not that successful, two reasons may explain it. The first one is the smaller number of samples. Totally, 30 samples were collected, and unbalanced samples may also influence the classification. Second, the sensitivity of gas sensors to NO_2 (oxidizing gas) is negative which is contrary to other five contaminants. Concluded from the digits in bold in Table 6.3, the HSVM classification with FLDA is always better than other models for HCHO, C_7H_8, NH_3, and CO. The 100% classification accuracy of C_7H_8, NH_3, and CO can be obtained on the testing samples by using the HSVM model. Also, the highest 94.23% classification accuracy of HCHO on the testing samples is obtained.

For visualization, Fig. 6.2 illustrates the classification accuracy of the test samples with three different proportions based on the presented six classifiers. We can see that HSVM model performs the best multi-class discrimination. Note that each node (from number 1 to 6) in Fig. 6.2 denotes one kind of classifier and three kinds of symbols ("square", "circle", and "triangle") represent three different proportions, respectively. The single SVM classifier performs the second best when the training–testing proportion is 50–50% and 80–20%. However, with 30% training samples SVM performs worse than SFAM, MLP, and FLDA models. The proposed HSVM is obviously superior to all the models considered. It also confirms that with small number of samples HSVM can still show the best classification performance.

To study the classification performance of different models using only four metal oxide semiconductor gas sensors without considering temperature and humidity in feature space, we perform all the classification procedure on the features with only four variables on the training and testing samples with the proportion of 80–20%. Table 6.4 presents the classification accuracy of training and testing samples separately without temperature and humidity integration. We can find that HSVM still performs the best discrimination. Another finding is that the accuracy of C_6H_6 decreased for all the models, which means that temperature and humidity are also important as classification features of C_6H_6. Therefore, both temperature and humidity are key features in pattern recognition for improving the classification performance of E-nose. From the datasheets of the sensors, we can also find that

Table 6.3 Classification accuracy with the proportion 50%–50% of training and testing samples

Class	Classification accuracy (%)											
	Train set						Test set					
	EDC	SFAM	MLP	FLDA	SVM	HSVM	EDC	SFAM	MLP	FLDA	SVM	HSVM
HCHO	17.69	100.0	93.08	71.54	98.41	100.0	15.38	69.23	91.54	71.54	86.92	**89.23**
C_6H_6	42.68	100.0	80.49	64.63	93.90	100.0	57.32	71.95	69.51	79.27	82.93	80.48
C_7H_8	69.70	100.0	90.91	90.91	96.67	96.97	54.55	69.70	90.91	93.94	96.97	**100.0**
CO	68.97	100.0	82.76	89.66	86.21	89.66	65.52	79.31	86.21	100.0	89.66	**100.0**
NH_3	66.67	100.0	66.67	73.33	93.33	93.33	78.57	57.14	50.00	71.43	71.43	**92.86**
NO_2	26.67	100.0	73.33	86.67	86.67	86.67	26.67	80.00	46.67	46.67	66.67	53.33
Mean	48.73	100.0	81.21	79.46	92.53	94.44	49.67	71.22	72.47	77.14	82.43	**85.98**
Total	37.83	100.0	86.19	74.34	95.01	97.70	39.27	70.96	80.86	77.56	85.47	**87.46**

The boldface type denotes the best performance.

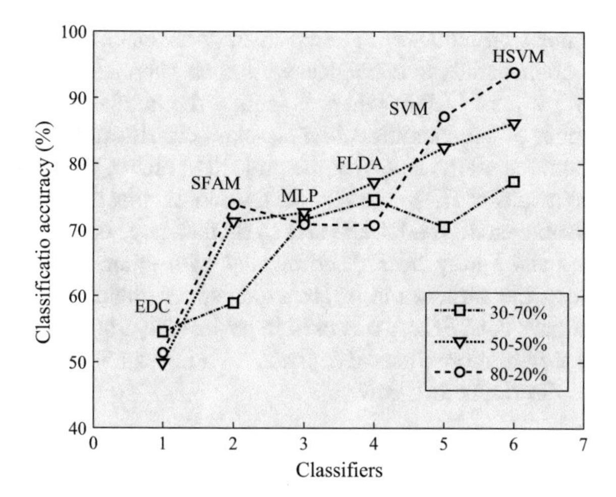

Fig. 6.2 Total classification accuracy on test samples with three proportions using six classifiers labeled as EDC, SFAM, MLP, FLDA, SVM, and HSVM (from number 1 to 6)

Table 6.4 Classification accuracy of the train and test sets without temperature and humidity integration

Class	Classification accuracy (%)											
	Train set						Test set					
	EDC	SFAM	MLP	FLDA	SVM	HSVM	EDC	SFAM	MLP	FLDA	SVM	HSVM
HCHO	21.63	100.0	88.94	61.54	90.53	94.37	28.84	86.53	94.06	69.23	92.47	**94.23**
C_6H_6	45.03	100.0	78.63	60.30	65.65	63.71	51.51	75.75	49.77	66.67	64.35	**69.69**
C_7H_8	75.47	100.0	92.45	94.34	98.11	100.0	38.46	84.61	96.15	100.0	96.09	**100.0**
CO	78.26	100.0	80.43	73.91	93.47	100.0	91.67	91.67	97.77	91.67	95.53	**100.0**
NH_3	56.52	100.0	73.91	78.26	78.26	87.00	60.00	84.00	92.89	88.00	87.94	**96.00**
NO_2	50.00	100.0	95.83	62.50	95.83	81.40	50.00	66.67	66.57	66.67	67.02	**100.0**
Mean	54.49	100.0	85.03	71.81	86.97	87.75	53.41	81.54	82.87	80.37	83.90	**93.32**
Total	42.26	100.0	85.36	66.80	84.60	86.25	46.81	82.97	82.83	76.60	84.60	**90.07**

The boldface type denotes the best performance.

temperature and humidity have a great influence on the sensitivity of metal oxide semiconductor sensors as the key environmental elements. Therefore, it is better to integrate both two variables in data treatments for classification. More experimental results can be referred to as [25].

6.4.2 Discussion

This chapter mainly investigated a multi-class problem of E-nose data using linear and nonlinear classification methods. In the hybrid HSVM discrimination model, the FLDA is developed as a pre-classification which uses a transformation matrix to reconstruct new patterns with more variables associated with each sub-classifier while not for dimension reduction referred in the previous study. Concretely, this work aims to obtain variables correlated with each sub-classifier through a projection matrix $\mathbf{W}_{6 \times 15}$ of FLDA, where 6 denotes the number of variables and 15 denotes the number of sub-classifiers in a six-class classification problem in terms of "one-against-one" strategy. Note that the projection matrix $\mathbf{W}$ should be obtained offline. From the results of PCA (see Fig. 6.1), we know that the E-nose data shows a linearly inseparable case. The HCHO and C_6H_6 data have completely overlapped in the data space and hardly been discriminated with other odorants. Due to the correlations among the variables in original data space, the classification task will also become difficult; thus, FLDA was used to project the original data space onto a new feature space with more linear independent variables related to each classifier and enhance the discriminatory power.

6.5 Summary

In this chapter, we studied the potential applicability of an E-nose in the classification of air contaminants indoor using different classification methods. Six classification models, such as the EDC, SFAM, MLP, individual FLDA, single SVM, and the HSVM, have been used for multi-class odor recognition. The experimental results demonstrate that the proposed HSVM shows the best classification performance compared with other classifiers.

References

1. M. Peris, L. Escuder-Gilabert, A 21st century technique for food control: electronics noses. Anal. Chim. Acta **638**, 1–15 (2009)
2. S.J. Dixon, R.G. Brereton, Comparison of performance of five common classifiers represented as boundary methods: euclidean distance to centroids, linear discriminant analysis, quadratic discriminant analysis, learning vector quantization and support vector machines, as dependent on data structure. Chemom. Intell. Lab. Syst. **95**, 1–17 (2009)
3. E. Llobet, E.L. Hines, J.W. Gardner, P.N. Bartlett, Fuzzy ARTMAP based electronic nose data anaylsis. Sens. Actuators, B Chem. **61**, 183–190 (1999)
4. Z. Xu, X. Shi, L. Wang, J. Luo, C.J. Zhong, S. Lu, Pattern recognition for sensor array signals using Fuzzy ARTMAP. Sens. Actuators, B Chem. **141**, 458–464 (2009)
5. P. Ciosek, W. Wroblewski, The analysis of sensor array data with various pattern recognition techniques. Sens. Actuators, B Chem. **114**, 85–93 (2006)
6. Q. Chen, J. Zhao, Z. Chen, H. Lin, D.A. Zhao, Discrimination of green tea quality using the electronic nose technique and the human panel test, comparison of linear and nonlinear classification tools. Sens. Actuators, B Chem. **159**, 294–300 (2011)
7. B. Debska, B. Guzowska-Swider, Application of artificial neural network in food classification. Anal. Chim. Acta **705**, 283–291 (2011)
8. W. Wu, Y. Mallet, B. Walczak, W. Penninckx, D.L. Massart, S. Heuerding, F. Erni, Comparison of regularized discriminant analysis, linear discriminant analysis and quadratic discriminant analysis, applied to NIR data. Anal. Chim. Acta **329**, 257–265 (1996)

9. K. Brudzewski, S. Osowski, T. Markiewicz, Classification of milk by means of an electronic nose and SVM neural network. Sens. Actuators, B Chem. **98**, 291–298 (2004)
10. L.H. Chiang, M.E. Kotanchek, A.K. Kordon, Fault diagnosis based on fisher discriminant analysis and support vector machines. Comp. Chem. Eng. **28**, 1389–1401 (2004)
11. G.A. Carpenter, S. Grossberg, N. Marcuzon, J.H. Reinolds, D.B. Rosen, Fuzzy ARTMAP: a neural network architecture for incremental supervised learning of analog multidimensional maps. IEEE. Trans. Neural. Netw. **3**, 698–713 (1992)
12. G.A. Carpenter, S. Grossberg, N. Marcuzon, D.B. Rosen, Fuzzy ART: fast stable learning and categorization of analogue patterns by an adaptive resonance system. Neural Netw. **4**, 759–771 (1991)
13. J.W. Gardner, E.L. Hines, M. Wilkinson, The application of artificial neural networks in an electronic nose. Meas. Sci. Technol. **1**, 446–451 (1990)
14. E. Llobet, J. Brezmes, X. Vilanova, J.E. Sueiras, X. Correig, Qualitative and quantitative analysis of volatile organic compounds using transient and steady-state responses of a thick film tin oxide gas sensor array. Sens. Actuators, B Chem. **41**, 13–21 (1997)
15. C. Maugis, G. Celeux, M.L. Martin-Magniette, Variable selection in model-based discriminant analysis. J. Multivar. Anal. **102**, 1374–1387 (2011)
16. V. Vapnik, *Statistical Learning Theory* (Wiley, New York, 1998)
17. V. Vapnik, *The Nature of Statistical Learning Theory* (Springer, New York, 1995)
18. H.L. Chen, D.Y. Liu, B. Yang, J. Liu, G. Wang, A new hybrid method based on local fisher discriminant analysis and support vector machines for hepatitis disease diagnosis. Expert Syst. Appl. **38**, 11796–11803 (2011)
19. C.K. Loo, A. Law, W.S. Lim, M.V.C. Rao, Probabilistic ensemble simplified fuzzy ARTMAP for sonar target differentiation. Neural Comput. Appl. **15**, 79–90 (2006)
20. S. Haykin, *Neural Networks, A Comprehensive Foundation* (Macmillan, New York, 2002)
21. L. Zhang, F.C. Tian, C. Kadri, B. Xiao, H. Li, L. Pan, H. Zhou, On-line sensor calibration transfer among electronic nose instruments for monitoring volatile organic chemicals in indoor air quality. Sens. Actuators, B Chem. **160**, 899–909 (2011)
22. C.W. Hsu, C.J. Lin, A comparison of methods for multiclass support vector machines. IEEE. Trans. Neural Netw. **13**, 415–425 (2002)
23. F. Sales, M.P. Callao, F.X. Rius, Multivariate standardization for correcting the ionic strength variation on potentiometric sensor arrays. Analyst **125**, 883–888 (2000)
24. J. Karhunen, Generalization of principal component analysis, optimization problems and neural networks. Neural Netw. **8**, 549–562 (1995)
25. L. Zhang et al., Classification of multiple indoor air contaminants by an electronic nose and a hybrid support vector machine. Sens. Actuators, B Chem. **174**, 114–125 (2012)

Chapter 7
Local Kernel Discriminant Analysis-Based Odor Recognition

Abstract This chapter proposes a new discriminant analysis framework (NDA) for dimension reduction and recognition. In the NDA, the between-class and the within-class Laplacian scatter matrices are designed from sample to sample, respectively, to characterize the between-class separability and the within-class compactness. Then, a discriminant projection matrix is solved by simultaneously maximizing the between-class Laplacian scatter and minimizing the within-class Laplacian scatter. Benefiting from the linear separability of the kernelized mapping space and the dimension reduction of principal component analysis (PCA), an effective kernel PCA plus NDA method (KNDA) is proposed for rapid detection of gas mixture components. In this chapter, the NDA framework is derived with specific implementations. Experimental results demonstrate the superiority of the proposed KNDA method in multi-class recognition.

Keywords Electronic nose · Discriminant analysis · Dimension reduction
Feature extraction · Multi-class recognition

7.1 Introduction

The classification methodologies have been widely studied in e-Nose applications. Artificial neural networks (ANNs), based on empirical risk minimization, are widely used for qualitative and quantification analysis, such as multilayer perceptron (MLP) neural network with back-propagation (BP) algorithm [1], RBF neural network [2], ARTMAP neural networks [3, 4]. Then, the decision tree method has also been proposed for classification [5]. Support vector machine (SVM) with complete theory, modeled with structural risk minimization, has also been widely studied in E-nose [6, 7]. The nonlinear mapping with different kernel functions (e.g., polynomial, Gaussian function) in SVM can make a linearly inseparable classification problem in original data space linearly separable in a high-dimensional feature space. Besides, SVM is used to solve a convex quadratic programming problem and can promise global optimum, rather than a local optimum of ANN.

L. Zhang et al., *Electronic Nose: Algorithmic Challenges*,
https://doi.org/10.1007/978-981-13-2167-2_7

Besides the classification methodologies, data preprocessing methods like feature extraction and dimension reduction methods [8–11] including principal component analysis (PCA), independent component analysis (ICA), kernel PCA (KPCA), linear discriminant analysis (LDA), singular value decomposition (SVD), etc., have also been combined with ANNs or SVMs for improving the prediction accuracy of e-Nose. Both feature extraction and dimension reduction aim to obtain useful features for classification. Dimension reduction can reduce the redundant information like de-noising but may lose some useful information in original data. In addition, classification method has also ability to automatically depress the useless components in samples learning process.

The methodologies of PCA, KPCA, LDA and the combination of KPCA and LDA have also wide application in many fields, such as time series forecasting, novelty detection, scene recognition, and face recognition as feature extraction and dimension reduction. Cao et al. employed a comparison of PCA, KPCA, and ICA combined with SVM for time series forecasting and find that KPCA has the best performance in feature extraction [12]. Xiao et al. proposed a L1 norm-based KPCA algorithm for novelty detection and obtained satisfactory effect using simulation dataset [13]. Hotta proposed a local feature acquisition method using KPCA for scene classification, and the performance is superior to conventional methods based on local correlation features [14]. Lu et al. [15] and Yang et al. [16] also proposed kernel direct discriminant analysis and KPCA plus LDA algorithms in face recognitions and given a complete kernel Fisher discriminant framework for feature extraction and recognition. Dixon et al. [17] presented a PLS-DA method used in gas chromatography mass spectrometry. A kernel PLS algorithm was also discussed in [18].

Inspired by these works, l_2-norm between each two sample vectors x_i and x_j in between-class and within-class is considered with a similarity matrix calculated by the Gaussian function $\exp\left(-\|x_i - x_j\|^2/t\right)$. Through the construction of the between-class Laplacian scatter matrix and within-class Laplacian scatter matrix, the new discriminant analysis (NDA) framework is realized by solving an optimization problem which makes the samples between-class more separable and the samples within-class more compactable. Considering the characteristic of linearly separable in high-dimensional kernel space, the Gaussian kernel function is introduced for mapping the original data space into a high-dimensional space. To make the within-class Laplacian scatter matrix nonsingular in the calculation of eigenvalue problem in which an inverse operation of the within-class Laplacian scatter matrix is necessary, PCA is used to reduce the dimension of the kernel space. The contribution of this chapter can be concluded as the proposed new discriminant analysis framework (KNDA) based on KPCA and its application in electronic nose for rapid detection of multiple kinds of pollutant gas components.

It is worthwhile to highlight several aspects of the proposed KNDA framework. First, in the NDA framework, each sample vector in between-class and within-class has been used for Laplacian scatter matrix, while in LDA, only the centroid of between-class and within-class is used to calculate the scatter matrix in which each sample's information cannot be well represented. Second, a similarity matrix by a Gaussian function is used to measure the importance of each two samples x_i and x_j with respect to their distance $\|x_i - x_j\|_2$ which is not considered in LDA. Third, the projection vector can be obtained by maximizing the between-class Laplacian scatter matrix and minimizing the within-class Laplacian scatter matrix. Fourth, the NDA is a supervised discriminant analysis framework, and KNDA is the combined learning framework of unsupervised KPCA and supervised NDA for feature extraction and recognition. Fifth, the recognition in this chapter is an intuitive Euclidean distance-based method and promising the stability and reliability of the results.

7.2 Related Work

7.2.1 PCA

PCA [19] is an unsupervised method in dimension reduction by projecting correlated variables into another orthogonal feature space, and thus, a group of new variables with the largest variance (global variance maximization) were obtained. The PC coefficients can be obtained by calculating the eigenvectors of the covariance matrix of the original dataset.

7.2.2 KPCA

KPCA does the PCA process in kernel space which introduces the advantage of high-dimension mapping of original data using kernel trick on the basis of PCA, in which the original input vectors are mapped to a high-dimensional feature space F. The mapping from the original data space to high-dimensional feature space can be represented by calculating the symmetrical kernel matrix of input training pattern vectors using a Gaussian kernel function shown as

$$K(x, x_i) = \exp\left(-\|x - x_i\|^2/\sigma^2\right) \tag{7.1}$$

where x and x_i denote the observation vectors, and σ^2 denotes the width of Gaussian that is commonly called kernel parameter.

In general, KPCA is to perform PCA algorithm in the high-dimensional feature space K and extract nonlinear feature. The size of the dimension depends on the number of training vectors. The PCA process of kernel matrix K is to perform the following eigenvalue operation

$$K \cdot \alpha = \lambda \cdot \alpha \tag{7.2}$$

where α denotes the set of eigenvectors corresponding to d eigenvalues and λ denotes the diagonal matrix (λ_{ii} is the eigenvalue, i = 1,..., d). The set of eigenvectors $\{\alpha|\alpha_i, i = 1, \ldots, r\}$ corresponding to the first r largest eigenvalues ordered in such a way $\lambda_1 > \lambda_2 > \cdots > \lambda_r$ is the kernel principal component coefficients (projection vectors). Therefore, the kernel PC scores can be obtained by multiplying the kernel matrix K by the PC coefficients.

7.2.3 LDA

LDA aims to maximize the ratio of between-class variance to the within-class variance in any particular dataset through a transformation vector w and therefore promise the maximum separability. Finally, a linear decision boundary between the two classes will be produced for classification.

To a binary classification (two classes), assume the dataset for the two classes to be X_1 and X_2, respectively. We write $X_1 = \{x_1^1, x_1^2, \ldots, x_1^{N_1}\}$, and $X_2 = \{x_2^1, x_2^2, \ldots, x_2^{N_2}\}$, N_1 and N_2 denote the numbers of column vectors for X_1 and X_2, and $\{x_i^j, i = 1, 2; j = 1, \ldots, N_i\}$ denotes the column vector (observation sample). Then, we set the total dataset Z in R^d as $Z = \{X_1, X_2\}$.

Then, the within-class scatter matrix S_{w} and between-class scatter matrix S_{b} can be represented as

$$S_{\mathrm{w}} = \sum_{i=1}^{2} \sum_{j=1}^{N_i} \left(x_i^j - \mu_i\right)\left(x_i^j - \mu_i\right)^{\mathrm{T}} \tag{7.3}$$

$$S_{\mathrm{b}} = \sum_{i=1}^{2} N_i \cdot (\mu_i - \bar{Z})(\mu_i - \bar{Z})^{\mathrm{T}} \tag{7.4}$$

where μ_i denotes the centroid of the i-th class, $\bar{Z}$ denotes the centroid of the total dataset Z, and symbol T denotes transpose.

If S_{w} is nonsingular, the optimal projection matrix w can be obtained by solving the following maximization problem

$$w = \operatorname{argmax} \frac{w^{\mathrm{T}} S_{\mathrm{b}} w}{w^{\mathrm{T}} S_{\mathrm{w}} w} \tag{7.5}$$

$\{w|w_i, i = 1, \ldots, m\}$ is the set of the eigenvectors of the $S_{\mathrm{w}}^{-1} S_{\mathrm{b}}$ corresponding to the m largest eigenvalues.

7.3 The Proposed Approach

7.3.1 NDA Framework

Let $x_1 = \left[x_1^1, x_1^2, \ldots, x_1^{N_1}\right]$ be the training set of class 1 with N_1 samples, and $x_2 = \left[x_2^1, x_2^2, \ldots, x_2^{N_2}\right]$ be the training set of class 2 with N_2 samples. The proposed NDA framework aims to find the best projection basis W from the training sets between class 1 and class 2 to transform the training set into low-dimensional feature spaces. To our knowledge, the two classes would be more separable if the between-class scatter matrix is larger and the within-class scatter becomes more compact. The proposed NDA is a supervised dimension reduction algorithm from sample to sample; thus, the similarity matrices A and B are introduced to construct the between-class Laplacian scatter and within-class Laplacian scatter. The similarity matrices A and B can be calculated as follows

$$A^{ij} = \exp\left(- \| x_1^i - x_2^j \|^2 / t\right), \quad i = 1, \ldots, N_1;\ j = 1, \ldots, N_2 \tag{7.6}$$

$$\begin{gathered} B_k^{ij} = \exp\left(-\left\|x_k^i - x_k^j\right\|^2 / t\right), \quad i = 1, \ldots, N_k; \\ j = 1, \ldots, N_k;\ k = 1, \ldots, c;\ c = 2 \end{gathered} \tag{7.7}$$

where t represents the width of Gaussian which is an empirical parameter. In this chapter, $t = 100$.

Therefore, from the viewpoint of classification, we aim to maximize the ratio of the between-class Laplacian scatter matrix $J_1(W)$ and the within-class Laplacian scatter matrix $J_2(W)$. The specific algorithm derivation of the proposed NDA framework is shown as follows.

The between-class Laplacian scatter matrix can be represented as

$$\begin{aligned} J_1(W) &= \frac{1}{N_1 \cdot N_2} \sum_{i=1}^{N_1} \sum_{j=1}^{N_2} \left\| W^{\mathrm{T}} x_1^i - W^{\mathrm{T}} x_2^j \right\|^2 A^{ij} \\ &= \frac{1}{N_1 \cdot N_2} \sum_{i=1}^{N_1} \sum_{j=1}^{N_2} tr\left[\left(W^{\mathrm{T}} x_1^i - W^{\mathrm{T}} x_2^j\right)\left(W^{\mathrm{T}} x_1^i - W^{\mathrm{T}} x_2^j\right)^{\mathrm{T}}\right] A^{ij} \\ &= \frac{1}{N_1 \cdot N_2} \sum_{i=1}^{N_1} \sum_{j=1}^{N_2} tr\left[W^{\mathrm{T}}\left(x_1^i - x_2^j\right)\left(x_1^i - x_2^j\right)^{\mathrm{T}} W\right] A^{ij} \\ &= \frac{1}{N_1 \cdot N_2} \sum_{i=1}^{N_1} \sum_{j=1}^{N_2} tr\left\{W^{\mathrm{T}}\left[\left(x_1^i - x_2^j\right)\left(x_1^i - x_2^j\right)^{\mathrm{T}} A^{ij}\right] W\right\} \\ &= tr\left\{W^{\mathrm{T}}\left[\frac{1}{N_1 \cdot N_2} \sum_{i=1}^{N_1} \sum_{j=1}^{N_2} \left(x_1^i - x_2^j\right)\left(x_1^i - x_2^j\right)^{\mathrm{T}} A^{ij}\right] W\right\} \end{aligned} \tag{7.8}$$

Now, we let

$$H_1 = \frac{1}{N_1 \cdot N_2} \sum_{i=1}^{N_1} \sum_{j=1}^{N_2} \left(x_1^i - x_2^j\right)\left(x_1^i - x_2^j\right)^{\mathrm{T}} A^{ij} \tag{7.9}$$

Then, we get $J_1(W) = tr(W^{\mathrm{T}} H_1 W)$.

Similarly, the within-class Laplacian scatter matrix can be represented as

$$\begin{aligned} J_2(W) &= \sum_{k=1}^{c} \frac{1}{N_k^2} \sum_{i=1}^{N_k} \sum_{j=1}^{N_k} \left\| W^{\mathrm{T}} x_k^i - W^{\mathrm{T}} x_k^j \right\|^2 B_k^{ij} \\ &= \sum_{k=1}^{c} \frac{1}{N_k^2} \sum_{i=1}^{N_k} \sum_{j=1}^{N_k} tr\left[\left(W^{\mathrm{T}} x_k^i - W^{\mathrm{T}} x_k^j\right)\left(W^{\mathrm{T}} x_k^i - W^{\mathrm{T}} x_k^j\right)^{\mathrm{T}}\right] B_k^{ij} \\ &= \sum_{k=1}^{c} \frac{1}{N_k^2} \sum_{i=1}^{N_k} \sum_{j=1}^{N_k} tr\left\{W^{\mathrm{T}}\left[\left(x_k^i - x_k^j\right)\left(x_k^i - x_k^j\right)^{\mathrm{T}} B_k^{ij}\right] W\right\} \\ &= tr\left\{W^{\mathrm{T}}\left[\sum_{k=1}^{c} \frac{1}{N_k^2} \sum_{i=1}^{N_k} \sum_{j=1}^{N_k} \left(x_k^i - x_k^j\right)\left(x_k^i - x_k^j\right)^{\mathrm{T}} B_k^{ij}\right] W\right\} \end{aligned} \tag{7.10}$$

We let

$$H_2 = \sum_{k=1}^{c} \frac{1}{N_k^2} \sum_{i=1}^{N_k} \sum_{j=1}^{N_k} \left(x_k^i - x_k^j\right)\left(x_k^i - x_k^j\right)^{\mathrm{T}} B_k^{ij} \tag{7.11}$$

Then, we have $J_2(W) = tr(W^{\mathrm{T}} H_2 W)$.

In this chapter, the algorithm aims to solve a two-class problem, that is, $c = 2$. Therefore, we can rewrite the H_2 as

$$\begin{aligned} H_2 &= \sum_{k=1}^{2} \frac{1}{N_k^2} \sum_{i=1}^{N_k} \sum_{j=1}^{N_k} \left(x_k^i - x_k^j\right)\left(x_k^i - x_k^j\right)^{\mathrm{T}} B_k^{ij} \\ &= \frac{1}{N_1^2} \sum_{i=1}^{N_1} \sum_{j=1}^{N_1} \left(x_1^i - x_1^j\right)\left(x_1^i - x_1^j\right)^{\mathrm{T}} B_1^{ij} + \frac{1}{N_2^2} \sum_{i=1}^{N_2} \sum_{j=1}^{N_2} \left(x_2^i - x_2^j\right)\left(x_2^i - x_2^j\right)^{\mathrm{T}} B_2^{ij} \end{aligned} \tag{7.12}$$

From the angle of classification, to make class 1 and class 2 more separable, we formulate the discriminative analysis model as the following optimization problem

$$\max J(W) = \max \frac{J_1(W)}{J_2(W)} = \max \frac{W^{\mathrm{T}} H_1 W}{W^{\mathrm{T}} H_2 W} \tag{7.13}$$

H_1 and H_2 have been derived in analysis; thus, we can find the projection basis W by solving the following eigenvalue problem

$$H_1\varphi = \lambda H_2\varphi \tag{7.14}$$

Then, the optimization problem in Eq. 7.13 can be transformed into the following maximization problem

$$\max \frac{\varphi_i^{\mathrm{T}} H_1 \varphi_i}{\varphi_i^{\mathrm{T}} H_2 \varphi_i} \tag{7.15}$$

According to $H_1\varphi = \lambda H_2\varphi$, we have

$$H_1\varphi_1 = \lambda_1 H_2\varphi_1, H_1\varphi_2 = \lambda_2 H_2\varphi_2, \ldots, H_1\varphi_i = \lambda_i H_2\varphi_i \tag{7.16}$$

Then, the maximization problem (7.15) can be solved as

$$\max \frac{\varphi_i^{\mathrm{T}} H_1 \varphi_i}{\varphi_i^{\mathrm{T}} H_2 \varphi_i} = \max \frac{\lambda_i \varphi_i^{\mathrm{T}} H_2 \varphi_i}{\varphi_i^{\mathrm{T}} H_2 \varphi_i} = \max \lambda_i \tag{7.17}$$

Let φ_1 be the eigenvector corresponding to the largest eigenvalue $\lambda_1(\lambda_1 > \lambda_2 > \cdots > \lambda_d)$, then the optimal projection basis W between class 1 and class 2 can be represented as $W = \varphi_1$.

7.3.2 The KPCA Plus NDA Algorithm (KNDA)

In this chapter, the KPCA method is combined with the proposed NDA framework for feature extraction and recognition in e-Nose application. It is worthwhile to highlight the two reasons of KPCA in this work. First, the introduction of kernel function mapping is on the basis of the consideration that in a high-dimensional kernel space, the patterns would become more separable linearly than the original data space. Second, the PCA is used for dimension reduction of the kernel data space and makes the number of variables less than the number of training samples so that we can guarantee that the within-class Laplacian scatter matrix H_2 in Eq. 7.12 to be nonsingular in NDA framework.

The pseudocodes of the KNDA training algorithm have been described in Algorithm 7.1.

The proposed NDA framework considers the two-class condition. To a multi-class (k classes, $k > 2$) problem, the NDA can also be useful by decomposing the multi-class problem into multiple two-class (binary) problems. Generally, "one-against-all (OAA)" and "one-against-one (OAO)" are often used in classification [20]. A study in [21] demonstrates that the OAO strategy would be a better choice in the case of $k \le 10$, while this chapter studies the discrimination of $k = 6$ kinds of pollutant gases. Therefore, $k(k - 1)/2 = 15$ NDA models are designed in this chapter.

7.3.3 Multi-class Recognition

Given a test sample vector z, we first transform the sample vector z through the optimal projection vector W_i obtained using the proposed NDA method as follows

$$z' = W_i^{\mathrm{T}} z \tag{7.18}$$

The recognition of z in a two-class (class 1 and class 2) problem can be employed in terms of the Euclidean distance 2-norm calculated as follows

$$\text{if } \left\| z' - \mu_i^1 \right\|_2 > \left\| z' - \mu_i^2 \right\|_2, \quad z \in \{\text{class 1}\}; \text{ else}, z \in \{\text{class 2}\} \tag{7.19}$$

where symbol $\| \cdot \|_2$ denotes 2-norm, and μ_i^1 and μ_i^2 denote the centroid of class 1 and class 2 in the i-model, respectively.

For a multi-class recognition, a majority voting mechanism in decision level is used based on the OAO strategy. The statics of vote number V_j for class j can be shown by

$$V_j = \sum_{i=1}^{k \cdot (k-1)/2} I\left(c_i = T_j\right), \quad j = 1, \ldots, k \tag{7.20}$$

where $I(\cdot)$ denotes the binary indicator function, c_i denotes the predicted label of the i-th sub-classifier, and T_j denotes the true label of class j. The class label of class j with the largest vote number $\max V_j$ is the discriminated class of the test sample vector. The pseudocodes of the proposed KNDA recognition (testing) process have been described in Algorithm 7.2.

The diagram of the proposed KNDA electronic nose recognition method has been illustrated in Fig. 7.1, wherein two parts are included: KNDA training and KNDA recognition. The specific implementations of KNDA training and KNDA recognition have been illustrated in Algorithm 7.1 and Algorithm 7.2, respectively. All the algorithms in this chapter are implemented in the platform of MATLAB software (version 7.8) on a laptop with Intel i5 CPU and 2 GB RAM.

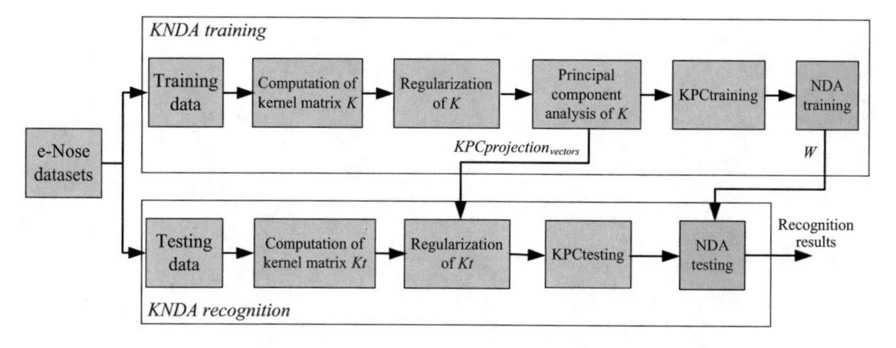

Fig. 7.1 Diagram of the proposed classification method

Algorithm 17.1. KNDA Training
Input: The training sets $x = [x_1, x_2, \ldots, x_m]$ of k classes, the parameter t, the kernel parameter σ of kernel space mapping, and the threshold of accumulated contribution rate of the kernel principal components.
Output: Projection basis W_i, $i = 1,\ldots, k(k-1)/2$, and the centroid set μ of $k \cdot (k-1)/2$ models.

Step 1 (kernel function mapping of the training sets):

1.1. Computation of the symmetrical kernel matrix $K_{m \times m}$ of the training sets x using Eq. 7.1.
1.2. Regularization of the kernel matrix K.

1.1.1. Centralization of K by using $K = K - \frac{1}{m}I \cdot K - \frac{1}{m}K \cdot I + \frac{1}{m}I \cdot K \cdot I$
1.1.2. Normalization by using $K = K/m$

Step 2 (kernel principal components analysis):

2.1. Eigenvalue decomposition of K by equation $K \cdot V = \lambda \cdot V$, where λ and V denote eigenvalues and eigenvectors, respectively.
2.2. Sort the eigenvalues in descending order $\lambda_1 > \lambda_1 > \cdots > \lambda_m$, and the sorted new_eigenvectors.
2.3. Calculate the accumulated contribution rate (ACR) which is shown by

$$\mathrm{ACR}_j = \frac{\sum_{k=1}^{j} \lambda_k}{\sum_{i=1}^{m} \lambda_i} \times 100, \quad j = 1, \ldots, m$$

2.4. Determine the number j of kernel principal components by using

$$j = \operatorname{argmin}\{\mathrm{ACR}_j \geq \text{threshold}\}$$

2.5. The $\mathrm{KPCprojection}_{\mathrm{vectors}}$ for kernel principal components projection is determined as

$$\text{KPCprojection}_{\text{vectors}} = \{\text{new\_eigenvectors}_i,\ i = 1,\ldots,j\}$$

2.6. Calculate the kernel principal components $\text{KernelPC}_{\text{training}}$ of the training sets by using

$$\text{KernelPC}_{\text{training}} = K \times \text{KPCprojection}_{\text{vectors}}$$

Step 3 (NDA framework):

For $i = 1, 2, \ldots, k(k-1)/2$, **repeat**

3.1. Take the training vectors of the i-th pair of classes from $\text{KernelPC}_{\text{training}}$, and calculate the between-class similarity matrix A and within-class similarity matrix B according to Eq. 7.6 and 7.7, respectively,
3.2. Calculate the between-class Laplacian scatter matrix H_1 and within-class Laplacian scatter matrix H_2 as shown in Eq. 7.9 and 7.11, respectively.
3.3. Solve the eigenvalue problem in Eq. 7.14 and get the eigenvector φ_1 corresponding to the largest eigenvalue.
3.4. Obtain the projection basis $W_i = \varphi_1$.
3.5. Calculate the i-th centroid pair $\mu_i = [\mu_i^1, \mu_i^2]$ of class 1 and class 2 in the i-th model.

End for

Step 4 (output the low-dimensional projection basis matrix *W*):
Output the projection basis matrix $W = \{W_i, i = 1,\ldots,k(k-1)/2\}$ and the centroid set $\mu = \{\mu_i, i = 1,\ldots,k(k-1)/2\}$.

Algorithm 2. KNDA Testing Input: The testing sets $z = [z_1, z_2, \cdots, z_n]$ of k classes, the kernel parameter σ of kernel space mapping, the $\text{KPCprojection}_{\text{vectors}}$, the projection basis matrix W, and the centroid $\mu = \{\mu_i^1, \mu_i^2\}$, $i = 1, \ldots, k(k-1)/2$ obtained in the KNDA training process.
Output: The predicted labels of testing samples.

Step 1 (kernel function mapping of the testing sets):

1.1. Computation of the symmetrical kernel matrix $Kt_{n\times n}$ of the testing sets z using Eq. 7.1.
1.2. Regularization of the kernel matrix Kt.

1.1.1. Centralization of Kt by using $Kt = Kt - \frac{1}{n}I \cdot Kt - \frac{1}{n}Kt \cdot I + \frac{1}{n}I \cdot Kt \cdot I$
1.1.2. Normalization by using $Kt = Kt/n$

Step 2 (kernel principal components projection of the testing sets):
Calculate the kernel principal components $\text{KernelPC}_{\text{testing}}$ of testing vectors

$$\text{KernelPC}_{\text{testing}} = Kt \times \text{KPCprojection}_{\text{vectors}}$$

Step 3 (multi-class recognition):

For $p = 1, 2, \ldots, n$, **repeat**

For $i = 1, \ldots, k(k-1)/2$, **repeat**

3.1. Low-dimensional projection of NDA.

$$\text{LowDim}^{i,p}_{\text{projection}} = W_i^{\text{T}} \cdot \text{KernelPC}^{p}_{\text{testing}}$$

3.2. Calculate the Euclidean distance between $\text{LowDim}^{i,p}_{\text{projection}}$ and the centroid μ_i, and discriminate the label c_i of the p-th sample in the i-th classifier according to Eq. 7.19.

End for

3.3. Compute the vote number V_j for class j of the p-th sample according to Eq. 7.20.

3.4. Predict the label of the p-th sample as $j = \arg\max V_j$.

End for

Step 4 (output the predicted labels of testing samples):
Output the predicted labels of the n testing samples.

7.4 Experiments

7.4.1 E-Nose System

Consider the selectivity, stability, reproducibility, sensitivity, and low cost of metal oxide semiconductor (MOS) gas sensors; our sensor array in e-Nose system consists of four metal oxide semiconductor gas sensors including TGS2602, TGS2620, TGS2201A, and TGS2201B. Moreover, a module with two auxiliary sensors for the temperature and humidity measurement is also used with the consideration that MOS gas sensors are sensitivity to environmental temperature and humidity. Therefore, six variables are contained in each observation. A 12-bit analog–digital converter is used as the interface between FPGA processor and sensor array for convenient digital signal processing. The e-Nose system can be connected to a PC via a Joint Test Action Group (JTAG) port for data storage and debugging programs. The e-Nose system and the experimental platform are illustrated in Chap. 3 in which the typical response of gas sensors in the sampling process (1. baseline, 2. transient response, 3. steady state response, 4. recover process) is also presented.

7.4.2 Dataset

In this chapter, the experiments were employed in a constant temperature and humidity chamber in a condition of room temperature (15–35 °C). The experimental process for each gas is similar in which three main steps are included. First, set the target temperature and humidity and collect the sensor baseline for 2 min. Second, inject the target gas by using a flowmeter with time controlled, and collect the steady state response of sensors for 8 min. Third, clean the chamber by air exhaust for 10 min and read the data for the sample by a laptop connected with the electronic nose through a JTAG.

For more information about all the samples, we have described the experimental temperature, relative humidity, and concentration for each sample of each gas in supplementary data. The number of formaldehyde, benzene, toluene, carbon monoxide, ammonia, and nitrogen dioxide samples are 188, 72, 66, 58, 60, and 38, respectively. In each sample, six variables (with six sensing units) are contained. All the experimental samples were obtained within two months by employing the e-Nose experiments continuously.

To determine the training sample index, we introduce the Kennard–Stone sequential (KSS) algorithm [22] based on Euclidean distance to select the most representative samples in the whole sample space for each gas. The selection starts by taking the pair of sample vectors (p_1, p_2) with the largest distance $d(p_1, p_2)$ among the samples for each gas. KSS follows a stepwise procedure that new selections are taken which would be the farthest from the samples already selected, until the number of training samples for each gas reaches. In this way, the most representative samples for each gas can be selected as training samples and guarantee the reliability of the learned model. The merit of KSS for training samples selection is to reduce the complexity of cross-validation in the performance evaluation. The remaining samples without being selected would be used for model testing. The specific number of training and testing samples after KSS selection for each gas is illustrated in Table 7.1.

Table 7.1 Statistic of the experimental data

Data	Formaldehyde	Benzene	Toluene	Carbon monoxide	Ammonia	Nitrogen dioxide
Training	125	48	44	38	40	25
Testing	63	24	22	20	20	13
Total	188	72	66	58	60	38

7.5 Results and Discussion

7.5.1 Contribution Rate Analysis

In the classification model, the kernel parameters σ^2 and the accumulated contribution rate (ACR) in KPCA are related to the actual classification performance. In experiments, six values {5, 6, 7, 8, 9, 10} of σ^2 and five values {95%, 96%, 97%, 98%, 99%} of the CR (the threshold of ACR) are selected for study and comparison, because these values have more positive effects in classifications than other values. Therefore, we do not use special optimizations to search the best parameters of KPCA. Totally, 30 kinds of combinations of (σ^2, CR) are studied in classification.

For KPCA analysis, we perform the KPCA algorithm on the total training samples (320 samples) of all gases. The size of the kernel matrix K should be 320 multiply 320. Table 7.2 presents the contribution rate analysis of KPCA including the number of kernel principal components (KPCs) with their ACR lower than the threshold CR. We can see that the number of KPCs (dimension) with ACR $<$ 95%, ACR $<$ 96%, ACR $<$ 97%, ACR $<$ 98%, and ACR $<$ 99% is 47, 53, 61, 73, and 95, respectively. From the contribution rate analysis, we can find that about 99% information can be obtained by the first 95 principal components which is much lower than 320, and about 95% information is obtained by only the first 47 principal components.

7.5.2 Comparisons

The classification performance of the proposed KNDA method can be shown by the average recognition rates and the total recognition rates of six kinds of chemicals. The average recognition rate denotes the average value of six recognition rates for

Table 7.2 Contribution rate analysis of KPCA

Threshold CR (%)	<95	<96	<97	<98	<99
Number of KPCs (dimension)	47	53	61	73	95
ACR (%)	94.85	95.91	96.96	97.96	98.99

six chemicals, and it can validate the balance of multi-class recognition. The low average recognition rate demonstrates that at least one class failed in recognition. The total recognition rate represents the ratio between the number of correctly recognized samples for six chemicals and the number of total testing samples of six chemicals.

To study the kernel parameter σ^2 and the CR mentioned in KPCA, an empirical way such that six values {5, 6, 7, 8, 9, 10} of σ^2 and five values {95%, 96%, 97%, 98%, 99%} of the CR (the threshold of ACR) are selected for study and comparison of KNDA, KLDA, KPLS-DA, and KSVM. In experiments, we find that the best classification performance of KNDA is in the case of $\sigma^2 = 7$ and CR = 97% with the average recognition rate as 94.14% and the total recognition rate as 95.06%. In KPCA plus LDA (KLDA), we can find that the best performance with average recognition rate as 92.94% and total recognition rate as 94.44% is obtained in the case of $\sigma^2 = 6$ and CR = 95%. In KPLS-DA, the best performance of the average recognition rate and total recognition rate is 89.96 and 93.21% in the case of $\sigma^2 = 6$ and CR = 99%. Through the comparison of the best classification performance, we can see that the proposed KNDA has a better performance than KLDA in feature extraction for multi-class recognition. Moreover, we have implemented the KPCA plus SVM (KSVM) method for multi-class classification. In experiments, we can find that the only one case with $\sigma^2 = 10$ and CR = 98% by using KSVM has an average recognition rate 92.97% that is higher than 90%, and the corresponding total recognition rate is 95.06%. Seen from the results, SVM has an equal performance with the proposed KNDA in total recognition rate, while the proposed KNDA has a better performance from the average recognition rate. The average recognition rate demonstrates that the proposed KNDA has a better balance of recognition than SVM.

For details of the results, we present variation curves of the total recognition rate and the average recognition rate of four kernel methods (KSVM, KLDA, KPLS-DA, and KNDA) in Figs. 7.2 and 7.3, respectively, with $\sigma^2 = 5, 6, \ldots, 10$ and the ACR changes from 50 to 99% for each σ^2. The relation between the kernel parameters σ^2 and ACR of KPCA and the classification performance can be shown. From Figs. 7.2 and 7.3, we can find that the KSVM performs better in low ACR, and the KLDA, KPLS-DA, and KNDA perform as good as KSVM with the increasing of ACR. When the ACR reaches 97%, the KNDA performs the best among KSVM, KLDA, and KPLS-DA. In contrast, KPLS-DA shows the worst performance among the four kernel-based methods. It is worth noting that the 97% is an inflection point, and there is an obvious reduction of recognition rate when ACR is 98 and 99% for KLDA and KNDA. This may be explained that the most useful information of the original data is the 97% principal components, and the remaining 3% is the redundant information which is not useful for recognition. We can see that the proposed KNDA is sensitive to the noise which is similar to KLDA, while SVM is a nonlinear classification model and noise insensitive. However, considering that smaller number of principal components is also desirable on the basis of the high recognition rate, the proposed KNDA framework is superior to SVM-based methods.

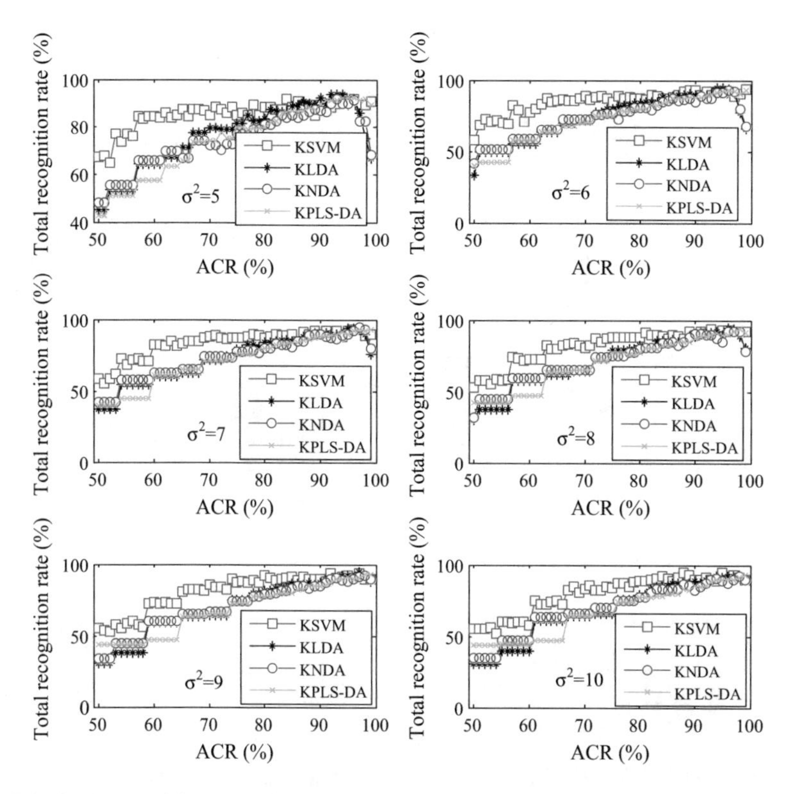

Fig. 7.2 Total recognition rates of all gases for four kernel trick-based methods

For completeness of comparisons, we have also studied the original SVM (SVM), PCA-based SVM (PCA-SVM), original LDA (LDA), PCA-based LDA (PCA-LDA), PLS-DA, kernel PLS-DA (KPLS-DA), and the proposed NDA and KNDA methods. The best classification performance of each method has been presented in Table 7.3 which shows the recognition rate for each gas, the average recognition rate, and the total recognition rate. It is worthwhile noting that several facets should be highlighted in Table 7.3. First, from the comparison of LDA and the proposed NDA framework, NDA has higher recognition rates than LDA. Second, as seen from the results of KSVM, the recognition rates of benzene and nitrogen dioxide are 87.50 and 76.92%, respectively. While the recognition rates of benzene and nitrogen dioxide are 100 and 84.62% which have been much improved by using KNDA. Thus, the average recognition rate of KNDA is higher than KSVM that also demonstrates that the proposed KNDA can effectively improve the imbalance of E-nose data in SVM classification. Note that no specific method is used to improve the imbalance of experimental samples for each class in this work. Third, it is worth noting that both PLS-DA and KPLS-DA recognition methods have the same flaw of over-training as PLS, due to the unpredictable number of

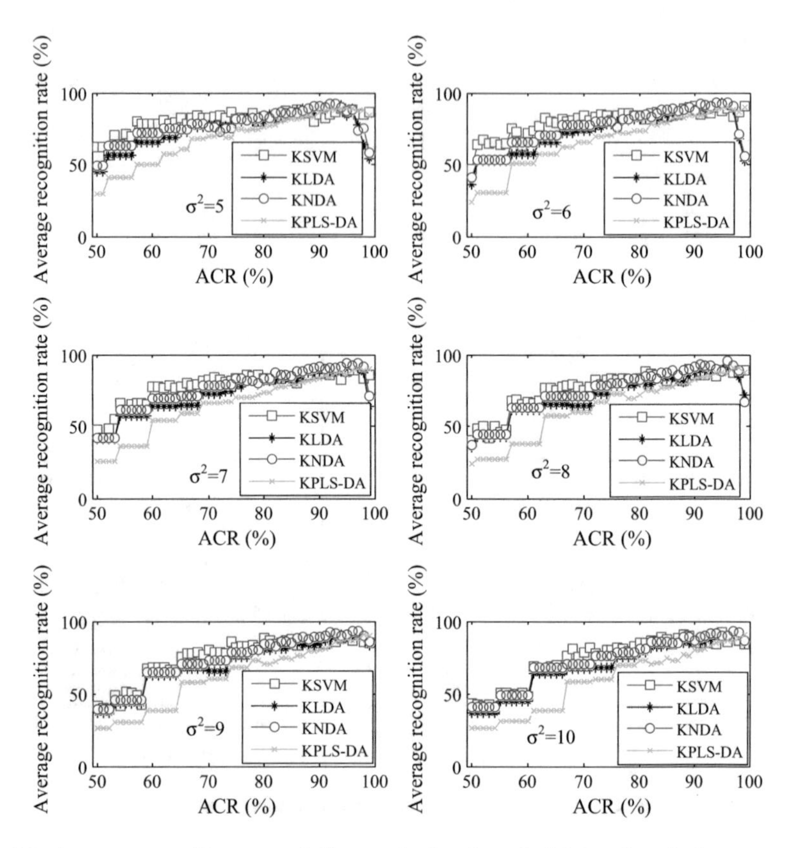

Fig. 7.3 Average recognition rates of all gases for four kernel trick-based methods

Table 7.3 Comparisons of classification accuracies with state-of-the-art methods

Methods	Recognition accuracy (%)							
	HCHO	C_6H_6	C_7H_8	CO	NH_3	NO_2	Average	Total
SVM	98.41	79.17	100.0	100.0	90.00	69.23	89.47	92.59
PCA-SVM	98.41	91.67	100.0	65.00	100.0	30.77	80.97	88.27
KSVM	98.41	87.50	100.0	100.0	95.00	76.92	92.97	95.06
LDA	88.89	66.67	90.91	100.0	90.00	30.77	77.87	82.72
PCA-LDA	82.54	58.33	86.36	90.00	90.00	30.77	73.00	77.16
PLS-DA	93.65	45.83	68.18	75.00	70.00	23.08	62.62	72.22
NDA	87.30	66.67	100.0	100.0	95.00	30.77	79.96	83.95
KLDA	95.24	100.0	95.45	95.00	95.00	76.92	92.94	94.44
KPLS-DA	98.41	91.67	95.45	95.00	90.00	69.23	89.96	93.21
KNDA	95.24	100.0	100.0	95.00	90.00	84.62	94.14	95.06

components in regression. However, to present the best results of each method for fair comparison, the results of PLS-DA and KPLS-DA in this chapter may be over-learned.

In E-nose, the generalization capability is a very important factor in a system. The KNDA is actually a feature extraction method, and the recognition is based on an intuitive Euclidean distance method. Instead, SVM recognition is to solve an optimization problem. Therefore, the generalization capability of SVM in recognition depends on the optimization effect including the parameter selection in training. That is, the recognition results would also be different with different SVM parameter selection or encountered with overfitting. In general, the recognition results of KNDA should be more stable and reliable.

7.5.3 Computational Efficiency

From the theories of these methods, the proposed NDA framework belongs to a linear discrimination, while KNDA introduces the kernel PCA in the proposed NDA framework. SVM is a nonlinear classification, and SVM aims to solve a convex quadratic programming problem. Though SVM has been widely studied for its complete theory in mathematics, SVM-based classifiers have also a large computational burden which is related with the number of support vectors. For analysis of the computational efficiency of each method, the average running time including the training time and recognition time of each method for 10 times has been presented in Table 7.4. The consumed time in multi-class classification using SVM-based classifiers (SVM, PCA-SVM, and KSVM) is generally more than 30 s, the LDA- and PLS-based methods (LDA, PCA-LDA, KLDA, PLS-DA, and KPLS-DA) take less than 1 s, and the proposed NDA and KNDA methods take 1.811 and 2.221 s, respectively. Though the proposed NDA framework has a little

Table 7.4 Comparison of algorithms' running time (in seconds)

Methods	Training	Recognition	Total time
SVM	33.0	0.310	33.31
PCA-SVM	35.0	0.352	35.35
KSVM	40.0	0.620	40.62
LDA	0.144	0.011	0.155
PCA-LDA	0.194	0.011	0.205
PLS-DA	0.037	0.026	0.063
NDA	1.801	0.010	1.811
KLDA	0.425	0.092	0.517
KPLS-DA	0.280	0.091	0.371
KNDA	2.119	0.102	2.221

higher computational efficiency than LDA, from the angle of synthesized consideration of the recognition accuracy and computational efficiency, the proposed KNDA is more acceptable in a real application in terms of its best performance among state-of-the-art methods.

7.6 Summary

This chapter presents a KNDA method with between-class separability and within-class compactness for multi-class classification. In KNDA, the KPCA contains high-dimensional kernel space mapping and principal component analysis. There are two merits: first, the samples between classes become linearly separable in the high-dimensional kernel space; second, the PCA is used to extract the most important information, reduce the dimension of the kernel space and guarantee the within-class Laplacian scatter matrix nonsingular in NDA training. Extensive experiments show that the proposed KNDA outperforms other methods in recognition performance and computational efficiency.

References

1. L. Zhang, F. Tian, C. Kadri, G. Pei, H. Li, L. Pan, Gases concentration estimation using heuristics and bio-inspired optimization models for experimental chemical electronic nose. Sens. Actuators, B **160**, 760–770 (2011)
2. Z. Ali, D. James, W.T. O'Hare, F.J. Rowell, S.M. Scott, Radial basis neural network for the classification of fresh edible oils using an electronic nose. J. Therm. Anal. Calorim. **71**, 147–154 (2003)
3. E. Llobet, E.L. Hines, J.W. Gardner, P.N. Bartlett, T.T. Mottram, Fuzzy ARTMAP based electronic nose data analysis. Sensors and Actuators B **61**, 183–190 (1999)
4. Z. Xu, X. Shi, L. Wang, J. Luo, C.J. Zhong, S. Lu, Pattern recognition for sensor array signals using Fuzzy ARTMAP. Sens. Actuators, B **141**, 458–464 (2009)
5. J.H. Cho, P.U. Kurup, Decision tree approach for classification and dimensionality reduction of electronic nose data. Sens. Actuators, B **160**, 542–548 (2011)
6. K. Brudzewski, S. Osowski, T. Markiewicz, J. Ulaczyk, Classification of gasoline with supplement of bio-products by means of an electronic nose and SVM neural network. Sens. Actuators, B **113**, 135–141 (2006)
7. M. Pardo, G. Sberveglieri, Classification of electronic nose data with support vector machines. Sens. Actuators, B **107**, 730–737 (2005)
8. B. Ehret, K. Safenreiter, F. Lorenz, J. Biermann, A new feature extraction method for odour classification. Sens. Actuators, B **158**, 75–88 (2011)
9. Y.G. Martín, J.L.P. Pavón, B.M. Cordero, C.G. Pinto, Classification of vegetable oils by linear discriminant analysis of electronic nose data. Anal. Chim. Acta **384**, 83–94 (1999)
10. L. Nanni, A. Lumini, Orthogonal linear discriminant analysis and feature selection for micro-array data classification. Expert Syst. Appl. **37**, 7132–7137 (2010)
11. S.K. Jha, R.D.S. Yadava, Denoising by singular value decomposition and its application to electronic nose data processing. IEEE Sens. J. **11**, 35–44 (2011)

12. L.J. Cao, K.S. Chua, W.K. Chong, H.P. Lee, Q.M. Gu, A comparison of PCA, KPCA and ICA for dimensionality reduction in support vector machine. Neurocomputing **55**(1–2), 321–336 (2003)
13. Y. Xiao, H. Wang, W. Xu, J. Zhou, L1 norm based KPCA for novelty detection. Pattern Recogn. **46**(1), 389–396 (2013)
14. K. Hotta, Local co-occurrence features in subspace obtained by KPCA of local blob visual words for scene classification. Pattern Recogn. **45**(10), 3687–3694 (2012)
15. J. Lu, K.N. Plataniotis, A.N. Venetsanopoulos, Face recognition using kernel direct discriminant analysis algorithms. IEEE Trans. Neural Netw. **14**(1), 117–126 (2003)
16. J. Yang, A.F. Frangi, J.Y. Yang, D. Zhang, J. Zhong, KPCA plus LDA: a complete kernel Fisher discriminant frame work for feature extraction and recognition. IEEE Trans. Pattern Anal. Mach. Intell. **27**(2), 230–244 (2005)
17. S.J. Dixon, Y. Xu, R.G. Brereton, H.A. Soini, M.V. Novotny, E. Oberzaucher, K. Grammer, D.J. Penn, Pattern recognition of gas chromatography mass spectrometry of human volatiles in sweat to distinguish the sex of subjects and determine potential discriminatory marker peaks. Chemometr. Intell. Lab. Syst. **87**(2), 161–172 (2007)
18. F. Lindgren, P. Geladi, S. Wold, The kernel algorithm for PLS. J. Chemom. **7**(1), 45–59 (1993)
19. J. Karhunen, Generalization of principal component analysis, optimization problems and neural networks. Neural Netw. **8**, 549–562 (1995)
20. C.W. Hsu, C.J. Lin, A comparison of methods for multiclass support vector machine. IEEE Trans. Neural Netw. **13**, 415–425 (2002)
21. L.H. Chiang, M.E. Kotanchek, A.K. Kordon, Fault diagnosis based on Fisher discriminant analysis and support vector machines. Comput. Chem. Eng. **28**, 1389–1401 (2004)
22. F. Sales, M.P. Callao, F.X. Rius, Multivariate standardization for correcting the ionic strength variation on potentiometric sensor arrays. Analyst **125**, 883–888 (2000)

Chapter 8
Ensemble of Classifiers for Robust Recognition

Abstract This chapter presents a novel multiple classifiers system called as improved support vector machine ensemble (ISVMEN), which solves a multi-class recognition problem in E-nose and aims to improve classification accuracy and robustness. The contributions of this chapter are presented in two aspects: First, in order to improve the accuracy of base classifiers, kernel principal component analysis (KPCA) method is used for nonlinear feature extraction; second, in the process of establishing classifiers ensemble, a new fusion approach conducting an effective base classifier weighted method is proposed. Experimental results show that the average classification accuracy of the proposed method has been significantly improved over base classifiers and majority voting-based fusion strategy.

Keywords Classifiers ensemble · Electronic nose · Multi-class recognition
Feature extraction · Support vector machine

8.1 Introduction

The recognition performance of E-nose instruments depends extremely on the pattern recognition algorithms. An ideal classifier should not only be able to identify the targets accurately but also tolerate environmental noise (interferences). The general classification methods cannot meet these requirements very well. Fortunately, it has been shown that classifiers ensemble approach can improve both the prediction precision [1] and generalization performance of a recognition system. Moreover, such kinds of ensemble approaches have been widely used in the multi-class recognition problems [2, 3]. Ensemble method is a learning approach by combing many base models to solve a given problem. The provided base models should be accurate enough and error-independent (diverse) in their predictions.

Although ensemble method is one of the advanced pattern recognition techniques within the machine learning community, only a few studies on their applications in E-nose data processing have been reported in the literature. Bagging decision trees were used for E-nose applications, and their VLSI implementation using 3D chip

L. Zhang et al., *Electronic Nose: Algorithmic Challenges*,
https://doi.org/10.1007/978-981-13-2167-2_8

technology was reported [4]. Shi et al. [5, 6] used heterogeneous classifiers including density models, KNN, ANN, and SVM for odor discrimination. In [7], the authors showed how ensemble learning methods could be used in an array of chemical sensors (non-selective field transistors) to cope with the interference problem. Gao et al. [8, 9] used modular neural networks ensemble to predict simultaneously both the classes and concentrations of several kinds of odors; in the first approach, they used MLPs for base learners, and in their second approach, a module or panel comprises various predictors, namely MLPs, MVLR, QMVLR, and SVM were used. In [10], Hirayama et al. demonstrated that it was possible to detect liquefied petroleum gas (LPG) calorific power with high recognition rate (up to 99%) using an E-nose and a committee of machines, even with the failure (fault) of one sensor. In [11], Bona et al. used a hybrid algorithm to generate an ensemble of 100 multi-layer perceptrons (MLPs) for the classification of seven categories of coffee. Recently, Vergara et al. [12] proposed an ensemble method that used support vector machines as base classifiers to cope with the problem of drift in chemical gas sensors. Amini et al. [13] used an ensemble of classifiers on data from a single metal oxide gas sensor (SP3-AQ2, FIS Inc. from Japan) operated at six different rectangular heating voltage pulses (temperature modulation), to identify three gas analytes including methanol, ethanol, and 1-butanol at range of 100–2000 ppm.

Feature extraction is one of the key steps in pattern recognition systems. Principal component analysis (PCA) and independent component analysis (ICA) are among the most widely used feature extraction methods. However, both PCA and ICA are linear feature extraction methods; they may therefore become ineffective in case of nonlinear features. On the other hand, kernel principal component analysis (KPCA) is a nonlinear feature extraction method that integrates kernel trick into standard PCA [14]. It does feature extraction by mapping the original inputs into a high-dimensional feature space, and then the new features are analyzed by PCA in the high-dimensional feature space [15]. Among the above-mentioned three methods, KPCA was found to have a better performance in nonlinear feature extraction [16].

In real-time applications, due to varying atmospheric conditions (i.e., temperature, humidity, pressure), the sensors' responses also vary nonlinearly with gas concentration. The nonlinearity of the sensor array can adversely affect the precision and robustness of the classifier. Therefore, there is a need to come up with an alternative that takes this aspect into consideration. This chapter proposes a novel classifiers ensemble which combines KPCA (for feature extraction) and SVM for classification of multiple indoor air pollutants. Firstly, KPCA method was used to extract separable features from the E-nose data; then five SVM base classifiers were trained using different training samples. Finally, the outputs from these base classifiers were combined using an effective weight fusion method.

8.2 Data Acquisition

The datasets used in this chapter were obtained by our E-nose system. Detailed description of the E-nose system can be found in our previous publications [17, 18]. However, to make the chapter self-contained, we reproduce the system structure and describe briefly the experimental setup.

8.2.1 Electronic Nose System

The sensor array in our E-nose system comprises seven sensors: four metal oxide semiconductor gas sensors (TGS2620, TGS2602, TGS2201A, and TGS2201B), humidity, temperature, and oxygen. A 12-bit analog–digital converter (A/D) is used as interface between the sensor array and a Field-Programmable Gate Array (FPGA) processor. The A/D converts analog signals from sensor array into digital signals which are used by the FPGA for further processing. The FPGA also serves as a control unit. Data collected from the sensor array can be saved as ASCII text files on a PC through Joint Test Action Group (JTAG) port and related software.

8.2.2 Experimental Setup

The experimental platform mainly consists of E-nose system, PC, temperature–humidity-controlled chamber, humidifier, air sampler, flowmeter, air pump, standard instrument, and so on. The specific experimental setup [17] is shown in Chap. 3. The experimental setup has also been mentioned in [19, 20]. All experiments were carried out inside the chamber. The experimental procedure in this chapter can be summarized as follows. Firstly, set the required temperature and humidity control after placing the E-nose system in temperature–humidity chamber; then inject the target gas into the chamber using a pump; finally, the E-nose system will be exposed to the target gas and begin to collect data. It is worth noting that a single experiment consists of three stages: exposure to clean air for 2 min for baseline; exposure to each gas analyte for 8 min for response; and exposure to clean air for 2 min again to allow the sensors recover.

The sensor array in this work is composed of four metal oxide semiconductor gas sensors (i.e., TGS2602, TGS2201A, TGS2201B, and TGS2620) due to their high sensitivity and quick response to detectable gases. The sensing element of the sensors is comprised of a metal oxide semiconductor layer formed on an alumina substrate of a sensing chip together with an integrated heater. In the presence of a detectable gas, the sensor's conductivity increases depending on the gas concentration in the air. Figure 8.1 illustrates the sensor response process when exposed to target gas in an experiment. In our experiment, the sampling frequency is set to

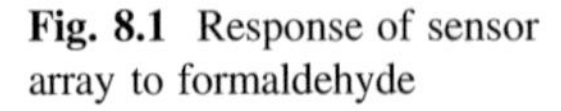

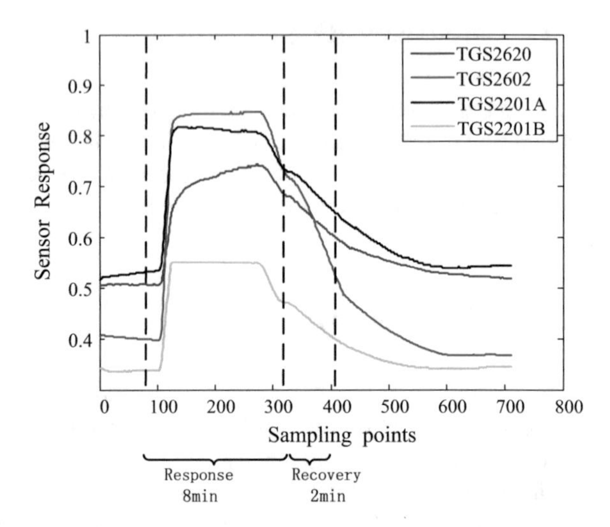

Fig. 8.1 Response of sensor array to formaldehyde

1/6 Hz, and thus, 20 observations would be obtained per minute. Then it is clear to see that the experiment involves 8 min of the exposure of the sensors to gas and 2 min for the sensors to recover.

8.2.3 E-Nose Data

This chapter contributes to the classification of six indoor air contaminants including formaldehyde (HCHO), benzene (C_6H_6), toluene (C_7H_8), carbon monoxide (CO), ammonia (NH_3), and nitrogen dioxide (NO_2). To simulate the indoor environments (e.g., temperature and humidity), the samples for each contaminant were obtained under certain temperature–humidity conditions: 15, 25, 30, and 35 °C for temperature, 40, 60, and 80% for relative humidity. Actually, it is nearly impossible to go through all of the temperature and humidity. How to ensure the precision of the product at a different temperature–humidity seems to be especially tricky. In other words, the generalization performance of the proposed classifier is of great importance. Ensemble method has been used to solve this problem in the current work. As to the pressure, all of the experiments were implemented at standard atmospheric pressure for that our product is used in indoor environment. In addition, with the limits of the experimental facilities, we do not change the air pressure conditions as the temperature–humidity does. The numbers of samples for HCHO, C_6H_6, C_7H_8, CO, NH_3, and NO_2 are 260, 164, 66, 58, 48, and 30, respectively. Detailed information on these samples is available in [17].

During the process of establishing ISVMEN model, the whole dataset was divided into three parts: training set, testing set, and validation set. Firstly, 20% of the samples were labeled as **dataset_va** which were obtained by using Kennard–Stone sequential (KSS) algorithm used in [21] and the 80% remaining samples were

Table 8.1 Distribution of training, test, and validation sets

Datasets	Number of samples in the subset					
	HCHO	C_6H_6	C_7H_8	CO	NH_3	NO_2
dataset_tr	156	99	40	35	29	18
dataset_te	52	32	13	11	9	6
dataset_va	52	33	13	12	10	6
Total	260	164	66	58	48	30

labeled as **dataset_learning**. Then we randomly select 75% samples from **dataset_learning** to form the training set labeled as **dataset_tr**, while the remaining 25% are left for test labeled as **dataset_te**. We should implement repeated random samplings for L times if there are L base classifiers in the ensemble model. Finally, **dataset_tr** and **dataset_te** were used to generate base classifiers and the corresponding weight of each classifier; **dataset_va** was used to validate the performance of the ensemble. The distribution of training set (**dataset_tr**), test set (**dataset_te**), and validation set (**dataset_va**) for each class is shown in Table 8.1.

Although the experimental setup does not change too much compared with the previous method [17], the classification model for indoor air pollutants proposed in this work is quite different. In [17], the proposed classification model HSVM is based on single classifier and linear feature extraction method. However, in this work, some innovations have been made to improve the prediction precision and generalization performance of the classifier. The contributions of this chapter are described as follows. First, in order to improve the accuracy of base classifier, KPCA method is used to extract nonlinear feature of six different indoor air pollutants; second, in the process of establishing classifiers ensemble, a new fusion approach that employs an effective weighted method for classifier ensemble is proposed.

8.3 Methods

8.3.1 KPCA-Based Feature Extraction

PCA and ICA have been successfully used for feature extraction in E-nose systems. In this section, a nonlinear formulation of PCA is described. For in-depth description of PCA and ICA, we refer the reader to [22, 23].

KPCA is one approach of generalizing linear PCA into nonlinear case using the kernel method. The key idea of KPCA is through the nonlinear transform $\Phi(\cdot)$ which maps the sample data from the input space to high-dimensional space F and performs PCA in the new feature space [24–27].

Suppose that there are M observation samples $\mathbf{X}_i$ with $\mathbf{x}_i \in R^N (i = 1, 2, \ldots, M)$, $\Phi(\mathbf{x}_i)$ represents a high-dimensional feature space and N denotes the length of the original $\mathbf{x}_i$, then if these vectors $\Phi(\mathbf{x}_i)$ meet the zero mean condition, the covariance matrix in the new feature space can be expressed as [28]:

$$\mathbf{C} = \frac{1}{M}\sum_{i=1}^{M} \Phi(\mathbf{x}_i)\Phi(\mathbf{x}_i)^{\mathrm{T}} \tag{8.1}$$

The eigenvalue decomposition of covariance matrix $\mathbf{C}$ is given by

$$\lambda \mathbf{\upsilon} = \mathbf{C}\upsilon \tag{8.2}$$

where λ and $\mathbf{\upsilon}$ denote the eigenvalue and eigenvector of $\mathbf{C}$, respectively.

The eigenvector $\mathbf{\upsilon}$ can be expressed in terms of linear combination of the samples

$$\mathbf{\upsilon} = \sum_{i=1}^{M} \beta_i \Phi(\mathbf{x}_i) \tag{8.3}$$

Taking into consideration our mapping $\Phi(\mathbf{x}_k)$ and the result in Eq. 8.2, we can write

$$\lambda(\Phi(\mathbf{x}_k)\cdot \mathbf{\upsilon}) = \Phi(\mathbf{x}_k)\cdot \mathbf{C}\upsilon \quad (k = 1, 2, \ldots, M) \tag{8.4}$$

Let us define a $M \times M$ matrix $\mathbf{K}$ as

$$\mathbf{K}_{ij} = \mathrm{K}(\mathbf{x}_i, \mathbf{x}_j) = \langle \Phi(\mathbf{x}_i), \Phi(\mathbf{x}_j)\rangle \quad (i, j = 1, 2, \cdots, M) \tag{8.5}$$

Combining Eqs. 8.3, 8.4, and 8.5, we can obtain the following equation

$$M\lambda\boldsymbol{\alpha} = \mathbf{K}\boldsymbol{\alpha} \tag{8.6}$$

where $M\lambda$ is the eigenvalue of $\mathbf{K}$ and coefficient vector $\boldsymbol{\alpha} = (\alpha_1, \alpha_2, \ldots, \alpha_M)^{\mathrm{T}}$ is the eigenvector. After normalization of the eigenvector $\mathbf{\upsilon}$, we get

$$\mathbf{\upsilon}^k \cdot \mathbf{\upsilon}^k = \langle \mathbf{\upsilon}^k, \mathbf{\upsilon}^k \rangle = 1, \quad k = 1, 2, \ldots, M \tag{8.7}$$

The principal components (scores) $\mathbf{t}_k$ $(k = 1, 2, \ldots, M)$ can be expressed as

$$\mathbf{t}_k = \langle \mathbf{\upsilon}^k, \Phi(\mathbf{x})\rangle = \sum_{i=1}^{M} \alpha_i^k \mathrm{K}(\mathbf{x}, \mathbf{x}_i) \tag{8.8}$$

8.3.2 Base Classifiers

In this chapter, KPCA and SVM were combined to generate required number of base classifiers. Support vector machines are formulated based on the framework of statistical learning theory. They involve minimization of structural risk. As a two-class classification problem, SVM could separate the datasets by searching for an optimal separating hyperplane between them [29, 30].

Given that the sample is repressed as $\mathbf{x}_i \in R^N, i = 1, 2, \ldots, M$, and each sample belongs to a class $y_i \in \{-1, 1\}$. For linear classification, we can identify two classes by an optimal separating hyperplane

$$\mathbf{w}^{\mathrm{T}}\mathbf{x} + b = 0 \tag{8.9}$$

We can obtain the optimal values for $\mathbf{W}$ and b by solving a constrained convex quadratic programming problem, using Lagrange multipliers $\alpha_i\,(i = 1, 2, \ldots, M)$. The decision function can be expressed as

$$f(x) = \mathrm{sgn}\left(\sum_{i=1}^{M} \alpha_i y_i \mathrm{K}(\mathbf{x}_i, \mathbf{x}) + b\right) \tag{8.10}$$

where those $\mathbf{X}_i$ with nonzero α_i are the "support vectors."

In most real-life applications, linear separation of datasets cannot be achieved successfully. To overcome this, an alternative method is proposed. It works by first projecting original data to a higher dimensional space, and then performing subsequent analysis in the new space. If the mapping function is $\varphi(\mathbf{x})$, then the optimal separating hyperplane function can be written as:

$$f(\mathbf{x}) = \mathbf{w} \cdot \varphi(\mathbf{x}) + b \tag{8.11}$$

We can get a more generalized form as

$$f(x) = \sum_{i=1}^{M} \alpha_i \cdot y_i \cdot (\varphi(\mathbf{x}_i) \cdot \varphi(\mathbf{x})) + b \tag{8.12}$$

In a high-dimensional space, determination of $\langle \varphi(\mathbf{x}_i), \varphi(\mathbf{y}_i) \rangle$ involves high computational cost. Fortunately, it turns out that this inner product can be evaluated using an appropriate kernel such that: $\mathrm{K}(\mathbf{x}_i, \mathbf{x}) = \langle \varphi(\mathbf{x}_i), \varphi(\mathbf{x}) \rangle$. In this chapter, we use Gaussian RBF kernel which is defined as follows

$$\mathrm{K}(\mathbf{x}_i, \mathbf{x}_j) = \exp\left(\frac{-\left\|\mathbf{x}_i - \mathbf{x}_j\right\|^2}{\sigma^2}\right) \tag{8.13}$$

where σ^2 is the kernel parameter which determines the bandwidth of RBF. The decision function can be expressed as

$$f(x) = \text{sgn}\left(\sum_{i=1}^{M} (\alpha_i \text{K}(\mathbf{x}_i, \mathbf{x}) + b\right) \tag{8.14}$$

where α_i and b are the optimal decision parameters.

8.3.3 Multi-class ISVMEN

An effective and promising alternative to build a classifier is to train many classifiers and combine their decisions, which is called ensemble classifier system (ECS) and is currently among some of the hot research areas. In the course of building ECS, there are several important issues that could be grouped into three main problems [31]: (1) selection of the topology of the ECS; (2) design of base classifiers; and (3) design of fusion (ensemble) method.

Usually, parallel topology is the most common topology because this structure has good methodological background [32]. The selection of base classifiers is also important. Indeed, ensemble learning will improve the generalization and prediction performance of individual classifiers only if these classifiers are accurate and diverse enough in their predictions. The fusion of base classifiers is the most important idea in classifiers ensemble. In this step, both the diversity among base classifiers and their accuracy are exploited to provide optimal combination (or fusion). Therefore, the choice of a collective decision making method is of paramount importance. There are two main fusion methods: majority voting and weight assignment. Majority voting utilizes the concept of democratic decision making. It states that the class label with more than half of base classifiers prediction is accepted as true [33]. The other approach is based on discriminant analysis. The main form of discrimination is a posterior probability typically associated with probabilistic pattern recognition model.

In this work, we focus on the issues of designing effective base classifiers and fusion method. A classifiers ensemble approach called as ISVMEN based on KPCA and SVM for classification of multiple indoor air pollutants using an E-nose was proposed. The architecture of ISVMEN model is depicted in Fig. 8.2.

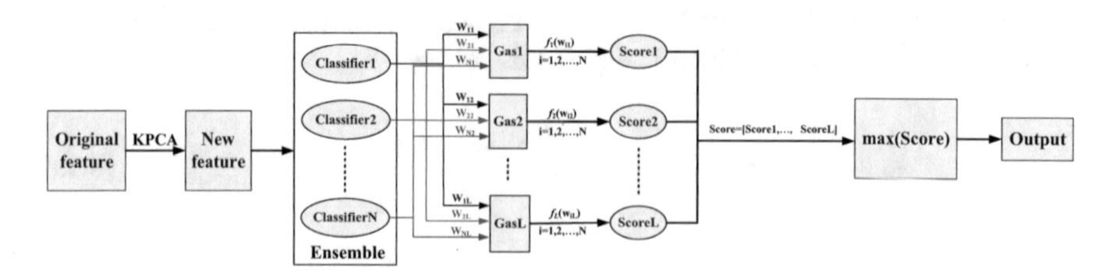

Fig. 8.2 Architecture of ISVMEN

In the proposed ISVMEN model, KPCA is first used for obtaining important nonlinear features from original data. KPCA results showed that the first 16 principal components accounted for 95% cumulative variance in original data. Thus, using these principal components, useful information from datasets of six kinds of air pollutants was extracted, which constitute the new features. Then, these new features were used as inputs of SVM multi-class classifier. In this chapter, five base classifiers based on SVM were designed. Finally, a weighted voting fusion was used to combine the classifiers. The implementation process of ISVMEN can be illustrated as follows.

Suppose that there are totally n training samples with p variables. Suppose also that the number of training samples for HCHO (formaldehyde), C_6H_6 (benzene), C_7H_8 (toluene), CO (carbon monoxide), NH_3 (ammonia), and NO_2 (nitrogen dioxide) is n_1, n_2, n_3, n_4, $n_{5,}$ respectively. Thus, the initial training dataset can be represented by an $n \times p$ matrix as follows

$$\mathbf{X}_{\text{old}} = \left\{\mathbf{X}_1, \mathbf{X}_2, \mathbf{X}_3, \mathbf{X}_{4,}, \mathbf{X}_5, \mathbf{X}_6\right\} = \begin{bmatrix} x_{11} & x_{12} & \cdots & x_{1p} \\ x_{21} & x_{22} & \cdots & x_{2p} \\ \vdots & \vdots & & \vdots \\ x_{n1} & x_{n2} & \cdots & x_{np} \end{bmatrix} \tag{8.15}$$

where the number of columns $p = 6$ and it represents the dimension of the original data; the number n of rows represents the total number of training samples; $\mathbf{X}_i (i = 1, 2, \ldots, 6)$ represents the training data of the i-th kind of gas.

It is worth noting that KPCA makes use of kernel trick to project original data $\mathbf{X}_{\text{old}}$ onto a higher dimensional feature space and gets the principal components as new features $\mathbf{X}_{\text{new}}$ by principal component analysis. After KPCA, we can obtain more linearly separable features, thereby getting better performance especially with limited training samples. Thus, the new training data matrix $\mathbf{X}_{\text{new}}$ can be expressed as

$$\mathbf{X}_{\text{new}} = \begin{bmatrix} x_{11} & x_{12} & \cdots & x_{1q} \\ x_{21} & x_{22} & \cdots & x_{2q} \\ \vdots & \vdots & & \vdots \\ x_{n1} & x_{n2} & \cdots & x_{nq} \end{bmatrix} \tag{8.16}$$

where q, the number of columns, depends on the number of principal components; the best value of q is obtained on the basis of cumulative variance (see Table 8.2) and prediction accuracy (see Fig. 8.3). Considering these two indicators, the best value of q was found to be 17.

The training goal is defined as

$$\mathbf{Y}_{\text{goal}} = [1, 2, 3, 4, 5, 6]^{\text{T}} \tag{8.17}$$

where 1, 2, 3, 4, 5, and 6 stand for the labels of HCHO, C_6H_6, C_7H_8, CO, NH_3, and $NO_{2,}$ respectively.

Table 8.2 Results of KPCA

Principle components	Accumulating contribution rate		
	Eigenvalue	Variance (%)	Cumulative variance (%)
PC1	68.07	29.95	29.95
PC2	53.78	23.66	53.61
PC3	22.23	9.78	63.40
PC4	16.99	7.47	70.88
PC5	13.33	5.86	76.75
PC6	11.12	4.89	81.64
PC7	7.60	3.34	84.99
PC8	4.66	2.05	87.04
PC9	3.75	1.65	88.69
PC10	2.89	1.27	89.97
PC11	2.87	1.26	91.23
PC12	2.41	1.06	92.29
PC13	2.23	0.98	93.28
PC14	1.92	0.84	94.13
PC15	1.66	0.73	94.85
PC16	1.39	0.61	95.47
PC17	0.98	0.43	95.90

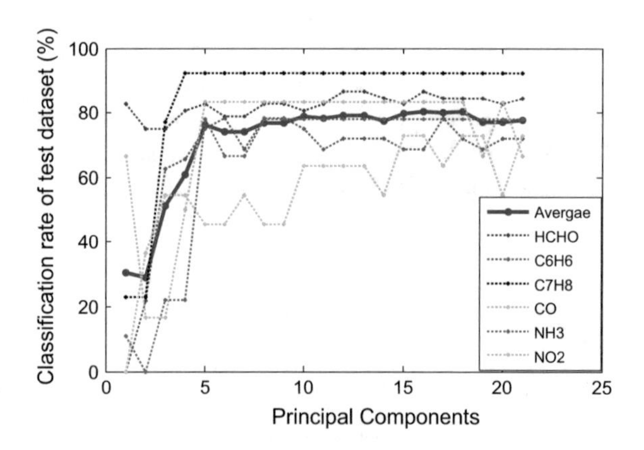

Fig. 8.3 Classification rate of test dataset in terms of number of principal components

Using these new features as input instances, a base classifier Ψ can classify any novel input instance as one of the predefined class labels $\mathbf{Y}_{\text{goal}}$. The base classifier Ψ is defined as [18]:

$$\Psi : \mathbf{X}_{\text{new}} \rightarrow \mathbf{Y}_{\text{goal}} \tag{8.18}$$

We use five classifiers $\Psi_1, \Psi_2, \Psi_3, \Psi_4, \Psi_5$. Each classifier as an expert makes decision according to its prediction accuracy on test dataset. The output from five base classifiers can be expressed as:

$$\hat{\boldsymbol{\Psi}} = [\Psi_1(\mathbf{x}), \Psi_2(\mathbf{x}), \ldots, \Psi_5(\mathbf{x})]^{\mathrm{T}} \tag{8.19}$$

where $\Psi_i(\mathbf{x}) \in \mathbf{Y}_{\text{goal}}, i = 1, 2, \ldots, 5$ means the output of the i-th base classifier on instance $\mathbf{x}$. To integrate decisions from base classifiers, we need to express the value $\Psi_i(\mathbf{x}) \in \mathbf{Y}_{\text{goal}}, i = 1, 2, \ldots, 5$ using binary encoding. The binary encoding method is shown as follows.

- If the output of the i-th (i = 1, 2, …, 5) base classifier is HCHO, namely $\Psi_i(\mathbf{x}) = 1$, then encoding it by $\boldsymbol{\Psi}_{codei}(\mathbf{x}) = [1 \quad 0 \quad 0 \quad 0 \quad 0 \quad 0]^{\mathrm{T}}$;
- Similarly, if $\Psi_i(\mathbf{x}) = 2$, then encoding it by $\boldsymbol{\Psi}_{codei}(\mathbf{x}) = [0 \quad 1 \quad 0 \quad 0 \quad 0 \quad 0]^{\mathrm{T}}$;
- if $\Psi_i(\mathbf{x}) = 3$, then encoding it by $\boldsymbol{\Psi}_{codei}(\mathbf{x}) = [0 \quad 0 \quad 1 \quad 0 \quad 0 \quad 0]^{\mathrm{T}}$;
- if $\Psi_i(\mathbf{x}) = 4$, then encoding it by $\boldsymbol{\Psi}_{codei}(\mathbf{x}) = [0 \quad 0 \quad 0 \quad 1 \quad 0 \quad 0]^{\mathrm{T}}$;
- if $\Psi_i(\mathbf{x}) = 5$, then encoding it by $\boldsymbol{\Psi}_{codei}(\mathbf{x}) = [0 \quad 0 \quad 0 \quad 0 \quad 1 \quad 0]^{\mathrm{T}}$;
- if $\Psi_i(\mathbf{x}) = 6$, then encoding it by $\boldsymbol{\Psi}_{codei}(\mathbf{x}) = [0 \quad 0 \quad 0 \quad 0 \quad 0 \quad 1]^{\mathrm{T}}$;

Then we have

$$\hat{\boldsymbol{\Psi}}_{code} = [\boldsymbol{\Psi}_{code1}(\mathbf{x}), \boldsymbol{\Psi}_{code2}(\mathbf{x}), \ldots, \boldsymbol{\Psi}_{code5}(\mathbf{x})]^{\mathrm{T}} = \begin{bmatrix} \hat{\Psi}_{code11} & \cdots & \hat{\Psi}_{code16} \\ \vdots & & \vdots \\ \hat{\Psi}_{code51} & \cdots & \hat{\Psi}_{code56} \end{bmatrix} \tag{8.20}$$

where $\widehat{\boldsymbol{\Psi}}_{code}$ is a matrix with 5×6 and the element $\hat{\Psi}_{codeij}(i = 1, 2, \ldots, 5; j = 1, 2\ldots, 6)$ is the result of the i-th classifier on the j-th gas after binary encoding; specifically, each row represents a classifier while each column represents a target gas.

Suppose that the total weight matrix of five base classifiers is expressed by

$$\mathbf{W} = [\mathbf{w}_1, \mathbf{w}_2, \mathbf{w}_3, \mathbf{w}_4, \mathbf{w}_5]^{\mathrm{T}} = \begin{vmatrix} w_{11} & w_{12} & w_{13} & w_{14} & w_{15} & w_{16} \\ w_{21} & w_{22} & w_{23} & w_{24} & w_{25} & w_{26} \\ w_{31} & w_{32} & w_{33} & w_{34} & w_{35} & w_{36} \\ w_{41} & w_{42} & w_{43} & w_{44} & w_{45} & w_{46} \\ w_{51} & w_{52} & w_{53} & w_{54} & w_{55} & w_{56} \end{vmatrix} \tag{8.21}$$

where w_{ij} $(i = 1, 2, \ldots, 5;\ j = 1, 2\ldots, 6)$ means the weight of the i-th base classifier on the j-th gas. The entries in each row represent the weights assigned to each base classifier. It is worth mentioning that the same classifier may have different weights for the six gases. The weights of each base classifier were computed as follows [24]

$$w_{ij} = \frac{\log \frac{p_{ij}}{1-p_{ij}}}{\sum_{i=1}^{5} \log \frac{p_{ij}}{1-p_{ij}}}, \quad i = 1, 2, \ldots, 5; \; j = 1, 2, \ldots, 6 \tag{8.22}$$

where p_{ij} is the accuracy of the i-th classifier on the j-th gas and it was obtained from test dataset. It has been proved that in literature [34], the ensemble classifier system's classification accuracy could reach the maximum and take full advantage of the prior information of base classifiers if one uses Eq. 8.22 to calculate the weights.

The classifiers ensemble system makes final decision based on Eq. 8.23 and Eq. 8.24,

$$\text{score}_j = f_j(w_{ij}, \hat{\Psi}_{codeij}) = \sum_{i=1}^{5} w_{ij} \cdot \hat{\Psi}_{codeij}, j = 1, 2, \ldots, 6 \tag{8.23}$$

where score_j is the score of the j-th gas. That is, each gas has its own score, and the label of the gas can be represented by the maximum score,

$$\text{Gas\_label} = \max(\text{score}_1, \text{score}_2, \text{score}_3, \text{score}_4, \text{score}_5, \text{score}_6) \tag{8.24}$$

8.4 Results and Discussion

All the computations are carried out in MATLAB R2011b software. In order to make an impartial comparison between base classifiers and the proposed ensemble classifier, we evaluated in terms of the classification accuracy on validation dataset (see Table 8.1) which was mainly used to validate the performance of ensemble classifier. The specific process and approach of obtaining the training set, test set, validation set have been described in Sect. 2.3. The training set was used to generate base classifiers, and we assess the classification accuracy of base classifiers using homologous testing set. Based on base classifiers' classification accuracy, we can obtain the weights for each base classifier using Eq. 8.22. From Table 8.1, it is easy to find that the dataset for six kinds of contaminants is quite imbalanced. The total numbers of HCHO and C_6H_6 are 260 and 164, respectively; however, the total number of the remaining four gases is far less than them. Especially, NO_2 has only 30 samples. In this case, a single classifier could hardly obtain good classification results for all gases. That is, the class with fewer samples is easier to be classified as the class with more samples.

KPCA results of the training samples are presented in Table 8.2. Table 8.2 shows the eigenvalues and the cumulative variance of the first 17 principal components. From this table, we can see that the variance contribution of the first 16 components reaches 95.47% which means that the sample information can be represented fully by using the first 16 principal components. In the implementation of KPCA, there are two major issues to be solved: one is to search for the optimal number of principal components, and the other is to find the best value of kernel parameter σ. In the beginning, the first principal component was used as input of SVM. Then the number of inputs (i.e., the number of principal components) of SVM increased gradually, up to 22 inputs. The reason why we stop at 22 principal components is that the first 16 principal components have explained more than 95% of the total variance. Then it is unnecessary to increase the principal components continually. Figure 8.3 illustrates the variation trend of classification accuracy versus the number of principal components. From this figure, one can notice that the best average accuracy was obtained at point 17 and point 19, as indicated by the blue line. However, considering the dimension aspect, we finally adopted 17 principal components. In addition, we have also compared KPCA with two linear feature extraction methods, namely PCA and ICA. Experimental results in Table 8.3 showed the superiority of KPCA over the linear feature extraction methods. In Table 8.3, KPCA obtained the best classification accuracy of 81.7%; however, the mean accuracy of raw feature, PCA, and ICA are only 76.8, 78.1, and 51.3%, respectively. Besides, for a single gas, KPCA had a good classification accuracy of 100, 100, and 92.3% for CO, NH_3, and HCHO, respectively. By taking into account the mean recognition rate and the single gas recognition rate, finally, we used KPCA as the feature extraction method to improve the base classifier's accuracy.

Table 8.4 shows the classification accuracy of five base classifiers for training and test samples. The five base classifiers were labeled SVM1, SVM2, SVM3, SVM4, and SVM5, respectively. From this table, we can see that the training results are desirable and all the mean accuracy exceeded 93%; however, the test results are not uniformly good. For SVM1 and SVM5, SVM1 has a good accuracy of 100.0% on toluene but a poor accuracy of 77.7% on ammonia; the situation for SVM5 is reversed. Besides, for the five base classifiers, the best classification accuracy on NO_2 and C_6H_6 is only 66 and 75%, respectively; the worst classification accuracy on NO_2 and C_6H_6 is low to 50 and 59%, respectively. Therefore, the single classifiers are not uniformly good for all gases. More specifically, the classification precision and generalization ability of single classifier can hardly meet our requirement.

Table 8.3 Comparison of classification accuracy on test dataset for four feature extraction methods

Class	Classification accuracy (%) on test set			
	Raw	PCA	ICA	KPCA
HCHO	90.3	94.2	66.5	92.3
C_6H_6	84.8	75.7	59.5	63.6
C_7H_8	100.0	92.3	51.2	84.6
CO	58.3	83.3	28.8	100.0
NH_3	60.0	40.0	63.3	100.0
NO_2	66.6	83.3	40.0	50.0
Average	76.8	78.1	51.3	**81.7**

The boldface type denotes the best performance.

Table 8.4 Classification accuracy of five base classifiers for training and testing samples

Class	Classification accuracy (%)									
	Train set					Test set				
	SVM1	SVM2	SVM3	SVM4	SVM5	SVM1	SVM2	SVM3	SVM4	SVM5
HCHO	94.8	98.0	92.9	96.1	94.2	73.0	82.6	84.6	76.9	86.5
C_6H_6	95.9	95.9	88.8	88.8	93.9	65.5	75.0	75.0	65.6	59.3
C_7H_8	100.0	100.0	100.0	97.5	100.0	100.0	92.3	100.0	100.0	84.6
CO	100.0	97.1	97.1	97.1	100.0	81.8	90.9	81.8	100.0	100.0
NH_3	93.1	100.0	100.0	96.5	100.0	77.7	100.0	88.8	88.8	100.0
NO_2	94.4	94.4	94.4	83.3	94.4	66.6	66.6	66.6	50.0	50.0
Average	96.3	97.6	95.5	93.2	97.1	77.4	84.5	82.8	80.2	78.5

The performance of the proposed multiple classifiers system and that of the base classifiers was evaluated on a validation dataset which has never been used (see Table 8.5). In addition, the performance of another fusion mechanism, majority voting, was evaluated. Majority voting (labeled as MV, see Table 8.5) uses the concept of democratic decision making wherein all base classifiers have the same weight which is the reciprocal of the number of base classifiers.

From Table 8.5, it can be seen that the classification accuracy of the proposed method (ISVMEN) is higher than that of any of the base classifiers. Of course one base classifier may have good performance on several gases. On the other hand, the proposed method, ISVMEN, can obtain the best discrimination accuracy for almost all gases. For C_7H_8, CO, and NH_3, almost 100% classification accuracy was obtained using our method. Due to limited number of samples, four out of five base classifiers had classification accuracy less than 70% for NO_2. However, ISVMEN model, the proposed method, can overcome sample unbalanced problem and classify each gas more accurately. Also, the average classification accuracy has been improved from less than 86 to 92.58%. At the same time, compared to MV fusion method (majority voting), the proposed fusion method can obtain better result, and average recognition accuracy has achieved 92.58% which is higher than 90.1% obtained using MV with an increment of two percent in classification accuracy. In conclusion, the ensemble classifier proposed in current work is superior to base classifiers for both classification accuracy and generation ability.

Table 8.5 Comparison of classification accuracy for six models using validation dataset

Class	Classification accuracy (%) of validation set						
	SVM1	SVM2	SVM3	SVM4	SVM5	MV	ISVMEN
HCHO	84.6	88.4	86.5	92.3	92.3	92.3	90.3
C_6H_6	78.7	81.8	81.8	78.7	63.6	81.8	81.8
C_7H_8	100.0	100.0	100.0	84.6	84.6	100.0	100.0
CO	83.3	100.0	91.6	91.6	100.0	83.3	100.0
NH_3	60.0	100.0	90.0	90.0	100.0	100.0	100.0
NO_2	83.3	50.0	66.6	50.0	50.0	83.3	83.3
Average	81.6	86.7	86.1	81.2	81.7	90.1	92.5

Table 8.6 Multi-class classification results on validation set using ISVMEN model

Class	*N*	Classified as					
		HCHO	C_6H_6	C_7H_8	CO	NH_3	NO_2
HCHO	52	**47**	5	0	0	0	0
C_6H_6	33	5	**22**	0	1	0	0
C_7H_8	13	0	0	**13**	0	0	0
CO	12	0	0	0	**12**	0	0
NH_3	10	0	0	0	0	**10**	0
NO_2	6	1	0	0	0	0	**5**

The boldface type denotes the best performance.

Table 8.6 presents the classification results of validation samples using ISVMEN model. The digits in diagonal line denote the number of correctly classified samples, and others mean the number of misclassified samples. It is clear to find that almost all the gases can be classified correctly.

According to the front analysis, it is easy to reach the conclusion that the proposed method ISVMEN is better favorably with other methods. Two main reasons can explain it: (1) using nonlinear KPCA to extract features and improve the precision of base classifiers. In [9], it shows that the six gases are linearly inseparable, and then linear feature extraction methods (i.e., PCA, ICA) are unable to obtain helpful nonlinear feature. In Table 8.3, it is easy to find that the classification accuracy of base classifier SVM has been greatly improved after KPCA. (2) The use of improved weighted fusion approach results in highly diverse base classifiers. The prediction accuracy of each base classifier depends on the gases to be predicted. For example, for classifier 1 which refers to SVM1 in Table 8.4, the prediction accuracy on C_7H_8 is very high, but it has a poor accuracy on C_6H_6. If uniform weights are used for all gases just as MV wherein all base classifiers have the same weight which is the reciprocal of the number of base classifiers, then diversity among base classifiers will not be fully exploited. As to the common fusion method MV, all the base classifiers would be assigned an equal weight, and then one cannot take full advantage of their diversity. Therefore, we adopt improved weighted approach in which weights assignment is done based on the predictive accuracy of each base classifier on each gas. After the weights for each base classifier have been determined, we used Eq. 8.23 to combine the decisions from all base classifiers thereby obtaining final score for each gas. In decision level, the gas with the highest score was considered as the winner.

8.5 Summary

A novel multi-class recognition approach ISVMEN is proposed in this chapter for classification of multiple indoor air contaminants. It is based on multiple classifier systems, in which KPCA and SVM were combined to build the base classifiers.

Then, an effective fusion strategy is presented to integrate the decisions from the base classifiers. The purpose of this work is to improve the prediction accuracy and robustness of the pattern recognition scheme in E-nose. The performance of the proposed method was compared with that of base classifiers, and also that of standard majority voting. Experimental results show that the proposed ISVMEN model can achieve better performance in both recognition accuracy and generalization compared with that of base classifiers, and also that of standard majority voting.

References

1. L.I. Kuncheval, Combining pattern classifiers: Methods and algorithms. IEEE Tran. Neural Netw. **18**, 964 (2007)
2. Y. Su, S. Shan, X. Chen, W. Gao, Hierarchical ensemble of global and local classifiers for face recognition. IEEE Trans. Image Process. **18**, 1885–1896 (2009)
3. D. Tao, X. Tang et al., Asymmetric bagging and random subspace for support vector machines-based relevance feedback in image retrieval. IEEE Trans. Pattern Anal. Mach. Intell. **28**, 1088–1099 (2006)
4. A. Bermak, D. Martinez, A compact 3D VLSI classifier using bagging threshold network ensembles. IEEE Trans. Neural Netw. **14**(5), 1097–1109 (2003)
5. M. Shi, S. Brahim-Belhouari, A. Bermak, Quantization errors in committee machine for gas sensor application. IEEE Int. Symp. Circ. Syst. ISCAS **3**, 1911–1914 (2005)
6. M. Shi, S. Brahim-Belhouari, A. Bermark, D. Martinez, Committee machine for odor discrimination in gas sensor array, in *Proceedings of the 11th International Symposium on olfaction and electronic nose (ISOEN)* (2005), pp. 74–76
7. S. Bermejo, J. Cabestany, Ensemble learning for chemical sensor arrays. Neural Process. Lett. **19**, 25–35 (2004)
8. D.Q. Gao, M.M. Chen, J. Yan, Simultaneous estimation of classes and concentrations of odors by an electronic nose using combinative and modular multilayer perceptrons. Sens. Actuators, B **107**, 773–781 (2005)
9. D.Q. Gao, W. Chen, Simultaneous estimation of odor classes and concentrations using an electronic nose with function approximation model ensembles. Sens. Actuators, B **120**, 584–594 (2007)
10. V. Hirayama, F.J. Ramirez-Fernandez, W.J. Salcedo, Committee machine for LPG calorific power classification. Sens. Actuators, B **116**, 62–65 (2006)
11. E. Bona, R. Silva, D. Borsato, D.G. Bassoli, Neural network for instant coffee classification through an electronic nose. Int. J. Food Eng. **7** (2011). https://doi.org/10.2202/1556-3785.2002
12. A. Vergara, S. Vembu, T. Ayhan, A.R. Margaret, L.H. Margie, H. Ramón, Chemical gas sensor drift compensation using classifier ensembles. Sens. Actuators, B **166–167**, 320–329 (2012)
13. A. Amini, M.A. Bagheri, G. Montazer, Improving gas identification accuracy of a temperature-modulated gas sensor using an ensemble of classifiers. Sens. Actuators, B **187**, 241–246 (2013)
14. B. Scholkopf, A. Smola, K.R. Muller, Nonlinear component analysis as a Kernel eigenvalue problem. Neural Comput. **10**, 1299–1319 (1998)
15. Z.L. Sun, D.S. Huang, Y.M. Cheun, Extracting nonlinear features for multispectral images by FCMC and KPCA. Digit. Signal Proc. **15**, 331–346 (2005)
16. L.J. Cao, K.S. Chua, W.K. Chong et al., A comparison of PCA, KPCA and ICA for dimensionality reduction in support vector machine. Neurocomputing **55**, 321–336 (2003)

17. L. Zhang, F.C. Tian, H. Nie et al., Classification of multiple indoor air contaminants by an electronic nose and a hybrid support vector machine. Sens. Actuators, B **174**, 114–125 (2012)
18. L. Zhang, F.C. Tian, S. Liu et al., Chaos based neural network optimization for concentration estimation of indoor air contaminants. Sens. Actuators, A **189**, 161–167 (2013)
19. C. Kadri, F.C. Tian, L. Zhang et al., Neural network ensembles for online gas concentration estimation using an electronic nose. Int. J. Comput. Sci. Issue **10**, 129–135 (2013)
20. F.C. Tian, H.J. Li, L. Zhang et al., A denoising method based on PCA and ICA in electronic nose for gases quantification. J. Comput. Inf. Syst. **8**, 5005–5015 (2012)
21. L. Zhang, F. Tian, C. Kadri, B. Xiao, H. Li, L. Pan, H. Zhou, On-line sensor calibration transfer among electronic nose instruments for monitoring volatile organic chemicals in indoor air quality. Sens. Actuators, B **160**, 899–909 (2011)
22. I.T. Jolliffe, *Principal Component Analysis*, 2nd edn. (Springer, New York, 2002)
23. A. Hyvärinen, E. Oja, Independent component analysis: algorithms and applications. Neural Netw. **13**, 411–430 (2000)
24. S. Mika, B. Scholkopf, A.J. Smola, Kernel PCA and de-nosing in feature spaces. Adv. Neural. Inf. Process. Syst. **11**, 536–542 (1999)
25. J.H. Cho, J.M. Lee, S.W. Choi, D. Lee, I.B. Lee, Fault identification for process monitoring using kernel principal component analysis. Chem. Eng. Sci. **60**(1), 279–288 (2005)
26. S.W. Choi, C. Lee, J.M. Lee, J.H. Park, I.B. Lee, Fault detection and identification of nonlinear process based on kernel PCA. Chemometr. Intell. Lab Syst **75**(1), 55–67 (2005)
27. J.D. Shao, G. Rong, J.M. Lee, Learning a data-dependent kernel function for KPCA-based nonlinear process monitoring. Chem. Eng. Res. Des. **87**, 1471–1480 (2009)
28. Y.P. Zheng, L.P. Zhang, Fault diagnosis of wet flue gas desulphurization system based on KPCA, in *The 19th International Conference on Industrial Engineering and Engineering Management* (2013), pp. 279–288
29. E. Gumus, N. Kilic, A. Sertbas et al., Evaluation of face recognition techniques using PCA, wavelets and SVM. Expert Syst. Appl. **37**, 6404–6408 (2010)
30. N. Cristianini, J. Taylor, *An Introduction to Support Vector Machines and Other Kernel-based Learning Methods* (Cambridge University Press, Cambridge, 2000)
31. M. Wozniak, M. Zmyslony, Designing combining classifier with trained fuser–analytical and experimental. Neural Netw. Word **20**, 807–978 (2010)
32. L.I. Kuncheva, Combining pattern classifiers: methods and algorithms. IEEE Trans. Neural Netw. **18**, 964 (2004)
33. A. Szczurek, B. Krawczyk, M. Maciejewska et al., VOCs classification based on the committee of classifiers coupled with single sensor signals. Chemometr. Intell. Lab. Syst. **125**, 1–166 (2013)
34. L.I. Kuncheva, C.J. Whitaker, C.A. Shipp et al., Limits on the majority vote accuracy in classifier fusion. Pattern Anal. Appl. **6**, 22–31 (2003)

Part III
E-Nose Drift Compensation: Challenge II

Chapter 9
Chaotic Time Series-Based Sensor Drift Prediction

Abstract Chemical sensor drift shows a chaotic behavior and unpredictability in long-term observation, such that constructing an appropriate sensor drift treatment is difficult. This chapter introduces a new methodology for chaotic time series modeling of chemical sensor observations in embedded phase space. This method realizes a long-term prediction of sensor baseline and drift based on phase space reconstruction (PSR) and radial basis function (RBF) neural network. PSR can memory all of the properties of a chaotic attractor and clearly show the motion trace of a time series; thus, PSR makes the long-term drift prediction using RBF neural network become possible. Experimental observation data of three metal oxide semiconductor sensors in a year demonstrates the obvious chaotic behavior through the Lyapunov exponents. Results demonstrate that the proposed model can make long-term and accurate prediction of time series chemical sensor baseline and drift.

Keywords Sensor drift · Chaotic time series · Long-term prediction
Phase space reconstruction · Radial basis function neural network

9.1 Introduction

Sensor drift is caused by unknown dynamic process in the sensor system including poisoning, aging of sensors, or environmental variations [1–3]. The drift will decrease the selectivity and sensitivity of sensors, and an E-nose consisting of drifted sensors will lose its effectiveness as a usable monitor because the pattern recognition model cannot fit the drifted data. Therefore, sensor drift has been recognized as the most challenging issue in E-nose research [4].

In recent years, many drift counteraction methods have been proposed by researchers. The most commonly used methods are multivariate component correction [5], principal component analysis [6, 7], adaptive drift correction based on evolutionary algorithm [8], orthogonal signal correction [9], and drift compensation based on estimation theory [10]. However, all the methods assume that the long-term sensor drift trend can be tracked through the drift direction in principal

L. Zhang et al., *Electronic Nose: Algorithmic Challenges*,
https://doi.org/10.1007/978-981-13-2167-2_9

component space or the samples' distribution. Unfortunately, the long-term sensor drift has no regular trend, and the sensor drift is not always positive or negative from a long-term view [4]. In addition, previous methods can only be analyzed off-line in computer, and it is difficult to be used in a real-life application. Therefore, through a long-term observation of sensor response within one year, we studied the chaotic characteristic of long-term sensor response using Lyapunov exponents in embedded phase space and proposed a novel drift prediction model using chaotic time series prediction method based on phase space reconstruction and RBF neural network.

Chaos is a universal phenomenon of nonlinear dynamic systems. Chaos is an irregular motion and seemingly unpredictable random behavior exhibited by a deterministic nonlinear system under deterministic conditions. Chaotic time series prediction is based on phase space reconstruction (PSR) and aims to find the embedding way, dimensions, and time delay of the attractor in terms of the present observation sequence [11, 12]. The time series will reconstruct the attractor of system in a high dimension without changing its topology, and an appropriate time delay will transform the prediction problem into a short evolutionary process in phase space. Phase space reconstruction preserves the attractor's properties. Therefore, to an unknown drift model, chaotic attractor theory makes drift prediction in embedded phase space become reasonable.

9.2 Data

9.2.1 Long-Term Sensor Data

The electronic nose system has been presented in [13–15]. The picture of our E-nose system and the internal circuit of print circuit board (PCB) have been illustrated in Chap. 3. The Joint Test Action Group (JTAG) is used for communication between PC and the PCB, such as write programs from PC to the CPU, and read data from the CPU to PC. Three metal oxide semiconductor sensors were studied in this work for sensor drift. They are TGS2620 and TGS2201 with dual outputs (TGS2201A/B). The sensors started work on September 04, 2011, and stopped on July 06, 2012. The operation surrounding sensors is inside the room. The heater voltage is 5 V. The total number of sampling points for each sample is set as 3000. The data of 3 days is collected, and totally, 136 samples within one year were obtained. Consider that the redundancy of 3000 sampling points in each sample around 3 days will cause the complexity of analysis; thus, we extract 100 points uniformly from each sample with the interval of 30 points. Totally, 13,600 points of sensor baseline within one year were obtained. Figure 9.1 illustrates the extracted observations of each sensor within one year. We can find from Fig. 9.1 that there is a reduction of sensor response at the approximated position of 6725. It is due to that the temperature indoor descends when the air conditioner was turned

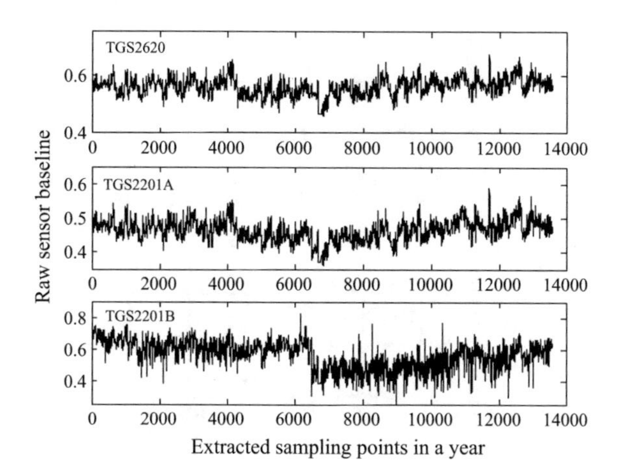

Fig. 9.1 Raw observation points of sensors within one year

on and results in a reduction of sensor response. Note that the unit of sensors' raw outputs is voltage (V); however, the responses in Fig. 9.1 is the output of 12 bit-ADC (analog-to-digital converter) after 0–1 normalization by dividing the maximum 4095 ($2^{12} - 1$).

9.2.2 Discrete Fourier Transform (DFT)-Based Feature Extraction

Although sensor drift is difficult to describe, some properties of drift are known. The drift has a low-frequency behavior, and it can also be considered as a band-limited signal [16]. The sensor baseline is extremely related to environmental factors (e.g., temperature, humidity, and pressure). Therefore, DFT is used for spectrum analysis of the long-term sensor signal. The representation of DFT is shown as

$$X(k) = \frac{1}{N}\sum_{n=0}^{N-1} x(n)\mathrm{e}^{-j\cdot(2\pi/N)\cdot k\cdot n}, \quad k = 0, 1, \ldots, N-1 \tag{9.1}$$

where $\{x(n), n = 0, \ldots, N - 1\}$ denotes the sensor baseline signal vector, N is the length of vector $\mathbf{x}$, and $\{X(k), k = 0, \ldots, N - 1\}$ denotes the DFT.

Figure 9.2 presents the spectrum plots of three sensors through the DFT operation. In this work, the frequency segment between 0.001 and 0.01 Hz is selected as desired spectrum of drift signal, and the power magnitude is set to 0 for other frequencies. Generally, the frequency components of environmental factors should be lower than drift. Then, an inverse DFT (IDFT) is employed for drift signal in time domain. The representation of IDFT is shown as

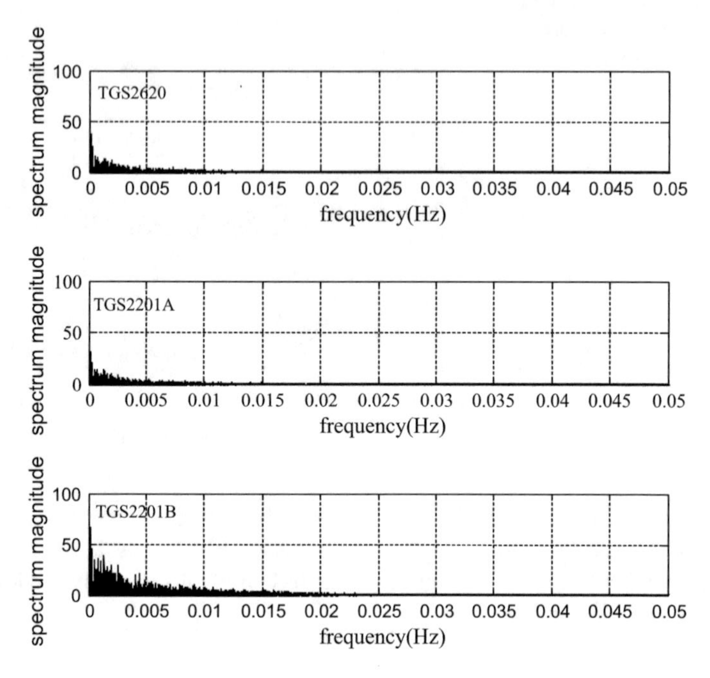

Fig. 9.2 Spectrum plots of DFT for three sensors

$$\tilde{x}(n) = \sum_{k=0}^{N-1} X(k)\mathrm{e}^{j\cdot(2\pi/N)\cdot k\cdot n}, \quad n = 0, 1, \ldots, N-1 \tag{9.2}$$

where $\{\tilde{x}(n), n = 0, \ldots, N-1\}$ denotes the reconstructed drift signal in time domain. Figure 9.3 presents the reconstructed drift signal of three sensors using IDFT operations.

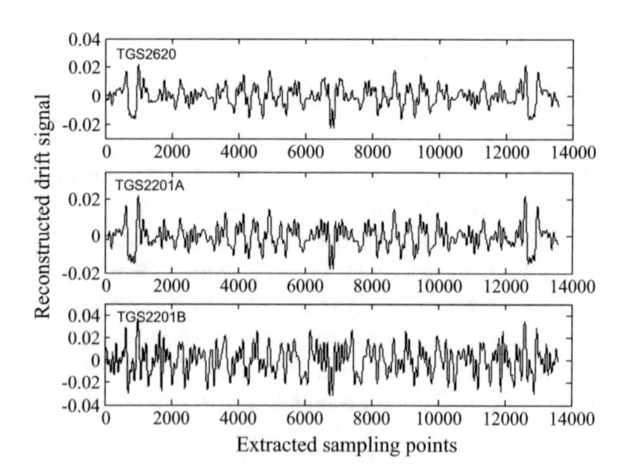

Fig. 9.3 Reconstructed sensor drift signal using IDFT

It is worth noting that this work is realized with two potential assumptions: First, the sensor drift shows a sufficiently low frequency; second, the drift from the long-term data time series should not change during the practical lifetime of the E-nose.

9.3 Chaotic Time Series Prediction

9.3.1 Phase Space Reconstruction

Phase space reconstruction (PSR) is the basic theory in the analysis of chaotic dynamic systems, founded by Taken [17]. Taken's theorem proves that we can reconstruct a phase space from a one-dimensional chaotic time series which has an equal phase space with dynamic system in topology. The discrimination, analysis, and prediction are employed in the reconstructed phase space; thus, PSR is the key point in chaotic time series study. The embedding dimension m and time delay τ are the most important variables in PSR. The selections of m and τ have been well studied by researchers in chaos [18–23]. In this chapter, false nearest neighbor (FNN) method [18] is used to determine the embedding dimension m. Auto-correlated function is used to calculate the time delay τ [18, 19].

Given a chaotic time series $\{x(t), t = 1, \ldots, N\}$, the reconstructed phase space can be represented by

$$\mathbf{X}(t) = \{x(t), x(t+\tau), \ldots, x(t+(m-1)\cdot\tau)\} \tag{9.3}$$

where $t = 1, \ldots, N - (m - 1)\tau$, and m ($m = 2, 3, \ldots$) is the embedding dimension, provided that $m \geq 2D + 1$, and D is the fractal dimension of attractor.

For clarity, we expand the Eq. 9.3 as follows

$$\begin{cases} \mathbf{X}(1) = [x(1), x(1+\tau), \ldots, x(1+(m-1)\cdot\tau)]^{\mathrm{T}} \\ \mathbf{X}(2) = [x(2), x(2+\tau), \ldots, x(2+(m-1)\cdot\tau)]^{\mathrm{T}} \\ \mathbf{X}(3) = [x(3), x(3+\tau), \ldots, x(3+(m-1)\cdot\tau)]^{\mathrm{T}} \\ \cdots \\ \mathbf{X}(N-(m-1)\cdot\tau-1) = [x(N-(m-1)\cdot\tau-1), x(N-(m-2)\cdot\tau), \ldots, x(N-1)]^{\mathrm{T}} \end{cases} \tag{9.4}$$

where symbol ^{T} denotes the transpose of a vector.

Lyapunov exponent is a useful tool to characterize a chaotic attractor quantitatively, which effectively measures the sensitivity of the chaotic orbit to its initial conditions and quantizes the attractor's dynamics of a complex system. When the Lyapunov exponent λ of a time series is positive, then the time series will become chaotic [12]. The computation of Lyapunov exponent in this work is employed using classical Wolf method [24].

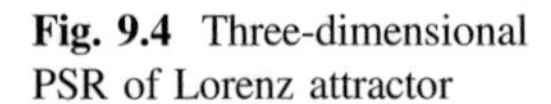
Fig. 9.4 Three-dimensional PSR of Lorenz attractor

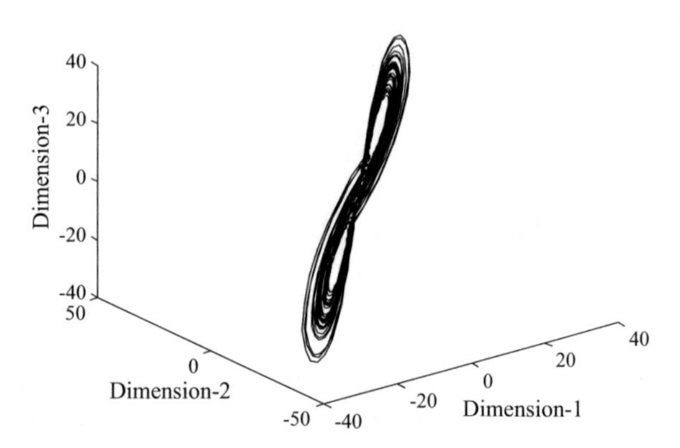

For PSR analysis, we generate a time series by numerically integrating the Lorenz system which consists of three ordinary differential equations [25]. The representation of the Lorenz system is shown as

$$\begin{cases} \dfrac{\mathrm{d}x}{\mathrm{d}t} = \sigma \cdot (y - x) \\ \dfrac{\mathrm{d}y}{\mathrm{d}t} = r \cdot x - y - x \cdot z \\ \dfrac{\mathrm{d}z}{\mathrm{d}t} = -b \cdot z + x \cdot y \end{cases} \tag{9.5}$$

where x is proportional to the intensity of convective motion; y is proportional to the horizontal temperature variation; z is proportional to the vertical temperature variation; σ, r, and b are constants. For analysis, the parameters σ, r, and b are set as 25, 3, and 50, respectively; the step size of integration is set to 0.01; the initial value of y is $[-1, 0, 1]^{\mathrm{T}}$. Figure 9.4 illustrates the PSR of Lorenz attractor in which the embedding dimension $m = 3$ and time delay $\tau = 1$.

9.3.2 Prediction Model

In chaotic time series prediction, radial basis function (RBF) neural network is used to trace the attractor in embedded phase space for each sensor. The use of RBF to model strange attractors representing time series data has been referred in [26]. RBF neural network [27] is a forward-feedback artificial neural network composed of three layers. The number of input nodes depends on the embedding dimension m. The input signal is just the reconstructed $\mathbf{X}(t)$ ($t = 1, \ldots, N - (m - 1)\tau$) in phase space. In the hidden layer, a Gaussian function is used as activation function. The structure of RBF neural network prediction in PSR is shown in Fig. 9.5.

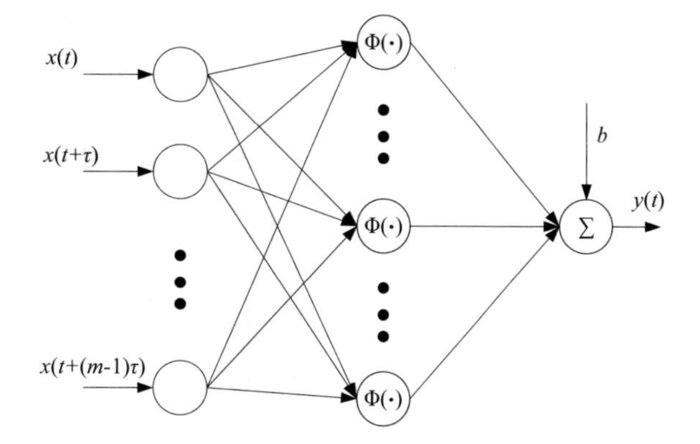

Fig. 9.5 Structure of RBF neural network for prediction in PSR

Combined with Eq. 9.4, the training input matrix can be shown by

$$\mathbf{P} = [\mathbf{X}(1), \mathbf{X}(2), \ldots, \mathbf{X}(N-(m-1)\cdot\tau-1)] \tag{9.6}$$

After expansion of **P** using Eqs. 9.4 and 9.6, we obtain

$$\mathbf{P} = \begin{bmatrix} x(1) & x(2) & \cdots & x(N-(m-1)\cdot\tau-1) \\ x(1+\tau) & x(2+\tau) & \cdots & x(N-(m-2)\cdot\tau-1) \\ \vdots & \vdots & \vdots & \vdots \\ x(1+(m-1)\cdot\tau) & x(2+(m-1)\cdot\tau) & \cdots & x(N-1) \end{bmatrix} \tag{9.7}$$

Then, the vector of training goal is represented as

$$\mathbf{T} = [x(2+(m-1)\cdot\tau), x(3+(m-1)\cdot\tau), \ldots, x(N)] \tag{9.8}$$

In this chapter, three models are constructed for TGS2620, TGS2201A, and TGS2201B, respectively. The chaotic time series prediction can be concluded as two steps: phase space reconstruction and RBF neural network learning. We apply the method to Lorenz attractor prediction for example analysis. We select 5000 points in Lorenz sequence for prediction, in which the first 1000 points are used for RBF neural network learning and the remaining 4000 points as test. The test results are presented in Fig. 9.6a, b, which illustrate the prediction of the Lorenz chaotic time series and the prediction residual error, respectively. We can see in Fig. 9.6 that the prediction of Lorenz chaotic time series is successful using the proposed model.

The well-known statistics root-mean-square error of prediction (RMSEP) [28] is also used to quantitatively measure the sensor baseline and drift prediction models in this work. Its representation can be expressed as

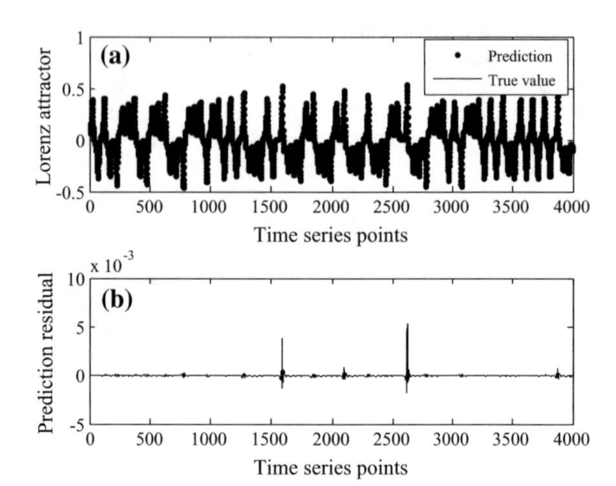

Fig. 9.6 Prediction of Lorenz time series using the proposed model

$$\text{RMSEP} = \sqrt{\frac{1}{N-1} \cdot \sum_{t=1}^{N} [s(t) - \hat{s}(t)]^2} \tag{9.9}$$

where N denotes the length of test sequence **s** which is the true sequence, and $\hat{\mathbf{s}}$ denotes the predicted output. In the ideal situation, if there is no error in prediction, then RMSEP= 0.

9.4 Results

9.4.1 Chaotic Characteristic Analysis

Chaotic characteristic of long-term sensor response has been confirmed in this work. Through analysis of $\{x(n), n = 1, \ldots, N\}$ of each sensor, the time delays calculated by auto-correlated function method for TGS2620, TGS2201A, and TGS2201B are 22, 24, and 23, respectively. The embedding dimension is calculated as $m = 3$. Three sub-figures in Fig. 9.7 illustrate the three-dimensional $[x(t), x(t + \tau), x(t + 2\tau)]$ PSR plots of the extracted 13,600 points for each sensor within a year which show the motion traces of sensors in PSR. Through the PSR plots in Fig. 9.7, we can find that the motion trace for each sensor has the characteristic of chaotic attractor.

In addition, for further confirmation, the Lyapunov exponents λ_1, λ_2, and λ_3 of time series calculated using Wolf method are 0.1412, 0.2435, and 0.2933 for TGS2620, TGS2201A, and TGS2201B sensor, respectively. The three Lyapunov exponents are positive, and thus, there is obviously chaotic behavior in long-term

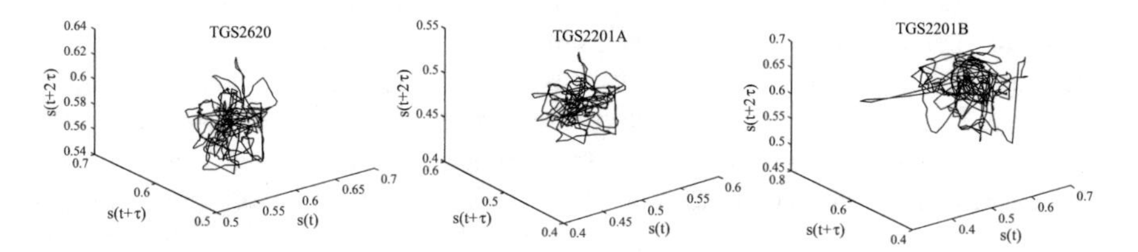

Fig. 9.7 PSR plots of sensor observation data

sensor observation in terms of chaotic attractor theory. This promotes us to study the sensor drift in a novel chaotic time series method.

9.4.2 *Drift Prediction*

The phase space reconstruction and RBF neural network are used for raw sensor baseline and reconstructed drift prediction in this work. The long-term observation data for each sensor includes 13,600 points (from September 04, 2011, to July 06, 2012). To validate the long-term predictability of each sensor, the first 1000 points in September, 2011 (from September 04, 2011, to September 25, 2011) are used to train a prediction model. The remaining 12600 points (from September 26, 2011, to July 06, 2012) are used for model test. Figure 9.8 presents the raw sensor baseline prediction results of three sensors. Figure 9.8a–c illustrates the predictions of TGS2620, TGS2201A, and TGS2201B, respectively. For more clear visualization of the prediction, we show the prediction results of detailed local segment from 12,000 to the end of the sequence in the Rectangular window (totally, 552 points) in Fig. 9.8a–c. The obvious results in Fig. 9.8a′–c′ correspond to the three rectangular windows of Fig. 9.8a–c, respectively. We can see in Fig. 9.8 that the prediction of the long-term baseline and drift signal reconstructed using DFT and IDFT is successful.

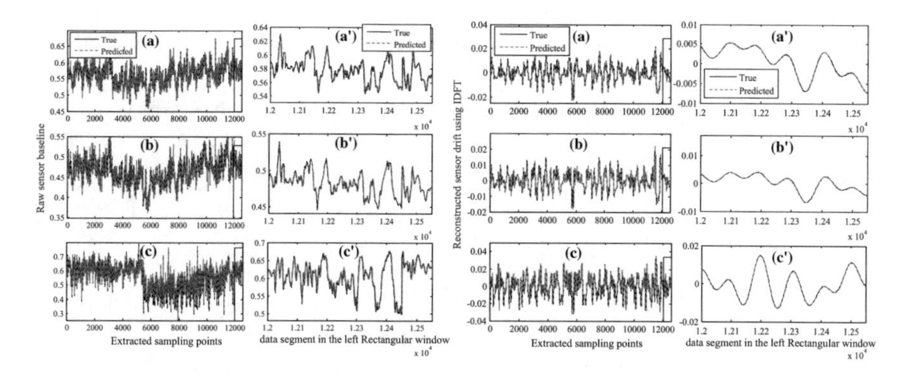

Fig. 9.8 Predictions of long-term sensor baseline using the proposed model

The prediction results shown in Fig. 9.8 are based on the RBF model trained by using the first 1000 data points of each sensor sequence. Due to that the number of training data points maybe useful to the potential researchers, we also experimentally show how many initial data points are required for prediction. Thus, the procedures of sensor baseline and drift predictions with smaller and larger numbers Ntr of training data points, such as 100, 250, 500, 750, 1250, and 1500, are performed, respectively. The predictions of TGS2620 sensor baseline with different numbers (Ntr) of training data points are shown in Fig. 9.9a. From the three figures, we can find that the prediction performance can be approved when Ntr is equal to 500, and there is little change when the number Ntr of training data points is larger than 1000. In addition, we also present the sensor drift predictions of TGS2620 with different numbers (Ntr) of training data points in Fig. 9.9b. We can also obtain from these figures that the smallest Ntr should be 500, and the modeling of RBF is enough when Ntr is not larger than 1000.

Besides the qualitative analysis, for quantitative evaluation of the prediction performance, Table 9.1 presents the RMSEPs of sensors' baseline and drift

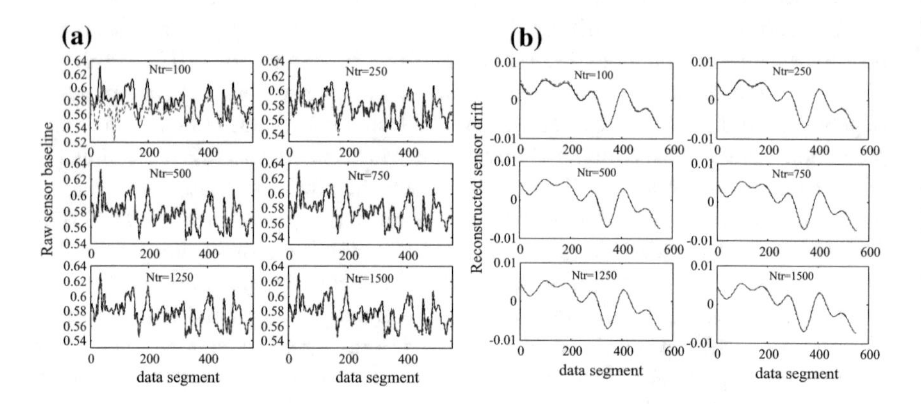

Fig. 9.9 TGS2620 sensor baseline predictions (**a**) and drift prediction (**b**) of the data segment

Table 9.1 RMSEP of sensor baseline and drift prediction with a different Ntr of training data points

Ntr	RMSEP of sensor baseline prediction			RMSEP of sensor drift prediction		
	TGS2620	TGS2201A	TGS2201B	TGS2620	TGS2201A	TGS2201B
100	0.0319	0.0330	0.5719	0.0050	0.0045	0.0299
250	0.0102	0.0121	0.0684	0.0021	0.0023	0.0098
500	0.0054	0.0060	0.0290	0.0011	0.0012	0.0035
750	0.0045	0.0044	0.0170	9.21e−04	0.0010	0.0030
1000	0.0045	0.0044	0.0179	9.19e−04	8.95e−04	0.0029
1250	0.0046	0.0045	0.0288	9.09e−04	8.93e−04	0.0027
1500	0.0046	0.0045	0.0177	9.12e−04	8.96e−04	0.0027

prediction modeling with a different number of training data points calculated by using the Eq. 9.9. From Table 9.1, we can find that the RMSEP reduces with increasing number of Ntr, and an appropriate value of Ntr should be between 500 and 1000.

The results presented in this work confirm that the long-term sensor baseline and drift signal show chaotic characteristic. Thus, it is difficult to find an effective drift compensation model in general ways. For example, the usual drift counteraction method is attributed to a fixed drift dataset. However, the drift model constructed using the fixed dataset may be not effective to a long-term drift. That is because the drift trend is uncertain, chaotic, and unpredictable. Chaotic time series prediction is developed based on chaotic attractor in embedded phase space. PSR remains the properties of raw chaotic attractor, and the prediction is feasible in PSR of a chaotic time series.

9.5 Summary

This chapter presents a novel methodology in the treatment of chemical sensor drift in embedded phase space. The chaotic characteristic of long-term sensor time series is studied in terms of chaotic attractor theory and Lyapunov exponents. In view of the uncertainty, unpredictability, and obvious chaotic characteristic, PSR and RBF neural network are combined together for long-term prediction of chaotic time series. In drift signal extraction, DFT technique is used for spectrum analysis and drift frequency selection for excluding the environmental noise frequencies. IDFT is then used for time-domain drift signal reconstruction. Qualitative and quantitative results of time series baseline and drift prediction demonstrate the effectiveness of the proposed method.

References

1. D. James, S.M. Scott, Z. Ali, W.T.O. Hare, Chemical sensors for electronic nose systems. Microchim. Acta **149**, 1–17 (2005)
2. S.M. Scott, D. James, Z. Ali, Data analysis for electronic nose systems. Microchim. Acta **156**, 183–207 (2007)
3. M. Holmberg, F.A.M. Davide, C. Di Natale, A.D. Amico, F. Winquist, I. Lundström, Drift counteraction in odour recognition applications: lifelong calibration method. Sens. Actuators B **42**, 185–194 (1997)
4. A.C. Romain, J. Nicolas, Long term stability of metal oxide-based gas sensors for e-nose environmental applications: an overview. Sens. Actuators B **146**, 502–506 (2010)
5. T. Artursson, T. Eklov, I. Lundström, P. Martensson, M. Sjostrom, M. Holmberg, Drift correction for gas sensors using multivariate methods. J. Chemometr. **14**, 711–723 (2000)
6. A. Ziyatdinov, S. Marco, A. Chaudry, K. Persaud, P. Caminal, A. Perera, Drift compensation of gas sensor array data by common principal component analysis. Sens. Actuators B **146**, 460–465 (2010)

7. H. Ding, J. Liu, Z. Shen, Drift reduction of gas sensor by wavelet and principal component analysis. Sens. Actuators B **96**, 354–363 (2003)
8. S. Di Carlo, M. Falasconi, E. Sanchez, A. Scionti, G. Squillero, A. Tonda, Increasing pattern recognition accuracy for chemical sensing by evolutionary based drift compensation. Pattern Recogn. Lett. **32**, 1594–1603 (2011)
9. M. Padilla, A. Perera, I. Montoliu, A. Chaudry, K. Persaud, S. Marco, Drift compensation of gas sensor array data by orthogonal signal correction. Chemometr. Intell. Lab. Syst. **100**, 28–35 (2010)
10. M.J. Wenzel, A.M. Brown, F. Josse, E.E. Yaz, Online drift compensation for chemical sensors using estimation theory. IEEE Sens. J. **11**(1), 225–232 (2011)
11. J. Zhang, K.C. Lam, W.J. Yan, H. Gao, Y. Li, Time series prediction using Lyapunov exponents in embedding phase space. Comput. Electr. Eng. **30**, 1–15 (2004)
12. M. Han, J. Xi, S. Xu, F.L. Yin, Prediction of chaotic time series based on the recurrent predictor neural network. IEEE Trans. Signal Process. **52**, 3409–3416 (2004)
13. L. Zhang, F. Tian, C. Kadri, G. Pei, H. Li, L. Pan, Gases concentration estimation using heuristics and bio-inspired optimization models for experimental chemical electronic nose. Sens. Actu. B Chem. **160**, 760–770 (2011)
14. L. Zhang, F. Tian, H. Nie, L. Dang, G. Li, Q. Ye, C. Kadri, Classification of multiple indoor air contaminants by an electronic nose and a hybrid support vector machine. Sens. Actuators B Chem. **174**, 114–125 (2012)
15. L. Zhang, F. Tian, S. Liu, J. Guo, B. Hu, Q. Ye, L. Dang, X. Peng, C. Kadri, J. Feng, Chaos based neural network optimization for concentration estimation of indoor air contaminants by an electronic nose. Sens. Actuators A Phys. (2012). https://doi.org/10.1016/j.sna.2012.10.023
16. D. Huang, H. Leung, Reconstruction of drifting sensor responses based on Papoulis-Gerchberg method. IEEE Sens. J. **9**, 595–604 (2009)
17. F. Taken, Detecting strange attractors in turbulence, in *Dynamical Systems and Turbulence* (Springer, Berlin, 1980)
18. H.D.I. Abarbanel, R. Brown, J.J. Sidorowich, L.S. Tsimring, The analysis of observed chaotic data in physical systems. Rev. Mod. Phys. **65**, 1331–1392 (1993)
19. M.T. Rosenstein, J.J. Collins, C.J. De Luca, A practical method for calculating largest Lyapunov exponents from small data sets. Physica D **65**, 117–134 (1993)
20. L. Cao, Practical method for determining the minimum embedding dimension of a scalar time series. Physica D **110**, 43–50 (1997)
21. M. Lei, Z. Wang, Z. Feng, A method of embedding dimension estimation based on symplectic geometry. Phys. Lett. A **303**, 179–189 (1994)
22. M.T. Rosenstein, J.J. Collins, J. De Luca Carlo, Reconstruction expansion as a geometry based framework for choosing proper delay times. Physica D **73**, 82–98 (1994)
23. T. Buzug, G. Pfister, Optimal delay time and embedding dimension for delay time coordinates by analysis of the global and local dynamical behavior of strange attractors. Phys. Rev. A **45**, 7073–7084 (1992)
24. A. Wolf, J.B. Swift, H.L. Swinney, J.A. Vastano, Determining Lyapunov exponents from a time series. Physica D **16**, 285–317 (1985)
25. E.N. Lorenz, Deterministic nonperiodic flow. J. Atmos. Sci. **20**, 130–141 (1963)
26. M. Casdagli, Nonlinear prediction of chaotic time series. Physica D **35**, 335–356 (1989)
27. Y. Lu, N. Sundararajan, P. Saratchandran, Performance evaluation of a sequential minimal radial basis function (RBF) neural network learning algorithm. IEEE Trans. Neural Netw. **9**, 308–318 (1998)
28. C. Jiann Long, I. Shafiqul, B. Pratim, Nonlinear dynamics of hourly ozone concentrations: nonparametric short term prediction. Atmos. Environ. **32**, 1839–1848 (1998)

Chapter 10
Domain Adaptation Guided Drift Compensation

Abstract This chapter addresses the sensor drift issue in E-nose from the viewpoint of machine learning. Traditional methods for drift compensation are laborious and costly due to the frequent acquisition and labeling process for gases samples recalibration. Extreme learning machines (ELMs) have been confirmed to be efficient and effective learning techniques for pattern recognition and regression. However, ELMs primarily focus on the supervised, semi-supervised, and unsupervised learning problems in single domain (i.e., source domain). Drift data and non-drift data can be recognized as cross-domain data. Therefore, this chapter proposes a unified framework, referred to as domain adaptation extreme learning machine (DAELM), which learns a cross-domain classifier with drift compensation. Experiments on the popular sensor drift data of multiple batches clearly demonstrate that the proposed DAELM significantly outperforms existing drift compensation methods.

Keywords Drift compensation · Electronic nose · Extreme learning machine
Domain adaptation · Transfer learning

10.1 Introduction

Extreme learning machine (ELM), proposed for solving a single-layer feed-forward network (SLFN) by Huang et al. [1, 2], has been proven to be effective and efficient algorithms for pattern classification and regression in different fields. ELM can analytically determine the output weights between the hidden layer and output layer using Moore-Penrose generalized inverse by adopting the square loss of prediction error, which then turns into solving a regularized least square problem efficiently in closed form. The hidden layer output is activated by an infinitely differentiable function with randomly selected input weights and biases of the hidden layer. Huang [3] rigorously proved that the input weights and hidden layer biases can be randomly assigned if the activation function is infinitely differentiable, and also showed that single SLFN with randomly generated additive or RBF nodes with

L. Zhang et al., *Electronic Nose: Algorithmic Challenges*,
https://doi.org/10.1007/978-981-13-2167-2_10

such activation functions can universally approximate any continuous function on any compact subspace of Euclidean space [4].

In recent years, ELM has witnessed a number of improved versions in models, algorithms and real-world applications. ELM shows a comparable or even higher prediction accuracy than that of SVMs which solves a quadratic programming problem. In [3], their differences have been discussed. Some specific examples of improved ELMs have been listed as follows. As the output weights are computed with predefined input weights and biases, a set of non-optimal input weights and hidden biases may exist. Additionally, ELM may require more hidden neurons than conventional learning algorithms in some special applications. Therefore, Zhu et al. [5] proposed an evolutionary ELM for more compact networks that speed the response of trained networks. In terms of the imbalanced number of classes, a weighted ELM was proposed for binary/multi-class classification tasks with both balanced and imbalanced data distribution [6]. Due to that the solution of ELM is dense which will require longer time for training in large-scale applications. Bai et al. [7] proposed a sparse ELM for reducing storage space and testing time. Besides, Li et al. [8] also proposed a fast sparse approximation of ELM for sparse classifiers training at a rather low complexity without reducing the generalization performance. For all the versions of ELM mentioned above, supervised learning framework was widely explored in application which limits its ability due to the difficulty in obtaining the labeled data. Therefore, Huang et al. [9] proposed a semi-supervised ELM for classification, in which a manifold regularization with graph Laplacian was set, and an unsupervised ELM was also explored for clustering.

In the past years, the contributions to ELM theories and applications have been made substantially by researchers from various fields. However, with the rising of big data, the data distribution obtained in different stages with different experimental conditions may change, i.e., from different domains. It is also well know that E-nose data collection and data labeling is tedious and labor ineffective, while the classifiers trained by a small number of labeled data are not robust and therefore lead to weak generalization, especially for large-scale application. Though ELM performs better generalization when a number of labeled data from source domain is used in learning, the transferring capability of ELM is reduced with a limited number of labeled training instances from target domains. Domain adaptation methods have been proposed for robust classifiers learning by leveraging a few labeled instances from target domains [10–14] in machine learning community and computer vision [15]. It is worth noting that domain adaptation is different from semi-supervised learning which assumes that the labeled and unlabeled data are from the same domain in classifier training.

In this chapter, we extend ELMs to handle domain adaptation problems for improving the transferring capability of ELM between multiple domains with very few labeled guide instances in target domain, and overcome the generalization disadvantages of ELM in multi-domains application. Specifically, we address the problem of sensor drift compensation in E-nose by using the proposed cross-domain learning framework. Inspired by ELM and knowledge adaptation, a unified domain adaptation ELM framework is proposed for sensor drift compensation.

10.2 Related Work

10.2.1 Drift Compensation

Electronic nose is an intelligent multi-sensor system or artificial olfaction system, which is developed as instrument for gas recognition [16, 17], tea quality assessment [18, 19], medical diagnosis [20], environmental monitor, and gas concentration estimation [21, 22], etc., by coupling with pattern recognition and gas sensor array with cross-sensitivity and broad spectrum characteristics. An excellent overview of the E-nose and techniques for processing the sensor responses can be referred to as [23, 24].

However, sensors are often operated over a long period in real-world application and lead to aging that seriously reduces the lifetime of sensors. This is so-called sensor drift caused by unknown dynamic process, such as poisoning, aging, or environmental variations [25]. Sensor drift has deteriorated the performance of classifiers [26] used for gas recognition of chemosensory systems or E-noses and plagued the sensory community for many years. Therefore, researchers have to retrain the classifier using a number of new samples in a period regularly for recalibration. However, the tedious work for classifier retraining and acquisition of new labeled samples regularly seems to be impossible for recalibration, due to the complicated gaseous experiments of E-nose and labor cost.

The drift problem can be formulated as follows.

Suppose $\mathcal{D}_1, \mathcal{D}_2, \ldots, \mathcal{D}_K$ are gas sensor datasets collected by an E-nose with K batches ranked according to the time intervals, where $\mathcal{D}_i = \{\mathbf{x}_j^i\}_{j=1}^{N_i}$, $i = 1, \ldots, K$, $\mathbf{x}_j^i$ denotes a feature vector of the j-th sample in batch i, and N_i is the number of samples in batch i. The sensor drift problem is that the feature distributions of $\mathcal{D}_2, \ldots, \mathcal{D}_K$ do not obey the distribution of $\mathcal{D}_1$. As a result, the classifier trained using the labeled data of $\mathcal{D}_1$ has degraded performance when tested on $\mathcal{D}_2, \ldots, \mathcal{D}_K$ due to the deteriorated generalization ability caused by drift. Generally, the mismatch of distribution between $\mathcal{D}_1$ and $\mathcal{D}_i$ becomes larger with increasing batch index i ($i > 1$) and aging. From the angle of domain adaptation, in this chapter, $\mathcal{D}_1$ is called source domain/auxiliary domain (without drift) with labeled data, $\mathcal{D}_2, \ldots, \mathcal{D}_K$ are referred to as target domain (drifted) in which only a limited number of labeled data is available.

Drift compensation has been studied for many years. Generally, drift compensation methods can be divided into three categories: component correction methods, adaptive methods, and machine learning methods. Specifically, multivariate component correction, such as CC-PCA [27] which attempts to find the drift direction using PCA and remove the drift component is recognized as a popular method in periodic calibration. However, CC-PCA assumes that the data from all classes behaves in the same way in the presence of drift that is not always the case. Additionally, evolutionary algorithm which optimizes a multiplicative correction factor for drift compensation [28] was proposed as an adaptive method. However, the generalization performance of the correction factor is limited for on-line use due to the nonlinear dynamic behavior of sensor drift. Classifier ensemble in machine learning was first proposed in [29] for drift compensation, which has shown improved gas recognition accuracy using the data with long-term drift. An overview of the drift compensation is referred to as [26]. Other recent methods to cope with drift can be referred to as [30–32].

Though researchers have paid more attention to sensor drift and aim to find some measures for drift compensation, sensor drift is still a challenging issue in machine olfaction community and sensory field. To our best knowledge, the existing methods are limited in dealing with sensor drift due to their weak generalization to completely new data in the presence of drift. Therefore, we aim to enhance the adaptive performance of classifiers to new drifting/drifted data using cross-domain learning with very low complexity. It would be very meaningful and interesting to train a classifier using very few labeled new samples (target domain) without giving up the recognized "useless" old data (source domain) and realize effective and efficient knowledge transfer (i.e., drift compensation) from source domain to multiple target domains.

10.2.2 Extreme Learning Machine

Given N samples $[\mathbf{x}_1, \mathbf{x}_2, \ldots, \mathbf{x}_N]$ and their corresponding ground truth $[\mathbf{t}_1, \mathbf{t}_2, \ldots, \mathbf{t}_N]$, where $\mathbf{x}_i = [x_{i1}, x_{i1}, \ldots, x_{\text{in}}]^{\text{T}} \in \mathbb{R}^n$ and $\mathbf{t}_i = [t_{i1}, t_{i1}, \ldots, t_{\text{im}}]^{\text{T}} \in \mathbb{R}^m$, n and m denote the number of input and output neurons, respectively. The output of the hidden layer is denoted as $\mathcal{H}(\mathbf{x}_i) \in \mathbb{R}^{1 \times L}$, where L is the number of hidden nodes and $\mathcal{H}(\cdot)$ is the activation function (e.g., RBF function, sigmoid function). The output weights between the hidden layer and the output layer being learned is denoted as $\boldsymbol{\beta} \in \mathbb{R}^{L \times m}$.

Regularized ELM aims to solve the output weights by minimizing the squared loss summation of prediction errors and the norm of the output weights for over-fitting control, formulated as follows

$$\begin{cases} \min\limits_{\boldsymbol{\beta}} \mathcal{L}_{\text{ELM}} = \frac{1}{2}\|\boldsymbol{\beta}\|^2 + C \cdot \frac{1}{2} \cdot \sum\limits_{i=1}^{N} \|\boldsymbol{\xi}_i\|^2 \\ \text{s.t.}\, \mathcal{H}(\mathbf{x}_i)\boldsymbol{\beta} = \mathbf{t}_i - \boldsymbol{\xi}_i, \quad i = 1, \ldots, N \end{cases} \tag{10.1}$$

where $\boldsymbol{\xi}_i$ denotes the prediction error w.r.t. to the i-th training pattern, and C is a penalty constant on the training errors.

By substituting the constraint term in Eq. 10.1 into the objective function, an equivalent unconstrained optimization problem can be obtained as follows

$$\min_{\boldsymbol{\beta} \in \mathbb{R}^{L \times m}} \mathcal{L}_{\text{ELM}} = \frac{1}{2}\|\boldsymbol{\beta}\|^2 + C \cdot \frac{1}{2} \cdot \|\mathbf{T} - \mathbf{H}\boldsymbol{\beta}\|^2 \tag{10.2}$$

where $\mathbf{H} = [\mathcal{H}(\mathbf{x}_1); \mathcal{H}(\mathbf{x}_2); \ldots; \mathcal{H}(\mathbf{x}_N)] \in \mathbb{R}^{N \times L}$ and $\mathbf{T} = [\mathbf{t}_1, \mathbf{t}_2, \ldots, \mathbf{t}_N]^{\mathrm{T}}$.

The optimization Eq. 10.2 is a well-known regularized least square problem. The closed-form solution of $\boldsymbol{\beta}$ can be easily solved by setting the gradient of the objective function Eq. 10.2 with respect to $\boldsymbol{\beta}$ to zero.

There are two cases when solving $\boldsymbol{\beta}$; i.e., if the number N of training patterns is larger than L, the gradient equation is overdetermined, and the closed-form solution can be obtained as

$$\boldsymbol{\beta}^* = \left(\mathbf{H}^{\mathrm{T}}\mathbf{H} + \frac{\mathbf{I}_L}{C} \right)^{-1} \mathbf{H}^{\mathrm{T}}\mathbf{T} \tag{10.3}$$

where $\mathbf{I}_{L \times L}$ denotes the identity matrix.

If the number N of training patterns is smaller than L, an under-determined least square problem would be handled. In this case, the solution of Eq. 10.2 can be obtained as

$$\boldsymbol{\beta}^* = \mathbf{H}^{\mathrm{T}} \left(\mathbf{H}\mathbf{H}^{\mathrm{T}} + \frac{\mathbf{I}_N}{C} \right)^{-1} \mathbf{T} \tag{10.4}$$

where $\mathbf{I}_{N \times N}$ denotes the identity matrix.

Therefore, in classifier training of ELM, the output weights can be computed by using Eq. 10.3 or Eq. 10.4 depending on the number of training instances and the number of hidden nodes. We refer interested readers to as [1] for more details of ELM theory and the algorithms.

10.3 Domain Adaptation Extreme Learning Machine

In this section, we present formulation of the proposed domain adaptation ELM framework, in which two methods referred to as source domain adaptation ELM (DAELM-S) and target domain adaptation ELM (DAELM-T) are introduced with their learning algorithms, respectively.

10.3.1 Source Domain Adaptation ELM (DAELM-S)

Suppose that the source domain and target domain are represented by "S" and "T". In this chapter, we assume that all the samples in the source domain are labeled data.

The proposed DAELM-S aims to learn a classifier $\boldsymbol{\beta}_{\mathrm{S}}$ using all labeled instances from the source domain by leveraging a limited number of labeled data from target domain. The DAELM-S can be formulated as

$$\min_{\boldsymbol{\beta}_\mathrm{S},\xi_\mathrm{S}^i,\xi_\mathrm{T}^i} \frac{1}{2}\|\boldsymbol{\beta}_\mathrm{S}\|^2 + C_\mathrm{S}\frac{1}{2}\sum_{i=1}^{N_\mathrm{S}}\|\xi_\mathrm{S}^i\|^2 + C_\mathrm{T}\frac{1}{2}\sum_{j=1}^{N_\mathrm{T}}\|\xi_\mathrm{T}^j\|^2 \tag{10.5}$$

$$\text{s.t.} \begin{cases} \mathbf{H}_\mathrm{S}^i\boldsymbol{\beta}_\mathrm{S} = \mathbf{t}_\mathrm{S}^i - \xi_\mathrm{S}^i, & i = 1,\ldots,N_\mathrm{S} \\ \mathbf{H}_\mathrm{T}^j\boldsymbol{\beta}_\mathrm{S} = \mathbf{t}_\mathrm{T}^j - \xi_\mathrm{T}^j, & j = 1,\ldots,N_\mathrm{T} \end{cases} \tag{10.6}$$

where $\mathbf{H}_\mathrm{S}^i \in \mathbb{R}^{1\times L}$, $\xi_\mathrm{S}^i \in \mathbb{R}^{1\times m}$, $\mathbf{t}_\mathrm{S}^i \in \mathbb{R}^{1\times m}$ denote the output of hidden layer, the prediction error and the label w.r.t. the i-th training instance $\mathbf{x}_\mathrm{S}^i$ from the source domain, $\mathbf{H}_\mathrm{T}^j \in \mathbb{R}^{1\times L}, \xi_\mathrm{T}^j \in \mathbb{R}^{1\times m}$, $\mathbf{t}_\mathrm{T}^j \in \mathbb{R}^{1\times m}$ denote the output of hidden layer, the prediction error and the label vector with respect to the j-th guide samples $\mathbf{x}_\mathrm{T}^j$ from the target domain, $\boldsymbol{\beta}_\mathrm{S} \in \mathbb{R}^{L\times m}$ is the output weights being solved, N_S and N_T denote the number of training instances and guide samples from the source domain and target domain, respectively, C_S and C_T are the penalty coefficients on the prediction errors of the labeled training data from source domain and target domain, respectively. In this chapter, $\mathbf{t}_\mathrm{S}^{i,j} = 1$ if pattern $\mathbf{x}_i$ belongs to the j-th class, and -1 otherwise. For example, $\mathbf{t}_\mathrm{S}^i = [1, -1, \ldots, -1]^\mathrm{T}$ if $\mathbf{x}_i$ belong to class 1.

From (10.5), we can find that the very few labeled guide samples from target domain can make the learning of $\boldsymbol{\beta}_\mathrm{S}$ "transferable" and realize the knowledge transfer between source domain and target domain by introducing the third term as regularization coupling with the second constraint in Eq. 10.6, which makes the feature mapping of the guide samples from target domain approximate the labels recognized with the output weights $\boldsymbol{\beta}_\mathrm{S}$. The structure of the proposed DAELM-S algorithm to learn M classifiers is illustrated in Fig. 10.1.

To solve the optimization Eq. 10.5, the Lagrangian multiplier equation is formulated as

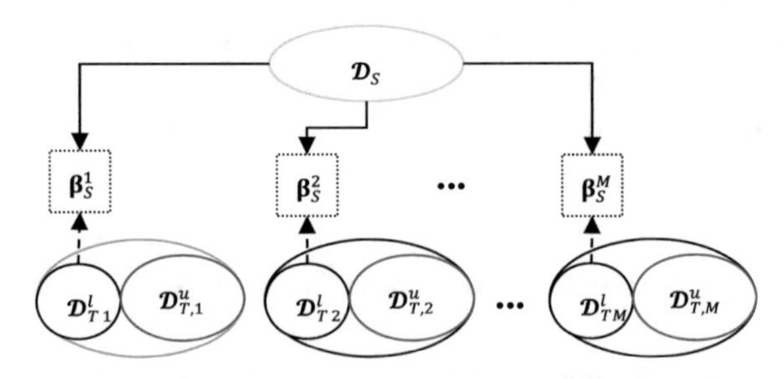

Fig. 10.1 Structure of DAELM-S algorithm with M target domains (M tasks). The solid arrow denotes the training data from source domain $\mathcal{D}_\mathrm{S}$ and the dashed arrow denotes the labeled data from target domain $\mathcal{D}_\mathrm{T}^l$ for classifier learning. The unlabeled data from target domain $\mathcal{D}_\mathrm{T}^u$ are not used in learning

$$L(\boldsymbol{\beta}_{\mathrm{S}}, \xi_{\mathrm{S}}^{i}, \xi_{\mathrm{T}}^{j}, \alpha_{\mathrm{S}}, \alpha_{\mathrm{T}}) = \frac{1}{2}\|\boldsymbol{\beta}_{\mathrm{S}}\|^2 + \frac{C_{\mathrm{S}}}{2}\sum_{i=1}^{N_{\mathrm{S}}}\|\xi_{\mathrm{S}}^{i}\|^2 + \frac{C_{\mathrm{T}}}{2}\sum_{j=1}^{N_{\mathrm{T}}}\|\xi_{\mathrm{T}}^{j}\|^2 - \boldsymbol{\alpha}_{\mathrm{S}}\left(\mathbf{H}_{\mathrm{S}}^{i}\boldsymbol{\beta}_{\mathrm{S}} - \mathbf{t}_{\mathrm{S}}^{i} + \xi_{\mathrm{S}}^{i}\right) - \boldsymbol{\alpha}_{\mathrm{T}}\left(\mathbf{H}_{\mathrm{T}}^{i}\boldsymbol{\beta}_{\mathrm{T}} - \mathbf{t}_{\mathrm{T}}^{i} + \xi_{\mathrm{T}}^{i}\right) \quad (10.7)$$

where $\boldsymbol{\alpha}_{\mathrm{S}}$ and $\boldsymbol{\alpha}_{\mathrm{T}}$ denote the multiplier vectors.

By setting the partial derivation with respect to $\boldsymbol{\beta}_{\mathrm{S}}, \xi_{\mathrm{S}}^{i}, \xi_{\mathrm{T}}^{j}, \boldsymbol{\alpha}_{\mathrm{S}}, \boldsymbol{\alpha}_{\mathrm{T}}$ as zero, we have

$$\begin{cases} \frac{\partial L}{\partial \boldsymbol{\beta}_{\mathrm{S}}} = 0 \to \boldsymbol{\beta}_{\mathrm{S}} = \mathbf{H}_{\mathrm{S}}^{\mathrm{T}}\boldsymbol{\alpha}_{\mathrm{S}} + \mathbf{H}_{\mathrm{T}}^{\mathrm{T}}\boldsymbol{\alpha}_{\mathrm{T}} \\ \frac{\partial L}{\partial \xi_{\mathrm{S}}} = 0 \to \boldsymbol{\alpha}_{\mathrm{S}} = C_{\mathrm{S}}\xi_{\mathrm{S}}^{\mathrm{T}} \\ \frac{\partial L}{\partial \xi_{\mathrm{T}}} = 0 \to \boldsymbol{\alpha}_{\mathrm{T}} = C_{\mathrm{T}}\xi_{\mathrm{T}}^{\mathrm{T}} \\ \frac{\partial L}{\partial \boldsymbol{\alpha}_{\mathrm{S}}} = 0 \to \mathbf{H}_{\mathrm{S}}\boldsymbol{\beta}_{\mathrm{S}} - \mathbf{t}_{\mathrm{S}} + \xi_{\mathrm{S}} = 0 \\ \frac{\partial L}{\partial \boldsymbol{\alpha}_{\mathrm{T}}} = 0 \to \mathbf{H}_{\mathrm{T}}\boldsymbol{\beta}_{\mathrm{S}} - \mathbf{t}_{\mathrm{T}} + \xi_{\mathrm{T}} = 0 \end{cases} \quad (10.8)$$

where $\mathbf{H}_{\mathrm{S}}$ and $\mathbf{H}_{\mathrm{T}}$ are the output matrix of hidden layer with respect to the labeled data from source domain and target domain, respectively. To analytically determine $\boldsymbol{\beta}_{\mathrm{S}}$, the multiplier vectors $\boldsymbol{\alpha}_{\mathrm{S}}$ and $\boldsymbol{\alpha}_{\mathrm{T}}$ should be solved first.

For the case that the number of training samples N_{S} is smaller than L ($N_{\mathrm{S}} < L$), then $\mathbf{H}_{\mathrm{S}}$ will have more columns than rows and be of full row rank, which leads to a under-determined least square problem, and infinite number of solutions may be obtained. To handle this problem, in Eq. 10.8, we substitute the 1st, 2nd, and 3rd equations into the 4th and 5th equations considering that $\mathbf{HH}^{\mathrm{T}}$ is invertible, and then there is

$$\begin{cases} \mathbf{H}_{\mathrm{T}}\mathbf{H}_{\mathrm{S}}^{\mathrm{T}}\boldsymbol{\alpha}_{\mathrm{S}} + \left(\mathbf{H}_{\mathrm{T}}\mathbf{H}_{\mathrm{T}}^{\mathrm{T}} + \frac{\mathbf{I}}{C_{\mathrm{T}}}\right)\boldsymbol{\alpha}_{\mathrm{T}} = \mathbf{t}_{\mathrm{T}} \\ \mathbf{H}_{\mathrm{S}}\mathbf{H}_{\mathrm{T}}^{\mathrm{T}}\boldsymbol{\alpha}_{\mathrm{T}} + \left(\mathbf{H}_{\mathrm{S}}\mathbf{H}_{\mathrm{S}}^{\mathrm{T}} + \frac{\mathbf{I}}{C_{\mathrm{S}}}\right)\boldsymbol{\alpha}_{\mathrm{S}} = \mathbf{t}_{\mathrm{S}} \end{cases} \quad (10.9)$$

Let $\mathbf{H}_{\mathrm{T}}\mathbf{H}_{\mathrm{S}}^{\mathrm{T}} = \mathcal{A}$, $\mathbf{H}_{\mathrm{T}}\mathbf{H}_{\mathrm{T}}^{\mathrm{T}} + \frac{\mathbf{I}}{C_{\mathrm{T}}} = \mathcal{B}, \mathbf{H}_{\mathrm{S}}\mathbf{H}_{\mathrm{T}}^{\mathrm{T}} = \mathcal{C}, \mathbf{H}_{\mathrm{S}}\mathbf{H}_{\mathrm{S}}^{\mathrm{T}} + \frac{\mathbf{I}}{C_{\mathrm{S}}} = \mathcal{D}$, then Eq. 10.9 can be written as

$$\begin{cases} \mathcal{A}\boldsymbol{\alpha}_{\mathrm{S}} + \mathcal{B}\boldsymbol{\alpha}_{\mathrm{T}} = \mathbf{t}_{\mathrm{T}} \\ \mathcal{C}\boldsymbol{\alpha}_{\mathrm{T}} + \mathcal{D}\boldsymbol{\alpha}_{\mathrm{S}} = \mathbf{t}_{\mathrm{S}} \end{cases} \to \begin{cases} \mathcal{B}^{-1}\mathcal{A}\boldsymbol{\alpha}_{\mathrm{S}} + \boldsymbol{\alpha}_{\mathrm{T}} = \mathcal{A}^{-1}\mathbf{t}_{\mathrm{T}} \\ \mathcal{C}\boldsymbol{\alpha}_{\mathrm{T}} + \mathcal{D}\boldsymbol{\alpha}_{\mathrm{S}} = \mathbf{t}_{\mathrm{S}} \end{cases} \quad (10.10)$$

Then $\boldsymbol{\alpha}_{\mathrm{S}}$ and $\boldsymbol{\alpha}_{\mathrm{T}}$ can be solved as

$$\begin{cases} \boldsymbol{\alpha}_{\mathrm{S}} = \left(\mathcal{C}\mathcal{B}^{-1}\mathcal{A} - \mathcal{D}\right)^{-1}\left(\mathcal{C}\mathcal{B}^{-1}\mathbf{t}_{\mathrm{T}} - \mathbf{t}_{\mathrm{S}}\right) \\ \boldsymbol{\alpha}_{\mathrm{T}} = \mathcal{B}^{-1}\mathbf{t}_{\mathrm{T}} - \mathcal{B}^{-1}\mathcal{A}\left(\mathcal{C}\mathcal{B}^{-1}\mathcal{A} - \mathcal{D}\right)^{-1}\left(\mathcal{C}\mathcal{B}^{-1}\mathbf{t}_{\mathrm{T}} - \mathbf{t}_{\mathrm{S}}\right) \end{cases} \quad (10.11)$$

Consider the 1st equation in Eq. 10.8, we obtain the output weights as

$$\begin{aligned}\boldsymbol{\beta}_{\mathrm{S}} &= \mathbf{H}_{\mathrm{S}}^{\mathrm{T}}\boldsymbol{\alpha}_{\mathrm{S}} + \mathbf{H}_{\mathrm{T}}^{\mathrm{T}}\boldsymbol{\alpha}_{\mathrm{T}} \\ &= \mathbf{H}_{\mathrm{S}}^{\mathrm{T}}\left(\mathcal{C}\mathcal{B}^{-1}\mathcal{A} - \mathcal{D}\right)^{-1}\left(\mathcal{C}\mathcal{B}^{-1}\mathbf{t}_{\mathrm{T}} - \mathbf{t}_{\mathrm{S}}\right) \\ &\quad + \mathbf{H}_{\mathrm{T}}^{\mathrm{T}}\left[\mathcal{B}^{-1}\mathbf{t}_{\mathrm{T}} - \mathcal{B}^{-1}\mathcal{A}\left(\mathcal{C}\mathcal{B}^{-1}\mathcal{A} - \mathcal{D}\right)^{-1}\left(\mathcal{C}\mathcal{B}^{-1}\mathbf{t}_{\mathrm{T}} - \mathbf{t}_{\mathrm{S}}\right)\right]\end{aligned} \tag{10.12}$$

where $\mathbf{I}$ is the identity matrix with size of N_{S}.

For the case that the number of training samples N_{S} is larger than L ($N_{\mathrm{S}} > L$), $\mathbf{H}_{\mathrm{S}}$ has more rows than columns and is of full column rank, which is an overdetermined least square problem. Then, we can obtain from the 1st equation in Eq. 10.8 that $\boldsymbol{\alpha}_{\mathrm{S}} = \left(\mathbf{H}_{\mathrm{S}}\mathbf{H}_{\mathrm{S}}^{\mathrm{T}}\right)^{-1}\left(\mathbf{H}_{\mathrm{S}}\boldsymbol{\beta}_{\mathrm{S}} - \mathbf{H}_{\mathrm{S}}\mathbf{H}_{\mathrm{T}}^{\mathrm{T}}\boldsymbol{\alpha}_{\mathrm{T}}\right)$, after which is substituted into the 4th and 5th equations, we can calculate the output weights $\boldsymbol{\beta}_{\mathrm{S}}$ as follows

$$\begin{aligned}&\begin{cases}\mathbf{H}_{\mathrm{S}}\boldsymbol{\beta}_{\mathrm{S}} + \boldsymbol{\xi}_{\mathrm{S}} = \mathbf{t}_{\mathrm{S}} \\ \mathbf{H}_{\mathrm{T}}\boldsymbol{\beta}_{\mathrm{S}} + \boldsymbol{\xi}_{\mathrm{T}} = \mathbf{t}_{\mathrm{T}}\end{cases} \rightarrow \begin{cases}\mathbf{H}_{\mathrm{S}}^{\mathrm{T}}\mathbf{H}_{\mathrm{S}}\boldsymbol{\beta}_{\mathrm{S}} + \frac{\mathbf{I}}{C_{\mathrm{S}}}\mathbf{H}_{\mathrm{S}}^{\mathrm{T}}\boldsymbol{\alpha}_{\mathrm{S}} = \mathbf{H}_{\mathrm{S}}^{\mathrm{T}}\mathbf{t}_{\mathrm{S}} \\ \mathbf{H}_{\mathrm{T}}\boldsymbol{\beta}_{\mathrm{S}} + \frac{\mathbf{I}}{C_{\mathrm{T}}}\boldsymbol{\alpha}_{\mathrm{T}} = \mathbf{t}_{\mathrm{T}}\end{cases} \\ &\rightarrow \begin{cases}\mathbf{H}_{\mathrm{S}}^{\mathrm{T}}\mathbf{H}_{\mathrm{S}}\boldsymbol{\beta}_{\mathrm{S}} + \frac{\mathbf{I}}{C_{\mathrm{S}}}\mathbf{H}_{\mathrm{S}}^{\mathrm{T}}\left(\mathbf{H}_{\mathrm{S}}\mathbf{H}_{\mathrm{S}}^{\mathrm{T}}\right)^{-1}\left(\mathbf{H}_{\mathrm{S}}\boldsymbol{\beta}_{\mathrm{S}} - \mathbf{H}_{\mathrm{S}}\mathbf{H}_{\mathrm{T}}^{\mathrm{T}}\boldsymbol{\alpha}_{\mathrm{T}}\right) = \mathbf{H}_{\mathrm{S}}^{\mathrm{T}}\mathbf{t}_{\mathrm{S}} \\ \boldsymbol{\alpha}_{\mathrm{T}} = C_{\mathrm{T}}\left(\mathbf{t}_{\mathrm{T}} - \mathbf{H}_{\mathrm{T}}\boldsymbol{\beta}_{\mathrm{S}}\right)\end{cases} \\ &\rightarrow \left(\mathbf{H}_{\mathrm{S}}^{\mathrm{T}}\mathbf{H}_{\mathrm{S}} + \frac{\mathbf{I}}{C_{\mathrm{S}}} + \frac{C_{\mathrm{T}}}{C_{\mathrm{S}}}\mathbf{H}_{\mathrm{T}}^{\mathrm{T}}\mathbf{H}_{\mathrm{T}}\right)\boldsymbol{\beta}_{\mathrm{S}} = \mathbf{H}_{\mathrm{S}}^{\mathrm{T}}\mathbf{t}_{\mathrm{S}} + \frac{C_{\mathrm{T}}}{C_{\mathrm{S}}}\mathbf{H}_{\mathrm{T}}^{\mathrm{T}}\mathbf{t}_{\mathrm{T}} \\ &\rightarrow \boldsymbol{\beta}_{\mathrm{S}} = \left(\mathbf{I} + C_{\mathrm{S}}\mathbf{H}_{\mathrm{S}}^{\mathrm{T}}\mathbf{H}_{\mathrm{S}} + C_{\mathrm{T}}\mathbf{H}_{\mathrm{T}}^{\mathrm{T}}\mathbf{H}_{\mathrm{T}}\right)^{-1}\left(C_{\mathrm{S}}\mathbf{H}_{\mathrm{S}}^{\mathrm{T}}\mathbf{t}_{\mathrm{S}} + C_{\mathrm{T}}\mathbf{H}_{\mathrm{T}}^{\mathrm{T}}\mathbf{t}_{\mathrm{T}}\right)\end{aligned} \tag{10.13}$$

where $\mathbf{I}$ is the identity matrix with size of L.

In fact, the optimization Eq. 10.5 can be reformulated as an equivalent unconstrained optimization problem in matrix form by substituting the constraints into the objective function,

$$\min_{\boldsymbol{\beta}_{\mathrm{S}}} L_{\mathrm{DAELM\text{-}S}}(\boldsymbol{\beta}_{\mathrm{S}}) = \frac{1}{2}\|\boldsymbol{\beta}_{\mathrm{S}}\|^2 + C_{\mathrm{S}}\frac{1}{2}\|\mathbf{t}_{\mathrm{S}} - \mathbf{H}_{\mathrm{S}}\boldsymbol{\beta}_{\mathrm{S}}\|^2 + C_{\mathrm{T}}\frac{1}{2}\|\mathbf{t}_{\mathrm{T}} - \mathbf{H}_{\mathrm{T}}\boldsymbol{\beta}_{\mathrm{S}}\|^2 \tag{10.14}$$

By setting the gradient of $L_{\mathrm{DAELM\text{-}S}}$ w.r.t. $\boldsymbol{\beta}_{\mathrm{S}}$ to be zero,

$$\nabla L_{\mathrm{DAELM\text{-}S}} = \boldsymbol{\beta}_{\mathrm{S}} - C_{\mathrm{S}}\mathbf{H}_{\mathrm{S}}^{\mathrm{T}}(\mathbf{t}_{\mathrm{S}} - \mathbf{H}_{\mathrm{S}}\boldsymbol{\beta}_{\mathrm{S}}) - C_{\mathrm{T}}\mathbf{H}_{\mathrm{T}}^{\mathrm{T}}(\mathbf{t}_{\mathrm{T}} - \mathbf{H}_{\mathrm{T}}\boldsymbol{\beta}_{\mathrm{S}}) = 0 \tag{10.15}$$

Then, we can easily solve Eq. 10.15 to obtain $\boldsymbol{\beta}_{\mathrm{S}}$ formulated in Eq. 10.13.

For recognition of the numerous unlabeled data in target domain, we calculate the output of DAELM-S network as

$$\mathbf{y}_{\mathrm{Tu}}^{k} = \mathbf{H}_{\mathrm{Tu}}^{k} \cdot \boldsymbol{\beta}_{\mathrm{S}}, \quad k = 1, \ldots, N_{\mathrm{Tu}} \tag{10.16}$$

where $\mathbf{H}_{\mathrm{Tu}}^{k}$ denote the hidden layer output with respect to the k-th unlabeled vector in target domain, and N_{Tu} is the number of unlabeled vectors in target domain.

The index corresponding to the maximum value in $\mathbf{y}_{\mathrm{Tu}}^{k}$ is the class of the k-th sample. For implementation, the DAELM-S algorithm is summarized as Algorithm 10.1.

Algorithm 10.1 DAELM-S

Input:

Training samples $\{\mathbf{X}_{\mathrm{S}}, \mathbf{t}_{\mathrm{S}}\} = \{\mathbf{x}_{\mathrm{S}}^{i}, \mathbf{t}_{\mathrm{S}}^{i}\}_{i=1}^{N_{\mathrm{S}}}$ of the source domain S;
Labeled guide samples $\{\mathbf{X}_{\mathrm{T}}, \mathbf{t}_{\mathrm{T}}\} = \{\mathbf{x}_{\mathrm{T}}^{j}, \mathbf{t}_{\mathrm{T}}^{j}\}_{j=1}^{N_{\mathrm{T}}}$ of the target domain T;
The tradeoff parameter C_{S} and C_{T} for source and target domain.

Output:

The output weights $\boldsymbol{\beta}_{\mathrm{S}}$;
The predicted output $\mathbf{y}_{\mathrm{Tu}}$ of unlabeled data in target domain.

Procedure:

1. Initialize the ELM network of L hidden neurons with random input weights $\mathbf{W}$ and hidden bias $\mathbf{B}$.
2. Calculate the output matrix $\mathbf{H}_{\mathrm{S}}$ and $\mathbf{H}_{\mathrm{T}}$ of hidden layer with source and target domains as $\mathbf{H}_{\mathrm{S}} = \mathcal{H}(\mathbf{W} \cdot \mathbf{X}_{\mathrm{S}} + \mathbf{B})$ and $\mathbf{H}_{\mathrm{T}} = \mathcal{H}(\mathbf{W} \cdot \mathbf{X}_{\mathrm{T}} + \mathbf{B})$.
3. **If** $N_{\mathrm{S}} < L$, compute the output weights $\boldsymbol{\beta}_{\mathrm{S}}$ using Eq. 10.12;

   Else, compute the output weights $\boldsymbol{\beta}_{\mathrm{S}}$ using Eq. 10.13.
4. Calculate the predicted output $\mathbf{y}_{\mathrm{Tu}}$ using Eq. 10.16.

Return The output weights $\boldsymbol{\beta}_{\mathrm{S}}$ and predicted output $\mathbf{y}_{\mathrm{Tu}}$.

10.3.2 Target Domain Adaptation ELM (DAELM-T)

In the proposed DAELM-S, the classifier $\boldsymbol{\beta}_{\mathrm{S}}$ is learned on the source domain with the very few labeled guide samples from the target domain as regularization. However, the unlabeled data is neglected which can also improve the performance of classification [33]. Different from DAELM-S, DAELM-T aims to learn a classifier $\boldsymbol{\beta}_{\mathrm{T}}$ on a very limited number of labeled samples from target domain, by leveraging numerous unlabeled data in target domain, into which a base classifier $\boldsymbol{\beta}_{B}$ trained by source data is incorporated. The proposed DAELM-T is formulated as

$$\min_{\boldsymbol{\beta}_{\mathrm{T}}} L_{\text{DAELM-T}}(\boldsymbol{\beta}_{\mathrm{T}}) = \frac{1}{2}\|\boldsymbol{\beta}_{\mathrm{T}}\|^2 + C_{\mathrm{T}}\frac{1}{2}\|\mathbf{t}_{\mathrm{T}} - \mathbf{H}_{\mathrm{T}}\boldsymbol{\beta}_{\mathrm{T}}\|^2 + C_{\mathrm{Tu}}\frac{1}{2}\|\mathbf{H}_{\mathrm{Tu}}\boldsymbol{\beta}_{B} - \mathbf{H}_{\mathrm{Tu}}\boldsymbol{\beta}_{\mathrm{T}}\|^2 \tag{10.17}$$

where $\boldsymbol{\beta}_{\mathrm{T}}$ denotes the learned classifier, $C_{\mathrm{T}}, \mathbf{H}_{\mathrm{T}}, \mathbf{t}_{\mathrm{T}}$ are the same as that in DAELM-S, $C_{\mathrm{Tu}}, \mathbf{H}_{\mathrm{Tu}}$ denote the regularization parameter and the output matrix of the hidden layer with respect to the unlabeled data in target domain. The first term is to reduce the risk of over-fitting, the second term is the least square loss function, and the third term is the regularization which means the domain adaptation between source domain and target domain. Note that $\boldsymbol{\beta}_B$ is a base classifier learned with source data. In this chapter, regularized ELM is used to train a base classifier $\boldsymbol{\beta}_B$ by solving

$$\min_{\boldsymbol{\beta}_B} L_{\mathrm{ELM}}(\boldsymbol{\beta}_B) = \frac{1}{2}\|\boldsymbol{\beta}_B\|^2 + C_{\mathrm{S}}\frac{1}{2}\|\mathbf{t}_{\mathrm{S}} - \mathbf{H}_{\mathrm{S}}\boldsymbol{\beta}_B\|^2 \tag{10.18}$$

where $C_{\mathrm{S}}, \mathbf{t}_{\mathrm{S}}, \mathbf{H}_{\mathrm{S}}$ denote the same meaning as that in DAELM-S.

The structure of the proposed DAELM-T is described in Fig. 10.2, from which we can see that the unlabeled data in target domain has also been exploited.

To solve the optimization Eq. 10.17, by setting the gradient of $L_{\mathrm{DAELM-T}}$ with respect to $\boldsymbol{\beta}_{\mathrm{T}}$ to be zero, we then have

$$\nabla L_{\mathrm{DAELM\text{-}T}} = \boldsymbol{\beta}_{\mathrm{T}} - C_{\mathrm{T}}\mathbf{H}_{\mathrm{T}}^{\mathrm{T}}(\mathbf{t}_{\mathrm{T}} - \mathbf{H}_{\mathrm{T}}\boldsymbol{\beta}_{\mathrm{T}}) - C_{\mathrm{Tu}}\mathbf{H}_{\mathrm{Tu}}^{\mathrm{T}}(\mathbf{H}_{\mathrm{Tu}}\boldsymbol{\beta}_B - \mathbf{H}_{\mathrm{Tu}}\boldsymbol{\beta}_{\mathrm{T}}) = 0 \tag{10.19}$$

If $N_{\mathrm{T}} > L$, then we can have from Eq. 10.19 that

$$\boldsymbol{\beta}_{\mathrm{T}} = \left(\mathbf{I} + C_{\mathrm{T}}\mathbf{H}_{\mathrm{T}}^{\mathrm{T}}\mathbf{H}_{\mathrm{T}} + C_{\mathrm{Tu}}\mathbf{H}_{\mathrm{Tu}}^{\mathrm{T}}\mathbf{H}_{\mathrm{Tu}}\right)^{-1}\left(C_{\mathrm{T}}\mathbf{H}_{\mathrm{T}}^{\mathrm{T}}\mathbf{t}_{\mathrm{T}} + C_{\mathrm{Tu}}\mathbf{H}_{\mathrm{Tu}}^{\mathrm{T}}\mathbf{H}_{\mathrm{Tu}}\boldsymbol{\beta}_B\right) \tag{10.20}$$

where I is the identity matrix with size of L.

If $N_{\mathrm{T}} < L$, we would like to obtain $\boldsymbol{\beta}_{\mathrm{T}}$ of the proposed DAELM-T according to the solving manner in DAELM-S. Let $\mathbf{t}_{\mathrm{Tu}} = \mathbf{H}_{\mathrm{Tu}}\boldsymbol{\beta}_B$, the model Eq. 10.17 can be rewritten as

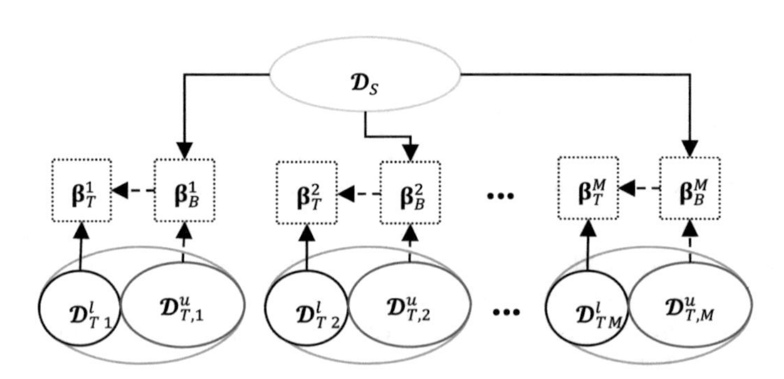

Fig. 10.2 Structure of DAELM-T algorithm with M target domains (M tasks). The solid arrow connected with $\boldsymbol{D}_{\mathrm{S}}$ denotes the training for base classifier $\boldsymbol{\beta}_B$, the dashed line connected with $\boldsymbol{D}_{\mathrm{T}}^u$ denotes the tentative test of base classifier using the unlabeled data from target domain, the solid arrow connected with $\boldsymbol{D}_{\mathrm{T}}^l$ denotes the terminal classifier learning of $\boldsymbol{\beta}_{\mathrm{T}}$, and the dashed arrow connected between $\boldsymbol{\beta}_B$ and $\boldsymbol{\beta}_{\mathrm{T}}$ denotes the regularization for learning $\boldsymbol{\beta}_{\mathrm{T}}$

$$\begin{aligned} &\min_{\boldsymbol{\beta}_{\mathrm{T}},\xi_{\mathrm{T}}^{i},\xi_{\mathrm{Tu}}^{i}} \frac{1}{2}\|\boldsymbol{\beta}_{\mathrm{T}}\|^{2}+C_{\mathrm{T}}\frac{1}{2}\sum_{i=1}^{N_{\mathrm{T}}}\left\|\xi_{\mathrm{T}}^{i}\right\|^{2}+C_{\mathrm{Tu}}\frac{1}{2}\sum_{j=1}^{N_{\mathrm{Tu}}}\left\|\xi_{\mathrm{Tu}}^{j}\right\|^{2} \\ &\text{s.t.} \begin{cases} \mathbf{H}_{\mathrm{T}}^{i}\boldsymbol{\beta}_{\mathrm{T}}=\mathbf{t}_{\mathrm{T}}^{i}-\xi_{\mathrm{T}}^{i}, & i=1,\ldots,N_{\mathrm{T}} \\ \mathbf{H}_{\mathrm{Tu}}^{j}\boldsymbol{\beta}_{\mathrm{T}}=\mathbf{t}_{\mathrm{Tu}}^{j}-\xi_{\mathrm{Tu}}^{j}, & j=1,\ldots,N_{\mathrm{Tu}} \end{cases} \end{aligned} \tag{10.21}$$

The Lagrange multiplier equation of Eq. 10.21 can be written as

$$\begin{aligned} L\left(\boldsymbol{\beta}_{\mathrm{T}},\xi_{\mathrm{T}}^{i},\xi_{\mathrm{Tu}}^{i},\boldsymbol{\alpha}_{\mathrm{T}},\boldsymbol{\alpha}_{\mathrm{Tu}}\right)=&\frac{1}{2}\|\boldsymbol{\beta}_{\mathrm{T}}\|^{2}+\frac{C_{\mathrm{T}}}{2}\sum_{i=1}^{N_{\mathrm{T}}}\left\|\xi_{\mathrm{T}}^{i}\right\|^{2}+\frac{C_{\mathrm{Tu}}}{2}\sum_{i=1}^{N_{\mathrm{Tu}}}\left\|\xi_{\mathrm{Tu}}^{j}\right\|^{2} \\ &-\boldsymbol{\alpha}_{\mathrm{T}}\left(\mathbf{H}_{\mathrm{T}}^{i}\boldsymbol{\beta}_{\mathrm{T}}-\mathbf{t}_{\mathrm{T}}^{i}+\xi_{\mathrm{T}}^{i}\right)-\boldsymbol{\alpha}_{\mathrm{Tu}}\left(\mathbf{H}_{\mathrm{Tu}}^{i}\boldsymbol{\beta}_{\mathrm{T}}-\mathbf{t}_{\mathrm{Tu}}^{i}+\xi_{\mathrm{Tu}}^{i}\right) \end{aligned} \tag{10.22}$$

By setting the partial derivation with respect to $\boldsymbol{\beta}_{\mathrm{T}}, \xi_{\mathrm{T}}^{i}, \xi_{\mathrm{Tu}}^{j}, \boldsymbol{\alpha}_{\mathrm{T}}, \boldsymbol{\alpha}_{\mathrm{Tu}}$ to be zero, we have

$$\begin{cases} \frac{\partial L}{\partial \boldsymbol{\beta}_{\mathrm{T}}}=0 \rightarrow \boldsymbol{\beta}_{\mathrm{T}}=\mathbf{H}_{\mathrm{T}}^{\mathrm{T}}\boldsymbol{\alpha}_{\mathrm{T}}+\mathbf{H}_{\mathrm{Tu}}^{\mathrm{T}}\boldsymbol{\alpha}_{\mathrm{Tu}} \\ \frac{\partial L}{\partial \xi_{\mathrm{T}}}=0 \rightarrow \boldsymbol{\alpha}_{\mathrm{T}}=C_{\mathrm{T}}\xi_{\mathrm{T}}^{\mathrm{T}} \\ \frac{\partial L}{\partial \xi_{\mathrm{Tu}}}=0 \rightarrow \boldsymbol{\alpha}_{\mathrm{Tu}}=C_{\mathrm{Tu}}\xi_{\mathrm{Tu}}^{\mathrm{T}} \\ \frac{\partial L}{\partial \boldsymbol{\alpha}_{\mathrm{T}}}=0 \rightarrow \mathbf{H}_{\mathrm{T}}\boldsymbol{\beta}_{\mathrm{T}}-\mathbf{t}_{\mathrm{T}}+\xi_{\mathrm{T}}=0 \\ \frac{\partial L}{\partial \boldsymbol{\alpha}_{\mathrm{Tu}}}=0 \rightarrow \mathbf{H}_{\mathrm{Tu}}\boldsymbol{\beta}_{\mathrm{T}}-\mathbf{t}_{\mathrm{Tu}}+\xi_{\mathrm{Tu}}=0 \end{cases} \tag{10.23}$$

To solve $\boldsymbol{\beta}_{\mathrm{T}}$, let $\mathbf{H}_{\mathrm{Tu}}\mathbf{H}_{\mathrm{T}}^{\mathrm{T}}=\mathcal{O}$, $\mathbf{H}_{\mathrm{Tu}}\mathbf{H}_{\mathrm{Tu}}^{\mathrm{T}}+\frac{\mathbf{I}}{C_{\mathrm{Tu}}}=\mathcal{P}$, $\mathbf{H}_{\mathrm{T}}\mathbf{H}_{\mathrm{Tu}}^{\mathrm{T}}=\mathcal{Q}$, and $\mathbf{H}_{\mathrm{T}}\mathbf{H}_{\mathrm{T}}^{\mathrm{T}}+\frac{\mathbf{I}}{C_{\mathrm{T}}}=\mathcal{R}$,

By calculating in the same way as Eqs. 10.9, 10.10, and 10.11, we get

$$\begin{cases} \boldsymbol{\alpha}_{\mathrm{T}}=\left(\mathcal{Q}\mathcal{P}^{-1}\mathcal{O}-\mathcal{R}\right)^{-1}\left(\mathcal{Q}\mathcal{P}^{-1}\mathbf{t}_{\mathrm{Tu}}-\mathbf{t}_{\mathrm{T}}\right) \\ \boldsymbol{\alpha}_{\mathrm{Tu}}=\mathcal{P}^{-1}\mathbf{t}_{\mathrm{Tu}}-\mathcal{P}^{-1}\mathcal{O}\left(\mathcal{Q}\mathcal{P}^{-1}\mathcal{O}-\mathcal{R}\right)^{-1}\left(\mathcal{Q}\mathcal{P}^{-1}\mathbf{t}_{\mathrm{Tu}}-\mathbf{t}_{\mathrm{T}}\right) \end{cases} \tag{10.24}$$

Therefore, when $N_{\mathrm{T}} < L$, the output weights can be obtained as

$$\begin{aligned} \boldsymbol{\beta}_{\mathrm{T}} &= \mathbf{H}_{\mathrm{T}}^{\mathrm{T}}\boldsymbol{\alpha}_{\mathrm{T}}+\mathbf{H}_{\mathrm{Tu}}^{\mathrm{T}}\boldsymbol{\alpha}_{\mathrm{Tu}} \\ &= \mathbf{H}_{\mathrm{T}}^{\mathrm{T}}\left(\mathcal{Q}\mathcal{P}^{-1}\mathcal{O}-\mathcal{R}\right)^{-1}\left(\mathcal{Q}\mathcal{P}^{-1}\mathbf{t}_{\mathrm{Tu}}-\mathbf{t}_{\mathrm{T}}\right) \\ &\quad +\mathbf{H}_{\mathrm{Tu}}^{\mathrm{T}}\left[\mathcal{P}^{-1}\mathbf{t}_{\mathrm{Tu}}-\mathcal{P}^{-1}\mathcal{O}\left(\mathcal{Q}\mathcal{P}^{-1}\mathcal{O}-\mathcal{R}\right)^{-1}\left(\mathcal{Q}\mathcal{P}^{-1}\mathbf{t}_{\mathrm{Tu}}-\mathbf{t}_{\mathrm{T}}\right)\right] \end{aligned} \tag{10.25}$$

where $\mathbf{t}_{\mathrm{Tu}}=\mathbf{H}_{\mathrm{Tu}}\boldsymbol{\beta}_{B}$, and $\mathbf{I}$ is the identity matrix with size of N_{T}.

For recognition of the numerous unlabeled data in target domain, we calculate the final output of DAELM-T as

$$\mathbf{y}_{\mathrm{Tu}}^{k} = \mathbf{H}_{\mathrm{Tu}}^{k} \cdot \boldsymbol{\beta}_{\mathrm{T}}, \quad k = 1, \ldots, N_{\mathrm{Tu}} \tag{10.26}$$

where $\mathbf{H}_{\mathrm{Tu}}^{k}$ denote the hidden layer output with respect to the k-th unlabeled sample vector in target domain, and N_{Tu} is the number of unlabeled vectors in target domain.

For implementation in experiment, the DAELM-T algorithm is summarized as Algorithm 10.2.

Algorithm 10.2 DAELM-T

Input:

Training samples $\{\mathbf{X}_{\mathrm{S}}, \mathbf{t}_{\mathrm{S}}\} = \left\{\mathbf{x}_{\mathrm{S}}^{i}, \mathbf{t}_{\mathrm{S}}^{i}\right\}_{i=1}^{N_{\mathrm{S}}}$ of the source domain S;
Labeled guide samples $\{\mathbf{X}_{\mathrm{T}}, \mathbf{t}_{\mathrm{T}}\} = \left\{\mathbf{x}_{\mathrm{T}}^{j}, \mathbf{t}_{\mathrm{T}}^{j}\right\}_{j=1}^{N_{\mathrm{T}}}$ of the target domain T;
Unlabeled samples $\{\mathbf{X}_{\mathrm{Tu}}\} = \left\{\mathbf{x}_{\mathrm{Tu}}^{k}\right\}_{k=1}^{N_{\mathrm{Tu}}}$ of the target domain T;
The tradeoff parameters C_{S}, C_{T} and C_{Tu}.

Output:

The output weights $\boldsymbol{\beta}_{\mathrm{T}}$;
The predicted output $\mathbf{y}_{\mathrm{Tu}}$ of unlabeled data in target domain.

Procedure:

1. Initialize the ELM network of L hidden neurons with random input weights $\mathbf{W}_1$ and hidden bias $\mathbf{B}_1$.
2. Calculate the output matrix $\mathbf{H}_{\mathrm{S}}$ of hidden layer with source domain as $\mathbf{H}_{\mathrm{S}} = \mathcal{H}(\mathbf{W}_1 \cdot \mathbf{X}_{\mathrm{S}} + \mathbf{B}_1)$.
3. **If** $N_{\mathrm{S}} < L$, compute the output weights $\boldsymbol{\beta}_B$ of the base classifier using Eq. 10.4;

   Else, compute the output weights $\boldsymbol{\beta}_B$ of the base classifier using Eq. 10.3.
4. Initialize the ELM network of L hidden neurons with random input weights $\mathbf{W}_2$ and hidden bias $\mathbf{B}_2$.
5. Calculate the hidden layer output matrix $\mathbf{H}_{\mathrm{T}}$ and $\mathbf{H}_{\mathrm{Tu}}$ of labeled and unlabeled data in target domains as $\mathbf{H}_{\mathrm{T}} = \mathcal{H}(\mathbf{W}_2 \cdot \mathbf{X}_{\mathrm{T}} + \mathbf{B}_2)$ and $\mathbf{H}_{\mathrm{Tu}} = \mathcal{H}(\mathbf{W}_2 \cdot \mathbf{X}_{\mathrm{Tu}} + \mathbf{B}_2)$.
6. **If** $N_{\mathrm{T}} < L$, compute the output weights $\boldsymbol{\beta}_{\mathrm{T}}$ using Eq. 10.25;

   Else, compute the output weights $\boldsymbol{\beta}_{\mathrm{T}}$ using Eq. 10.20.
7. Calculate the predicted output $\mathbf{y}_{\mathrm{Tu}}$ using Eq. 10.26.

Return The output weights $\boldsymbol{\beta}_{\mathrm{T}}$ and predicted output $\mathbf{y}_{\mathrm{Tu}}$.

Remark 1 From the algorithms of DAELM-S and DAELM-T, we observe that the same two stages as ELM are included: (1) feature mapping with randomly selected weights and biases; (2) output weights computation. For ELM, the algorithm is constructed and implemented in a single domain (source domain); as a result, the generalization performance is degraded in new domains. In the proposed DAELM framework, a limited number of labeled samples and numerous unlabeled data in target domain are exploited without changing the unified ELM framework, and the merits of ELM are inherited. The framework for DAELM might draw some new perspectives of domain adaptation for developing ELM theory.

Remark 2 We observe that the DAELM-S has similar structure in model and algorithm with DAELM-T. The essential difference lies in that numerous unlabeled data which may be useful for improving generalization performance are exploited in DAELM-T through a pre-learned base classifier. Specifically, DAELM-S learns a classifier using the labeled training data in source domain but draws some new knowledge by leveraging a limited number of labeled samples from target domain, such that the knowledge from target domain can be effectively transferred to source domain for generalization. While DAELM-T attempts to train a classifier using a limited number of labeled data from target domain as "main knowledge," but introduces a regularizer that minimizes the error between outputs of DAELM-T classifier $\boldsymbol{\beta}_{\mathrm{T}}$ and the base classifier $\boldsymbol{\beta}_{B}$ computed on the unlabeled input data.

10.4 Experiments

In this section, we will employ the sensor drift compensation experiment on the E-nose olfactory data by using the proposed DAELM-S and DAELM-T algorithms.

10.4.1 Description of Data

For verification of the proposed DAELM-S and DAELM-T algorithms, the long-term sensor drift big data of three years that was released in UCI Machine Learning Repository [34] by Vergara et al. [29, 35] is exploited and studied in this chapter.

The sensor drift big dataset was gathered during the period from January 2008 to February 2011 with 36 months in a gas delivery platform. Totally, this dataset contains 13,910 measurements (observations) from an electronic nose system with 16 gas sensors exposed to six kinds of pure gaseous substances including acetone, acetaldehyde, ethanol, ethylene, ammonia, and toluene at different concentration levels, individually. For each sensor, 8 features were extracted, and a 128-dimensional feature vector (8 features $\times$ 16 sensors) for each observation is formulated as a result. We refer readers to as [29] for specific technical details on

Table 10.1 Experimental data of sensor drift in electronic nose

Batch ID	Month	Acetone	Acetaldehyde	Ethanol	Ethylene	Ammonia	Toluene	Total
Batch 1	1, 2	90	98	83	30	70	74	445
Batch 2	3–10	164	334	100	109	532	5	1244
Batch 3	11, 12, 13	365	490	216	240	275	0	1586
Batch 4	14, 15	64	43	12	30	12	0	161
Batch 5	16	28	40	20	46	63	0	197
Batch 6	17, 18, 19, 20	514	574	110	29	606	467	2300
Batch 7	21	649	662	360	744	630	568	3613
Batch 8	22, 23	30	30	40	33	143	18	294
Batch 9	24, 30	61	55	100	75	78	101	470
Batch 10	36	600	600	600	600	600	600	3600

how to select the eight features for each sensor. In total, ten batches of sensor data that collected in different time intervals are included in the dataset. The details of the dataset are presented in Table 10.1.

For visualization of the drift behavior existing in the dataset, we first plot the sensor response before and after drifting. We view the data in batch 1 as non-drift, and select batch 2, batch 7, and batch 10 as drifted data, respectively, and the response is given in Fig. 10.3. It is known that sensor drift shows nonlinear behavior in a multi-dimensional sensor array, and it is impossible to intuitively and directly calibrate the sensor response using some linear or nonlinear transformation. Instead, we consider it as a space distribution adaptation using transfer learning and realize the drift compensation in decision level.

It is worth noting that sensor responses after drift cannot be calibrated directly due to the nonlinear dynamic behavior or chaotic behavior [31] of sensor drift. Therefore, drift compensation in decision level by data distribution adaptation and machine learning is more appealing.

Considering that a small number of labeled samples (guide samples) should be first selected from the target domains in the proposed DAELM-S and DAELM-T algorithms, while the labeled target data plays an important role in knowledge adaptation, we therefore adopt a representative labeled sample selection algorithm (SSA) based on the Euclidean distance $d(x_p, x_q)$ of a sample pair (x_p, x_q). For detail, the SSA algorithm is summarized as Algorithm 10.3.

The visual SSA algorithm in two-dimensional coordinate plane for selecting five guide samples from each target domain (batch) is shown in Fig. 10.4 as an example. The patterns marked as “1” denote the first two selected patterns (farthest distance) in Step 2. Then, the patterns marked as “2”, “3”, “4” denote the three selected patterns sequentially. The SSA is for the purpose that the labeled samples selected from target domains should be representative and global in the data space and promise the generalization performance of domain adaptation.

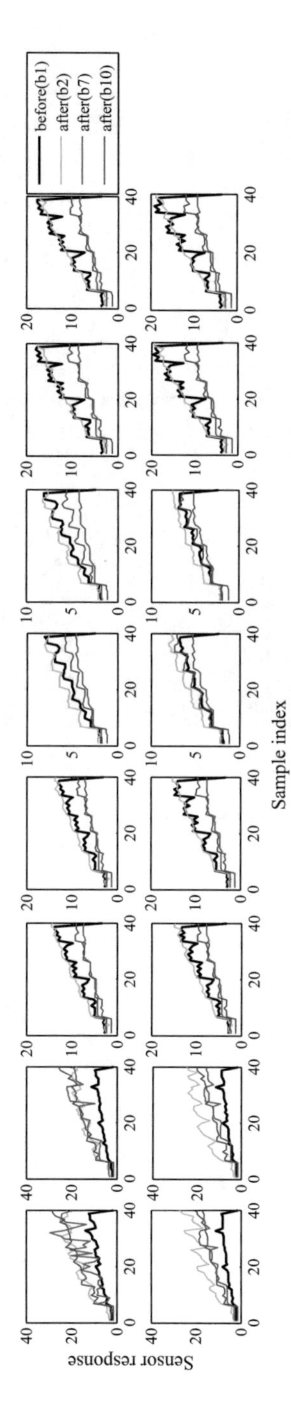

Fig. 10.3 The response of 16 sensors before (batch 1) and after drifting (batch 2, 7, and 10) under Acetone with various concentrations (i.e., 10, 50, 100, 150, 200, and 250 ppm). Totally, 40 samples including 6, 7, 7, 6, 7, and 7 samples for each concentration, respectively, are illustrated for visually shown the drift behavior

Algorithm 10.3 SSA algorithm

Input:

The data $\mathbf{X}_{\mathrm{T}}$ from target domain;
The predefined number k of labeled samples being selected.

Output:

The selected k labeled guide set $\mathcal{S}^l = \{s_1^l, s_2^l, \ldots, s_k^l\}$.

Procedure:

While the number of selected labeled instances does not reach k **do**

Step1: Calculate the Euclidean distance in pair-wise from each target domain, and select the farthest two patterns s_1^l and s_2^l as the labeled instances which is put into the guide set $\mathcal{S}^l = \{s_1^l, s_2^l\}$;

Step2: To a pattern x_i, calculate the Euclidean distances $d(x_i, \mathcal{S}^l)$, and the nearest distance in each pair for pattern x_i is put into the set $\mathcal{N}_d(x_i)$;

Step3: The pattern with the farthest distance in set $\mathcal{N}_d(x_i)$ is then selected as labeled sample s_3^l, and update selected labeled guide set $\mathcal{S}^l = \{s_1^l, s_2^l, s_3^l\}$;

Step4: **If** the size of the guide set $\mathcal{S}^l$ reaches the number k, **break**;

end while

Return $\mathcal{S}^l = \{s_1^l, s_2^l, \ldots, s_k^l\}$

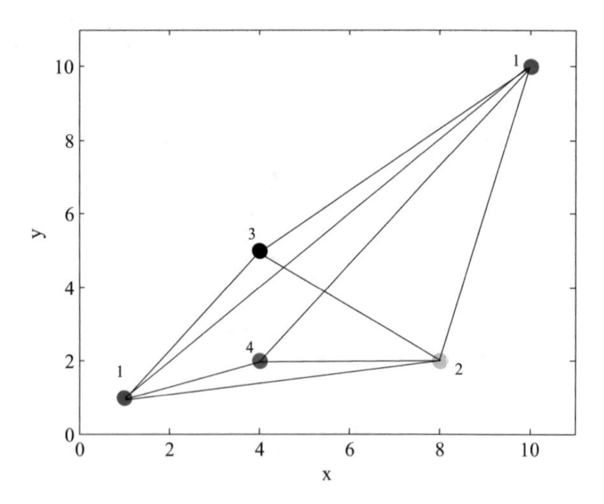

Fig. 10.4 Visual description of SSA Algorithm 10.3

10.4.2 Experimental Setup

We strictly follow the experimental setup in [29] to evaluate our DAELM framework. In default, the number of hidden neurons L is set as 1000, and The RBF function (i.e., *radbas*) with kernel width set as 1 is used as activation function (i.e., feature mapping function) in the hidden layer. The features are scaled appropriately to lie in interval $(-1, 1)$. In DAELM-S algorithm, the penalty coefficients C_S and C_T are empirically set as 0.01 and 10 throughout the experiments, respectively. In DAELM-T algorithm, the penalty coefficient C_S for base classifier is set as 0.001, C_T and C_{Tu} are set as 0.001 and 100 throughout the experiments, respectively. For effective verification of the proposed methods, two experimental settings according to [36] are given as follows.

- **Setting-1**: Take batch 1 (source domain) as fixed training set and tested on batch K, $K = 2, \ldots, 10$ (target domains);
- **Setting-2**: The training set (source domain) is dynamically changed with batch $K - 1$ and tested on batch K (target domain), $K = 2, \ldots, 10$.

Following the two settings, we realize our proposed DAELM framework and compare with multi-class SVM with RBF kernel (SVM-rbf), the geodesic flow kernel (SVM-gfk), and the combination kernel (SVM-comgfk). Besides, we also compared with the semi-supervised methods such as manifold regularization with RBF kernel (ML-rbf) and manifold regularization with combination kernel (ML-comgfk). The above machine learning-based methods have been reported for drift compensation [36] using the same dataset. The formulation of geodesic flow kernel as a domain adaptation method can be referred to as [37]. Additionally, the regularized ELM with RBF function in hidden layer (ELM-rbf) from [32] is also compared as baseline in experiments. The popular CC-PCA method [27] and classifier ensemble [29] for drift compensation are also reported in **Setting 1** and **Setting 2**.

Due to the random selection of input weights between input layer and hidden layer, and bias in hidden layer under ELM framework, in experiments, we run the ELM, DAELM-S, and DAELM-T for 10 times, and the average values are reported. Note that ELM is trained using the same labeled source data and target data as the proposed DAELM.

10.5 Results and Discussion

We conduct the experiments and discussion on **Setting 1** and **Setting 2**, respectively. The recognition results of nine batches for different methods under experimental setting 1 are reported in Table 10.2. We consider two conditions of DAELM-S with 20 labeled target samples and 30 labeled target samples, respectively. For DAELM-T, 40 and 50 labeled samples from the target domain are used, respectively, considering that DAELM-T trains a classifier only using a limited number of labeled samples from target domain. For visually observing the performance of all methods, we show the recognition accuracy on batches successively as Fig. 10.5a. From Table 10.2 and Fig. 10.5, we have the following observations:

1. SVM with the combined kernel of geodesic flow kernels (SVM-comgfk) performs better results than the popular CC-PCA method and other SVM-based methods in most batches, except the results of batch 4 and batch 8. It demonstrates that machine learning methods show more usefulness in drift compensation than traditional calibration.
2. Manifold learning with combined kernel (ML-comgfk) obtains an average accuracy of 67.3% and outperforms all baseline methods. It demonstrates that manifold regularization and combined kernel are more effective in semi-supervised learning with a limited number of samples.
3. The generalization performance and knowledge transfer capability of regularized ELM have been well improved by the proposed DAELM. The results of our DAELM-S and DAELM-T have an average improvement of about 30% in recognition accuracy than traditional ELM. The highest recognition accuracy of 91.86% under sensor drift is obtained using our proposed algorithm.
4. Both the proposed DAELM-S and DAELM-T significantly outperform all other existing methods including traditional CC-PCAM, SVM and manifold regularization-based machine learning methods. In addition, DAELM-T(50) has an obvious improvement than DAELM-T(40) and DAELM-S, which shows that more labeled target data is expected for DAELM-T. While DAELM-S can also perform well comparatively with fewer labeled target data. From the computations, due to that a base classifier is first trained in DAELM-T and more labeled target data need, DAELM-S may be a better choice in realistic applications.

From the experimental results in experimental **Setting 1**, the proposed methods outperform all other methods in drift compensation. We then follow the experimental **Setting 2**, i.e., trained on batch $K-1$ and tested on batch K, and report the results in Table 10.3. The performance variations of all methods are illustrated in Fig. 10.5b. From Table 10.3 and Fig. 10.5b, we have the following observations:

1. Manifold regularization-based combined kernel (ML-comgfk) achieves an average accuracy of 79.6% and outperforms other SVM-based machine learning algorithms and single kernel methods, which demonstrates that manifold learning and combined kernel can improve the classification accuracy, but limited capacity.
2. The classifier ensemble can improve the performance of the dataset with drift noise (an average accuracy of 80.0%). However, many base classifiers should be trained using the source data for ensemble, and it has no domain adaptability when tested on the data from target domains, which has been well referred in the proposed DAELM.

Table 10.2 Comparisons of recognition accuracy (%) under the experimental **setting 1**

Batch ID	Batch 2	Batch 3	Batch 4	Batch 5	Batch 6	Batch 7	Batch 8	Batch 9	Batch 10	Average
CC-PCA	67.00	48.50	41.00	35.50	55.00	31.00	56.50	46.50	30.50	45.72
SVM-rbf	74.36	61.03	50.93	18.27	28.26	28.81	20.07	34.26	34.47	38.94
SVM-gfk	72.75	70.08	60.75	75.08	73.82	54.53	55.44	69.62	41.78	63.76
SVM-comgfk	74.47	70.15	59.78	75.09	73.99	54.59	55.88	70.23	41.85	64.00
ML-rbf	42.25	73.69	75.53	66.75	77.51	54.43	33.50	23.57	34.92	53.57
ML-comgfk	80.25	74.99	78.79	67.41	77.82	71.68	49.96	50.79	53.79	67.28
ELM-rbf	70.63	66.44	66.83	63.45	69.73	51.23	49.76	49.83	33.50	57.93
Our DAELM-S(20)	87.57	96.53	82.61	81.47	84.97	71.89	78.10	87.02	57.42	**80.84**
Our DAELM-S(30)	87.98	95.74	85.16	95.99	94.14	83.51	86.90	100.0	53.62	**87.00**
Our DAELM-T(40)	83.52	96.34	88.20	99.49	78.43	80.93	87.42	100.0	56.25	**85.62**
Our DAELM-T(50)	97.96	95.34	99.32	99.24	97.03	83.09	95.27	100.0	59.45	**91.86**

The boldface type denotes the best performance.

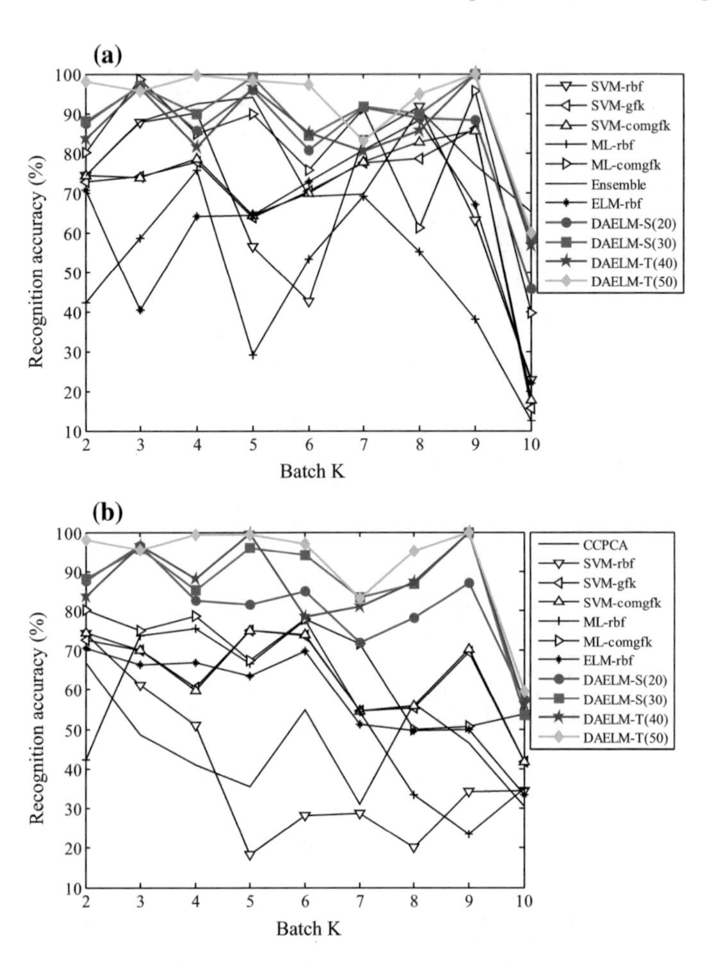

Fig. 10.5 Comparisons of different methods in experimental **Setting 1** (**a**) and **Setting 2** (**b**)

3. The proposed DAELM methods perform much better (91.82%) than all other existing methods for different tasks in recognition tested on drifted data. The robustness of the proposed methods with domain adaptability is proved for drift compensation in E-nose.

For studying the variations of recognition accuracy with the number *k* of labeled samples in target domain, different number *k* from the set of {5, 10, 15, 20, 25, 30, 35, 40, 45, 50} is explored by Algorithm 10.3 (SSA) and the proposed DAELM framework. Specifically, we present comparisons with different number of labeled samples selected from target domains. For fair comparison with ELM, the labeled target samples are feed into ELM together with the source training samples. The results for experimental **Setting 1** and **Setting 2** are shown in Fig. 10.6a, b, respectively, from which, we have:

Table 10.3 Comparisons of recognition accuracy (%) under the experimental **Setting 2**

Batch ID	$1 \rightarrow 2$	$2 \rightarrow 3$	$3 \rightarrow 4$	$4 \rightarrow 5$	$5 \rightarrow 6$	$6 \rightarrow 7$	$7 \rightarrow 8$	$8 \rightarrow 9$	$9 \rightarrow 10$	Average
SVM-rbf	74.36	87.83	90.06	56.35	42.52	83.53	91.84	62.98	22.64	68.01
SVM-gfk	72.75	74.02	77.83	63.91	70.31	77.59	78.57	86.23	15.76	68.56
SVM-comgfk	74.47	73.75	78.51	64.26	69.97	77.69	82.69	85.53	17.76	69.40
ML-rbf	42.25	58.51	75.78	29.10	53.22	69.17	55.10	37.94	12.44	48.17
ML-comgfk	80.25	98.55	84.89	89.85	75.53	91.17	61.22	95.53	39.56	79.62
Ensemble	74.40	88.00	92.50	94.00	69.00	69.50	91.00	77.00	65.00	80.04
ELM-rbf	70.63	40.44	64.16	64.37	72.70	80.75	88.20	67.00	22.00	63.36
Our DAELM-S(20)	87.57	96.90	85.59	95.89	80.53	91.56	88.71	88.40	45.61	**84.53**
Our DAELM-S(30)	87.98	96.58	89.75	99.04	84.43	91.75	89.83	100.0	58.44	**88.64**
Our DAELM-T(40)	83.52	96.41	81.36	96.45	85.13	80.49	85.71	100.0	56.81	**85.10**
Our DAELM-T(50)	97.96	95.62	99.63	98.17	97.13	83.10	94.90	100.0	59.88	**91.82**

The boldface type denotes the best performance.

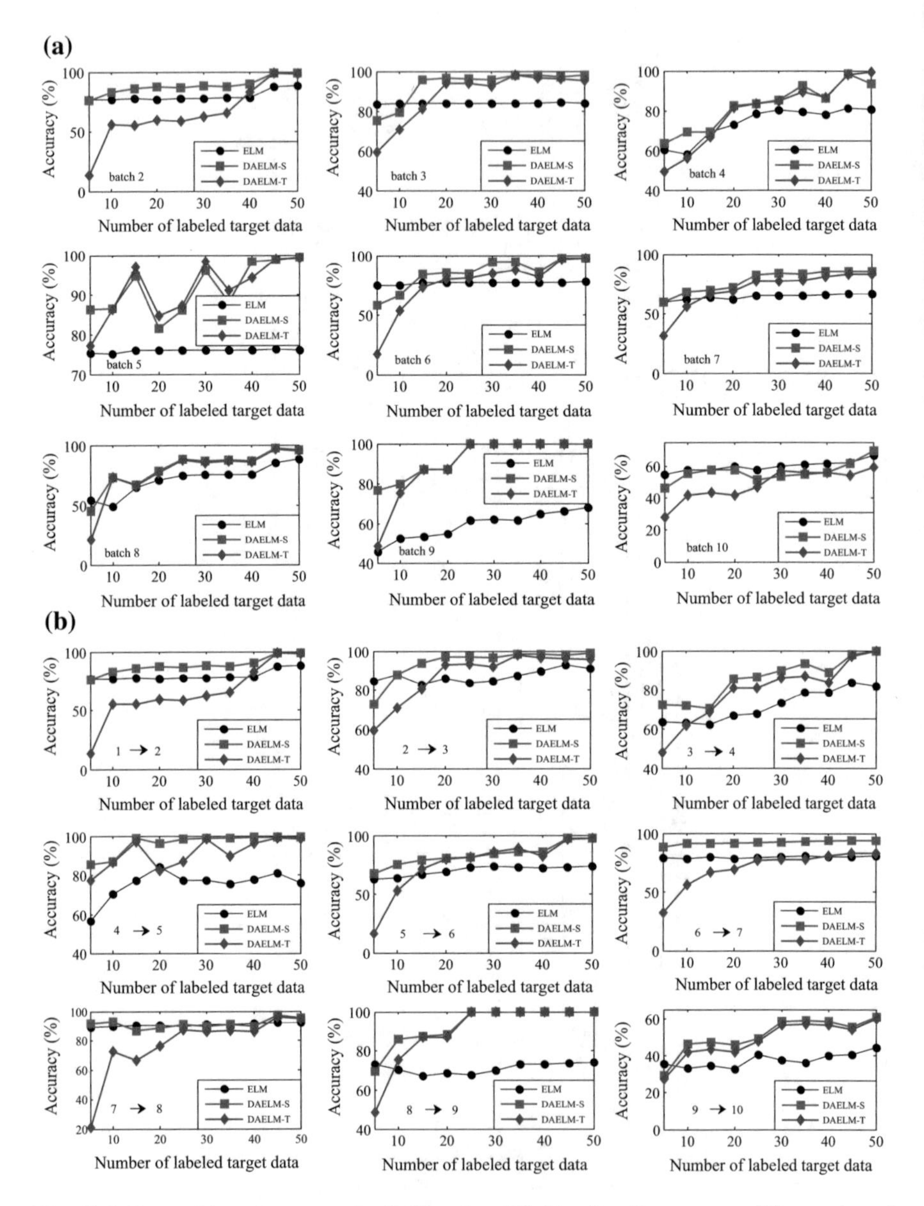

Fig. 10.6 Recognition accuracy under **Setting 1** and **Setting 2** with respect to different size of guide set (labeled samples from target domain)

1. The traditional ELM has little obvious improvement with the increase of the labeled samples from target domains, which clearly demonstrates that ELM has no the capability of knowledge adaptation.
2. Both DAELM-S and DAELM-T have significant enhancement in classification accuracy with increasing labeled data from target domain. Note that in batch 2 and batch 10 shown in Fig. 10.6a, our DAELM is comparative to ELM. The possible reason may be that little drift exist in batch 2 that leads to the small difference in classification task. While the data in batch 10 may be seriously noised by drift, the E-nose system may lose recognition ability only using batch 1 (Setting 1) for training. The proposed DAELM is still much better than ELM when tested on the seriously noised batch 10 in Setting 2 (Fig. 10.6b).
3. DAELM-S has superior performance to DAELM-T when the number k of labeled target samples used in knowledge adaptation is smaller, because DAELM-T does not consider the source data in classifier learning. Additionally, with the increase of the number k, DAELM-T has a comparative performance with DAELM-S which maybe a better choice when only a small number of labeled samples in target domain are available.

Throughout the chapter, the proposed DAELM framework is to cope with sensor drift in the perspective of machine learning in decision level, but not intuitively calibrate the single sensor response because the drift rules are difficult to be captured by some linear or nonlinear regression method due to its nonlinear/chaotic dynamic behavior. This work is to construct a learning framework with better knowledge adaptability and generalization capability to drift noise existing in dataset.

10.6 Summary

In this chapter, the sensor drift problem in electronic nose is addressed by using a knowledge adaptation-based machine learning approach, i.e., domain adaptation extreme learning machine (DAELM). From the angle of machine learning, the proposed methods provide new perspectives for exploring ELM theory, and also inherit the advantages of ELM including the feature mapping with randomly generated input weights and bias, the analytically determined solutions, and good generalization. Experiments on a long-term sensor drift dataset collected by E-nose clearly demonstrate the efficacy of our proposed method over other algorithms.

References

1. G.B. Huang, Q.Y. Zhu, C.K. Siew, Extreme learning machine: theory and applications. Neurocomputing **70**, 489–501 (2006)

2. G. Feng, G.B. Huang, Q. Lin, R. Gay, Error minimized extreme learning machine with growth of hidden nodes and incremental learning. IEEE. Trans. Neural Netw. **20**(8), 1352–1357 (2009)
3. G.B. Huang, H. Zhou, X. Ding, R. Zhang, Extreme learning machine for regression and multiclass classification. IEEE. Trans. Syst. Man Cybern Part B **42**(2), 513–529 (2012)
4. G.B. Huang, L. Chen, C.K. Siew, Universal approximation using incremental constructive feedforward networks with random hidden nodes. IEEE Trans. Neural Netw. **17**(4), 879–892 (2006)
5. Q.Y. Zhu, A.K. Qin, P.N. Suganthan, G.B. Huang, Evolutionary extreme learning machine. Pattern Recogn. **38**, 1759–1763 (2005)
6. W. Zong, G.B. Huang, Y. Chen, Weighted extreme learning machine for imbalance learning. Neurocomputing **101**, 229–242 (2013)
7. Z. Bai, G.B. Huang, D. Wang, H. Wang, M.B. Westover, Sparse extreme learning machine for classification. IEEE Trans. Cybern. (2014)
8. X. Li, W. Mao, W. Jiang, Fast sparse approximation of extreme learning machine. Neurocomputing **128**, 96–103 (2014)
9. G. Huang, S. Song, J.N.D. Gupta, C. Wu, Semi-supervised and unsupervised extreme learning machines. IEEE. Trans. Cybern. (2014)
10. J. Blitzer, R. McDonald, F. Pereira, Domain adaptation with structural correspondence learning, in *Proceedings of Conference on Empirical Methods in Natural Language Processing*, July 2006, pp. 120–128 (2006)
11. J. Yang, R. Yan, A.G. Hauptmann, Cross-domain video concept detection using adaptive SVMs, in *Proceedings of International Conference on Multimedia*, Sept 2007, pp. 188–197 (2007)
12. S.J. Pan, I.W. Tsang, J.T. Kwok, Q. Yang, Domain adaptation via transfer component analysis. IEEE Trans. Neural Netw. **22**(2), 199–210 (2011)
13. L. Duan, I.W. Tsang, D. Xu, T.S. Chua, Domain adaptation from multiple sources via auxiliary classifiers, in *Proceedings of International Conference on Machine Learning*, June 2009, pp. 289–296 (2009)
14. L. Duan, D. Xu, I.W. Tsang, Domain adaptation from multiple sources: domain-dependent regularization approach. IEEE Trans. Neural Netw. Learn. Syst. **23**(3), 504–518 (2012)
15. R. Gopalan, R. Li, R. Chellappa, Domain adaptation for object recognition: an unsupervised approach, in *Proceedings ICCV*, pp. 999–1006 (2011)
16. L. Zhang, F.C. Tian, A new kernel discriminant analysis framework for electronic nose recognition. Anal. Chim. Acta. **816**, 8–17 (2014)
17. L. Zhang, F. Tian, H. Nie, L. Dang, G. Li, Q. Ye, C. Kadri, Classification of multiple indoor air contaminants by an electronic nose and a hybrid support vector machine. Sens. Actuators B Chem. **174**, 114–125 (2012)
18. K. Brudzewski, S. Osowski, A. Dwulit, Recognition of coffee using differential electronic nose. IEEE. Trans. Instrum. Meas. **61**(6), 1803–1810 (2012)
19. B. Tudu, A. Metla, B. Das, N. Bhattacharyya, A. Jana, D. Ghosh, R. Bandyopadhyay, Towards versatile electronic nose pattern classifier for black tea quality evaluation: an incremental fuzzy approach. IEEE. Trans. Instrum. Meas. **58**(9), 3069–3078 (2009)
20. J.W. Garnder, H.W. Shin, E.L. Hines, An electronic nose system to diagnose illness. Sens. Actuators B Chem. **70**, 19–24 (2000)
21. L. Zhang, F. Tian, C. Kadri, G. Pei, H. Li, L. Pan, Gases concentration estimation using heuristics and bio-inspired optimization models for experimental chemical electronic nose. Sens. Actuators B Chem. **160**(1), 760–770 (2011)
22. L. Zhang, F. Tian, Performance study of multilayer perceptrons in a low-cost electronic. IEEE. Trans. Instrum. Meas. **63**(7), 1670–1679 (2014)
23. J.W. Gardner, P.N. Bartlett, *Electronic Noses: Principles and Applications* (Oxford University Press, Oxford, 1999)
24. R. Gutierrez-Osuna, Pattern analysis for machine olfaction: a review. IEEE Sens. J. **2**(3), 189–202 (2002)

25. M. Holmberg, F.A.M. Davide, C. Di Natale, A.D. Amico, F. Winquist, I. Lundström, Drift counteraction in odour recognition applications: lifelong calibration method. Sens. Actuators B Chem. **42**, 185–194 (1997)
26. S. Di Carlo, M. Falasconi, Drift correction methods for gas chemical sensors in artificial olfaction systems: techniques and challenges. Adv. Chem. Sens. 305–326 (2012)
27. T. Artursson, T. Eklov, I. Lundstrom, P. Martensson, M. Sjostrom, M. Holmberg, Drift correction for gas sensors using multivariate methods. J. Chemometr. **14**(5–6), 711–723 (2000)
28. S. Di Carlo, M. Falasconi, E. Sanchez, A. Scionti, G. Squillero, A. Tonda, Increasing pattern recognition accuracy for chemical sensing by evolutionary based drift compensation. Pattern Recogn. Lett. **32**(13), 1594–1603 (2011)
29. A. Vergara, S. Vembu, T. Ayhan, M.A. Ryan, M.L. Homer, R. Huerta, Chemical gas sensor drift compensation using classifier ensembles. Sens. Actuators B Chem. **166–167**, 320–329 (2012)
30. A.C. Romain, J. Nicolas, Long term stability of metal oxide-based gas sensors for e-nose environmental applications: an overview. Sens. Actuators B Chem. **146**, 502–506 (2010)
31. L. Zhang, F. Tian, S. Liu, L. Dang, X. Peng, X. Yin, Chaotic time series prediction of E-nose sensor drift in embedded phase space. Sens. Actuators B Chem. **182**, 71–79 (2013)
32. D.A.P. Daniel, K. Thangavel, R. Manavalan, R.S.C. Boss, ELM-based ensemble classifier for gas sensor array drift dataset, computational intelligence, cyber security and computational models. Adv. Intell. Syst. Comput. **246**, 89–96 (2014)
33. M. Belkin, P. Niyogi, V. Sindhwani, Manifold regularization: a geometric framework for learning from labeled and unlabeled examples. J. Mach. Learn. Res. **7**, 2399–2434 (2006)
34. http://archive.ics.uci.edu/ml/datasets/Gas+Sensor+Array+Drift+Dataset+at+Different +Concentrations
35. I.R. Lujan, J. Fonollosa, A. Vergara, M. Homer, R. Huerta, On the calibration of sensor arrays for pattern recognition using the minimal number of experiments. Chemometr. Intell. Lab. Syst. **130**, 123–134 (2014)
36. Q. Liu, X. Li, M. Ye, S. Sam Ge, X. Du, Drift compensation for electronic nose by semi-supervised domain adaptation. IEEE Sens. J. **14**(3), 657–665 (2014)
37. B. Gong, Y. Shi, F. Sha, K. Grauman, Geodesic flow kernel for unsupervised domain adaptation, in *Proceedings CVPR*, pp. 2066–2073 (2012)

Chapter 11
Domain Regularized Subspace Projection Method

Abstract This chapter addresses the time-varying drift with characteristics of uncertainty and unpredictability. Considering that drifted data is with different probability distribution from the regular data, we propose a machine learning-based subspace projection approach to project the data onto a new common subspace so that two clusters have similar distribution. Then drift can be automatically removed or reduced in the new common subspace. The merits are threefold: (1) the proposed subspace projection is unsupervised, without using any data label information; (2) a simple but effective domain distance is proposed to represent the mean distribution discrepancy metric; (3) the proposed anti-drift method can be easily solved by Eigen decomposition, and anti-drift is manifested with a well-solved projection matrix in real application. Experiments on synthetic data and real datasets demonstrate the effectiveness and efficiency of the proposed anti-drift method in comparison to state-of-the-art methods.

Keywords Anti-drift · Electronic nose · Subspace projection · Common subspace · Machine learning

11.1 Introduction

Drift is a well-known issue in instrumentation and measurement. In E-nose, sensor drift shows nonlinear dynamic behavior in a multi-dimensional sensor array [1], which is caused by many objective factors such as aging, poisoning, and the fluctuations of the ambient environmental variables (e.g., humidity, temperature) [2]. Due to the ambient variables, as a result, the instrument responds differently to a constant concentration of some contaminants at different ambient conditions. Drift effect can be understood as some uncertainty, which is therefore difficult to intuitively estimate and compensate by using some deterministic approaches. Drift is once thought to be an ill-posed problem due to the very irregular characteristics. Motivated by these findings, we aim to propose an instrumental drift adaptation

L. Zhang et al., *Electronic Nose: Algorithmic Challenges*,
https://doi.org/10.1007/978-981-13-2167-2_11

method from the angle of distribution consistency in this chapter. That is, we attempt to improve the distribution consistency of the observed data, rather than estimate the drift effect with uncertainty.

Although researchers have proposed many different methods for drift compensation from the angle of ensemble classifiers, semi-supervised learning, drift direction, and drift correction, the results are still unsatisfactory since the restricts of methods [3–10]. In this chapter, we propose a linear subspace alignment-based drift adaptation technique, which tends to achieve drift compensation by removing the distribution discrepancy between the normal E-nose data and the drifted E-nose data. From the angle of machine learning, a basic assumption in learning theory is that the probability distribution between the training data and the testing data should be consistent. Therefore, a statistical learning algorithm can effectively capture the implied rules in the training data, which can also be adapted to the testing data. However, once the data (e.g., drift effect) violates the basic assumption of distribution consistency, the learning model may fail. This chapter supposes that the drift effect seriously causes the distribution inconsistency of E-nose data. Therefore, our mission and idea is to improve the distribution consistency by subspace alignment between normal and drifted data. That is, we are not trying to calibrate the drift exactly or learn a drift-assimilation-based classifier, but attempting to enhance the statistical probability distribution consistency of the normal and drifted data.

To this end, under the construction of principal component analysis, we propose a domain regularized component analysis (DRCA) method, which aims at improving the distribution consistency and achieving drift adaptation in principal component subspaces of normal (source domain) and drifted data (target domain). Inspired by transfer learning and domain adaptation [6], in this chapter, we give that the normal data is from *source domain* and the drifted data is from *target domain*. The basic idea of the proposed DRCA method is illustrated vividly in Fig. 11.1.

11.2 Related Work

11.2.1 Existing Drift Compensation Approaches

Recently, many researchers have proposed different methods for handling drift issue of E-nose. Specifically, we present the existing work from classifier level and feature level. *In classifier level*, Martinelli et al. proposed an artificial immune system-based adaptive classifier for drift mitigation [6]. Vergara et al. proposed a classifier ensemble model for drift compensation, in which multiple SVMs with weighted ensemble are involved [4]. Also, a public long-term drift dataset was released and promotes the development in drift compensation. Liu and Tang [5] proposed a dynamic classifier ensemble for drift compensation in real application, in which the ensemble scheme is a dynamic weighted combination. Zhang et al. [6] proposed a domain adaptation extreme learning machine-based transfer learning

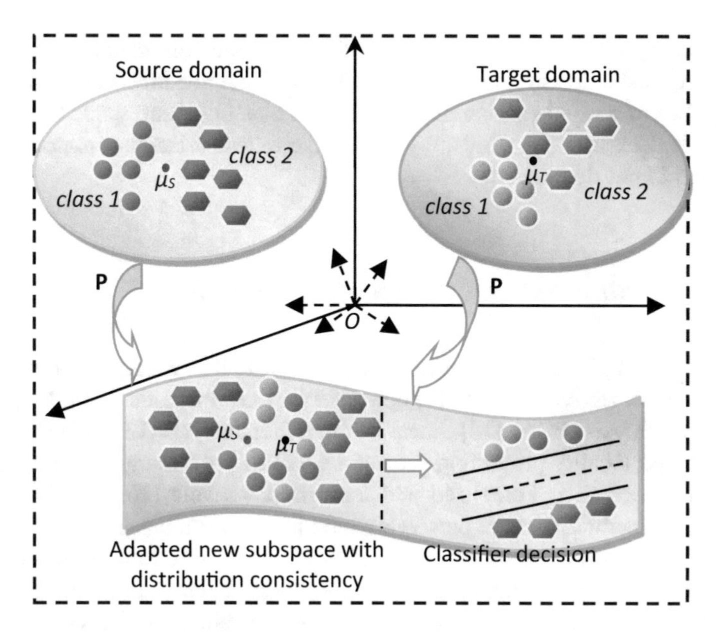

Fig. 11.1 Schematic diagram of the proposed DRCA method; after a subspace projection **P**, the source domain and target domain of different space distribution lie in a latent subspace with good distribution consistency (the centers of both domains become very close and drift is removed); in this latent subspace, the classification of two classes is successfully achieved

method for drift compensation, in which the classifier is learned with source and target domain simultaneously, by adaptive cross-domain learning. Liu et al. [7] also proposed a semi-supervised domain adaptation, in which manifold learning with affinity structure preservation is considered. These methods focus on the robust classifier learning, but neglect the important issue of data distribution mismatch. *In feature level*, Artursson et al. proposed a component correction-based principal component analysis method, which aims at finding the drift direction and then removes the drift directly from the data [8]. The residual is recognized to be the non-drifted data. Similarly, Padilla et al. [9] proposed an orthogonal signal correction method for drift compensation, in which the components orthogonal to the signal are recognized to be drift, and then removed directly. These methods suppose that the drift is some additive noise on the data, which may violate the basic property of drift. Di Carlo et al. [10] proposed an evolutionary drift compensation, which aims at learning some transformation corresponding to the best pattern recognition accuracy. However, the learned transformation is solved by taking the recognition accuracy as optimization objective, regardless of drift essence, such that the transformation is recognition accuracy oriented and overfitted. Ziyatdinov et al. [11] proposed a common principal component analysis (CPCA) method for drift compensation, in which the drift variance can be computed. Also, the CPCA method can find common components for all gases in feature space. In this chapter, the proposed DRCA method is completely different from the previous work. DRCA aims at finding a common subspace for both drifted and regular data, instead of the raw feature space. To our knowledge, it is the first time to address drift problem, by viewing drift to be some probability distribution bias in feature space between drift and regular data. In other words, we would like to find a latent common subspace where drift is not dominated.

11.2.2 Existing Subspace Projection Algorithms

Subspace learning aims at learning a low-dimensional subspace under different optimization objectives. Several popular subspace methods include principal component analysis (PCA) [12], linear discriminant analysis (LDA) [13], manifold learning-based locality preserving projections (LPP) [14], marginal fisher analysis (MFA) [15], and their kernelized and tensorized variants [16, 17]. PCA, as an unsupervised method, aims at preserving the maximum variance of data, such that the redundant information is removed. LDA is a supervised dimension reduction method, which target at maximizing the inter-class scatter matrix and minimizing the intra-class scatter matrix, such the linear separability is maximized in the subspace. LPP is an unsupervised dimension reduction technique with manifold assumption, which preserves the affinity structure during dimension reduction process from high to low dimension. MFA is recognized to be a comprehensive version of LDA and LPP, which integrates the intra-class compactness of LDA and the graph embedding of LPP. Therefore, MFA is a supervised subspace learning method with maximum separability and affinity structure preservation.

These subspace learning methods assume that the whole distribution of the data is consistent. However, the drift issue we are facing leads to much different distribution, so that these methods are no longer applicable. In this chapter, a PCA synthesis technique, a subspace projection with transfer capability and distribution alignment, i.e., DRCA, is proposed for handling the issue of distribution inconsistency and further achieving drift adaptation in E-noses.

11.3 Domain Regularized Component Analysis (DRCA)

11.3.1 Mathematical Notations

In this chapter, the source and target domain are defined by subscript "S" and "T", respectively. The training data of source and target domain is denoted as $\mathbf{X}_\mathrm{S} = \left[\mathbf{x}_\mathrm{S}^1, \ldots, \mathbf{x}_\mathrm{S}^{N_\mathrm{S}}\right] \in \Re^{D \times N_\mathrm{S}}$ and $\mathbf{X}_\mathrm{T} = \left[\mathbf{x}_\mathrm{T}^1, \ldots, \mathbf{x}_\mathrm{T}^{N_\mathrm{T}}\right] \in \Re^{D \times N_\mathrm{S}}$, respectively, where D is the number of dimensions, N_S and N_T are the number of training samples in both domains. Let $\mathbf{P} \in \Re^{D \times d}$ represents the basis transformation that maps the original space of source and target data to some subspace with dimension of d. $\|\cdot\|_\mathrm{F}$ and $\|\cdot\|_2$

denotes the Frobenius norm and l_2-norm. $\mathrm{Tr}(\cdot)$ denotes the trace operator and $(\cdot)^{\mathrm{T}}$ denotes the transpose operator. Throughout this chapter, matrix is written in capital boldface, vector is presented in lower boldface, and variable is in italics.

11.3.2 Problem Formulation

As illustrated in Fig. 11.2, we aim to learn a basis transformation **P** that maps the original space of source and target data to some subspace, where the feature distributions between the mapped source and target data $\mathbf{Y}_{\mathrm{S}} = \left[\mathbf{y}_{\mathrm{S}}^{1}, \ldots, \mathbf{y}_{\mathrm{S}}^{N_{\mathrm{S}}}\right] \in \Re^{d \times N_{\mathrm{S}}}$ and $\mathbf{Y}_{\mathrm{T}} = \left[\mathbf{y}_{\mathrm{T}}^{1}, \ldots, \mathbf{y}_{\mathrm{T}}^{N_{\mathrm{T}}}\right] \in \Re^{d \times N_{\mathrm{T}}}$ can be kept similar. Therefore, it is rational to have an idea that the mean distribution discrepancy (MDD) between $\mathbf{Y}_{\mathrm{S}}$ and $\mathbf{Y}_{\mathrm{T}}$ is minimized. Simply, the MDD concept is shown by the proposed domain distance, which is defined as the distance between the centers of the two domains. Therefore, the MDD minimization is formulated as

$$\min\|\boldsymbol{\mu}_{\mathrm{S}} - \boldsymbol{\mu}_{\mathrm{T}}\|_2^2 = \min\left\|\frac{1}{N_{\mathrm{S}}}\sum_{i=1}^{N_{\mathrm{S}}}\mathbf{y}_{\mathrm{S}}^{i} - \frac{1}{N_{\mathrm{T}}}\sum_{j=1}^{N_{\mathrm{T}}}\mathbf{y}_{\mathrm{T}}^{j}\right\|_2^2 \tag{11.1}$$

where $\boldsymbol{\mu}_{\mathrm{S}} = \frac{1}{N_{\mathrm{S}}}\sum_{i=1}^{N_{\mathrm{S}}}\mathbf{y}_{\mathrm{S}}^{i}$ and $\boldsymbol{\mu}_{\mathrm{T}} = \frac{1}{N_{\mathrm{T}}}\sum_{i=1}^{N_{\mathrm{T}}}\mathbf{y}_{\mathrm{T}}^{i}$ represents the centers in the new subspace.

According to the subspace projection, the new representation of source and target data in the lower-dimensional subspace can be formulated as

$$\mathbf{Y}_{\mathrm{S}} = \mathbf{P}^{\mathrm{T}}\mathbf{X}_{\mathrm{S}} = \mathbf{P}^{\mathrm{T}}\left[\mathbf{x}_{\mathrm{S}}^{1}, \ldots, \mathbf{x}_{\mathrm{S}}^{N_{\mathrm{S}}}\right] = \left[\mathbf{y}_{\mathrm{S}}^{1}, \ldots, \mathbf{y}_{\mathrm{S}}^{N_{\mathrm{S}}}\right] \tag{11.2}$$

$$\mathbf{Y}_{\mathrm{T}} = \mathbf{P}^{\mathrm{T}}\mathbf{X}_{\mathrm{T}} = \mathbf{P}^{\mathrm{T}}\left[\mathbf{x}_{\mathrm{T}}^{1}, \ldots, \mathbf{x}_{\mathrm{T}}^{N_{\mathrm{T}}}\right] = \left[\mathbf{y}_{\mathrm{S}}^{1}, \ldots, \mathbf{y}_{\mathrm{S}}^{N_{\mathrm{T}}}\right] \tag{11.3}$$

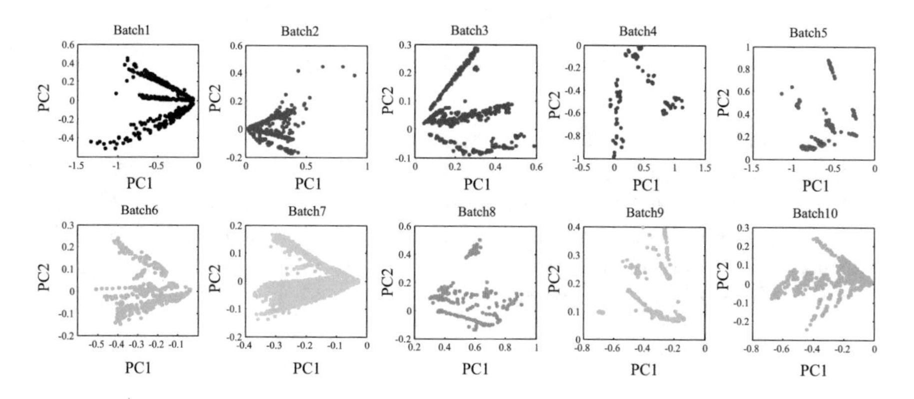

Fig. 11.2 Principal component subspace (PC1 vs. PC2) of the drifted data of 10 batches by PCA

Therefore, we have $\mathbf{y}_\mathrm{S}^i = \mathbf{P}^\mathrm{T}\mathbf{x}_\mathrm{S}^i$ and $\mathbf{y}_\mathrm{T}^j = \mathbf{P}^\mathrm{T}\mathbf{x}_\mathrm{T}^j$. By substituting Eqs. 11.2 and 11.3 into Eq. 11.1, the minimization problem Eq. 11.1 can be reformulated as

$$\min_{\mathbf{P}} \left\| \frac{1}{N_\mathrm{S}} \sum_{i=1}^{N_\mathrm{S}} \mathbf{P}^\mathrm{T}\mathbf{x}_\mathrm{S}^i - \frac{1}{N_\mathrm{T}} \sum_{j=1}^{N_\mathrm{T}} \mathbf{P}^\mathrm{T}\mathbf{x}_\mathrm{T}^j \right\|_2^2 \tag{11.4}$$

For learning such a basis transformation $\mathbf{P}$ that can minimize the mean distribution discrepancy in Eq. 11.4, we should also ensure that the projection does not distort the data itself, such that much more available information can be kept in the new subspace representation. Therefore, for source data, it is rational to maximize the following term:

$$\max_{\mathbf{P}} \mathrm{Tr}\left(\left(\mathbf{P}^\mathrm{T}\mathbf{X}_\mathrm{S}\right)\left(\mathbf{P}^\mathrm{T}\mathbf{X}_\mathrm{S}\right)^\mathrm{T} \right) = \max_{\mathbf{P}} \mathrm{Tr}\left(\mathbf{P}^\mathrm{T}\mathbf{X}_\mathrm{S}\mathbf{X}_\mathrm{S}^\mathrm{T}\mathbf{P}\right) \tag{11.5}$$

It can be seen that by solving Eq. 11.5, the variance (energy) of the source data in the new subspace is maximized, such that the data cannot be distorted and the most available information in the raw data can be remained.

Similarly, for target domain data, we aim at maximizing the following term

$$\max_{\mathbf{P}} \mathrm{Tr}\left(\left(\mathbf{P}^\mathrm{T}\mathbf{X}_\mathrm{T}\right)\left(\mathbf{P}^\mathrm{T}\mathbf{X}_\mathrm{T}\right)^\mathrm{T} \right) = \max_{\mathbf{P}} \mathrm{Tr}\left(\mathbf{P}^\mathrm{T}\mathbf{X}_\mathrm{T}\mathbf{X}_\mathrm{T}^\mathrm{T}\mathbf{P}\right) \tag{11.6}$$

In actual application, there is little data in target domain by comparing to the source domain. In order to learn an effective linear subspace $\mathbf{P}$ for drift adaptation, we propose a target domain regularized variance maximization idea and effectively avoid bias learning. Therefore, the variance maximization formulation shown in Eqs. 11.5 and 11.6 can be integrated together by using a trade-off parameter as follows

$$\begin{aligned} &\max_{\mathbf{P}} \mathrm{Tr}\left(\mathbf{P}^\mathrm{T}\mathbf{X}_\mathrm{S}\mathbf{X}_\mathrm{S}^\mathrm{T}\mathbf{P}\right) + \lambda \cdot \mathrm{Tr}\left(\mathbf{P}^\mathrm{T}\mathbf{X}_\mathrm{T}\mathbf{X}_\mathrm{T}^\mathrm{T}\mathbf{P}\right) \\ &= \max_{\mathbf{P}} \mathrm{Tr}\left(\mathbf{P}^\mathrm{T}\mathbf{X}_\mathrm{S}\mathbf{X}_\mathrm{S}^\mathrm{T}\mathbf{P} + \lambda \cdot \mathbf{P}^\mathrm{T}\mathbf{X}_\mathrm{T}\mathbf{X}_\mathrm{T}^\mathrm{T}\mathbf{P}\right) \\ &= \max_{\mathbf{P}} \mathrm{Tr}\left(\mathbf{P}^\mathrm{T}\left(\mathbf{X}_\mathrm{S}\mathbf{X}_\mathrm{S}^\mathrm{T} + \lambda \cdot \mathbf{X}_\mathrm{T}\mathbf{X}_\mathrm{T}^\mathrm{T}\right)\mathbf{P}\right) \end{aligned} \tag{11.7}$$

where λ denotes the trade-off (regularization) parameter.

The proposed target domain regularized component analysis (DRCA) model aims at minimizing the mean distribution discrepancy (MDD) in Eq. 11.4 and maximizing the regularized variance in Eq. 11.7 of source and target data, simultaneously. Therefore, by integrating Eqs. 11.4 and 11.7 together, the proposed DRSA model can be formulated as

$$\max_{\mathbf{P}} \frac{\mathrm{Tr}\left(\mathbf{P}^{\mathrm{T}}\left(\mathbf{X}_{\mathrm{S}}\mathbf{X}_{\mathrm{S}}^{\mathrm{T}} + \lambda \cdot \mathbf{X}_{\mathrm{T}}\mathbf{X}_{\mathrm{T}}^{\mathrm{T}}\right)\mathbf{P}\right)}{\left\| \frac{1}{N_{\mathrm{S}}}\sum_{i=1}^{N_{\mathrm{S}}} \mathbf{P}^{\mathrm{T}}\mathbf{x}_{\mathrm{S}}^{i} - \frac{1}{N_{\mathrm{T}}}\sum_{j=1}^{N_{\mathrm{T}}} \mathbf{P}^{\mathrm{T}}\mathbf{x}_{\mathrm{T}}^{j} \right\|_2^2} \tag{11.8}$$

Let $\boldsymbol{\mu}_{\mathrm{S}} = \frac{1}{N_{\mathrm{S}}}\sum_{i=1}^{N_{\mathrm{S}}} \mathbf{x}_{\mathrm{S}}^{i}$ and $\boldsymbol{\mu}_{\mathrm{T}} = \frac{1}{N_{\mathrm{T}}}\sum_{i=1}^{N_{\mathrm{T}}} \mathbf{x}_{\mathrm{T}}^{i}$ be the centers of source and target domain, then the maximization problem in Eq. 11.8 can be finally written as

$$\begin{aligned}
&\max_{\mathbf{P}} \frac{\mathrm{Tr}\left(\mathbf{P}^{\mathrm{T}}\left(\mathbf{X}_{\mathrm{S}}\mathbf{X}_{\mathrm{S}}^{\mathrm{T}} + \lambda \cdot \mathbf{X}_{\mathrm{T}}\mathbf{X}_{\mathrm{T}}^{\mathrm{T}}\right)\mathbf{P}\right)}{\left\| \mathbf{P}^{\mathrm{T}}\left(\frac{1}{N_{\mathrm{S}}}\sum_{i=1}^{N_{\mathrm{S}}} \mathbf{x}_{\mathrm{S}}^{i}\right) - \mathbf{P}^{\mathrm{T}}\left(\frac{1}{N_{\mathrm{T}}}\sum_{j=1}^{N_{\mathrm{T}}} \mathbf{x}_{\mathrm{T}}^{j}\right) \right\|_2^2} \\
&= \max_{\mathbf{P}} \frac{\mathrm{Tr}\left(\mathbf{P}^{\mathrm{T}}\left(\mathbf{X}_{\mathrm{S}}\mathbf{X}_{\mathrm{S}}^{\mathrm{T}} + \lambda \cdot \mathbf{X}_{\mathrm{T}}\mathbf{X}_{\mathrm{T}}^{\mathrm{T}}\right)\mathbf{P}\right)}{\left\| \mathbf{P}^{\mathrm{T}}\boldsymbol{\mu}_{\mathrm{S}} - \mathbf{P}^{\mathrm{T}}\boldsymbol{\mu}_{\mathrm{T}} \right\|_2^2} \\
&= \max_{\mathbf{P}} \frac{\mathrm{Tr}\left(\mathbf{P}^{\mathrm{T}}\left(\mathbf{X}_{\mathrm{S}}\mathbf{X}_{\mathrm{S}}^{\mathrm{T}} + \lambda \cdot \mathbf{X}_{\mathrm{T}}\mathbf{X}_{\mathrm{T}}^{\mathrm{T}}\right)\mathbf{P}\right)}{\mathrm{Tr}\left(\mathbf{P}^{\mathrm{T}}\boldsymbol{\mu}_{\mathrm{S}} - \mathbf{P}^{\mathrm{T}}\boldsymbol{\mu}_{\mathrm{T}}\right)\left(\mathbf{P}^{\mathrm{T}}\boldsymbol{\mu}_{\mathrm{S}} - \mathbf{P}^{\mathrm{T}}\boldsymbol{\mu}_{\mathrm{T}}\right)^{\mathrm{T}}} \\
&= \max_{\mathbf{P}} \frac{\mathrm{Tr}\left(\mathbf{P}^{\mathrm{T}}\left(\mathbf{X}_{\mathrm{S}}\mathbf{X}_{\mathrm{S}}^{\mathrm{T}} + \lambda \cdot \mathbf{X}_{\mathrm{T}}\mathbf{X}_{\mathrm{T}}^{\mathrm{T}}\right)\mathbf{P}\right)}{\mathrm{Tr}\left(\mathbf{P}^{\mathrm{T}}(\boldsymbol{\mu}_{\mathrm{S}} - \boldsymbol{\mu}_{\mathrm{T}})(\boldsymbol{\mu}_{S} - \boldsymbol{\mu}_{\mathrm{T}})^{\mathrm{T}}\mathbf{P}\right)}
\end{aligned} \tag{11.9}$$

11.3.3 Model Optimization

In the maximization problem Eq. 11.9, there are many possible solutions of $\mathbf{P}$ (i.e., non-unique solutions). To guarantee the unique property of solution, we impose an equality constraint on the optimization problem, and then, Eq. 11.9 can be written as

$$\begin{aligned}
&\max_{\mathbf{P}} \mathrm{Tr}\left(\mathbf{P}^{\mathrm{T}}\left(\mathbf{X}_{\mathrm{S}}\mathbf{X}_{\mathrm{S}}^{\mathrm{T}} + \lambda \cdot \mathbf{X}_{\mathrm{T}}\mathbf{X}_{\mathrm{T}}^{\mathrm{T}}\right)\mathbf{P}\right) \\
&\text{s.t. } \mathrm{Tr}\left(\mathbf{P}^{\mathrm{T}}(\boldsymbol{\mu}_{\mathrm{S}} - \boldsymbol{\mu}_{\mathrm{T}})(\boldsymbol{\mu}_{\mathrm{S}} - \boldsymbol{\mu}_{\mathrm{T}})^{\mathrm{T}}\mathbf{P}\right) = \eta
\end{aligned} \tag{11.10}$$

where η is a positive constant value.

To solve Eq. 11.10, the Lagrange multiplier function is written as

$$\begin{aligned}
L(\mathbf{P}, \rho) = &\mathrm{Tr}\left(\mathbf{P}^{\mathrm{T}}\left(\mathbf{X}_{\mathrm{S}}\mathbf{X}_{\mathrm{S}}^{\mathrm{T}} + \lambda \cdot \mathbf{X}_{\mathrm{T}}\mathbf{X}_{\mathrm{T}}^{\mathrm{T}}\right)\mathbf{P}\right) \\
&- \rho\left(\mathrm{Tr}\left(\mathbf{P}^{\mathrm{T}}(\boldsymbol{\mu}_{\mathrm{S}} - \boldsymbol{\mu}_{\mathrm{T}})(\boldsymbol{\mu}_{\mathrm{S}} - \boldsymbol{\mu}_{\mathrm{T}})^{\mathrm{T}}\mathbf{P}\right) - \eta\right)
\end{aligned} \tag{11.11}$$

where ρ denotes the Lagrange multiplier coefficient.

By setting the partial derivation of $L(\mathbf{P}, \rho)$ with respect to $\mathbf{P}$ to be 0, we have

$$\frac{\partial L(\mathbf{P}, \rho)}{\partial \mathbf{P}} = 0$$
$$\rightarrow \left((\boldsymbol{\mu}_S - \boldsymbol{\mu}_T)(\boldsymbol{\mu}_S - \boldsymbol{\mu}_T)^T\right)^{-1}(\mathbf{X}_S\mathbf{X}_S^T + \lambda \cdot \mathbf{X}_T\mathbf{X}_T^T)\mathbf{P} = \rho\mathbf{P} \tag{11.12}$$

From Eq. 11.12, we can observe that the $\mathbf{P}$ can be obtained by solving the following eigenvalue decomposition problem of matrix $\mathbf{A}$,

$$\mathbf{AP} = \rho\mathbf{P} \tag{11.13}$$

where $\mathbf{A} = \left((\boldsymbol{\mu}_S - \boldsymbol{\mu}_T)(\boldsymbol{\mu}_S - \boldsymbol{\mu}_T)^T\right)^{-1}(\mathbf{X}_S\mathbf{X}_S^T + \lambda \cdot \mathbf{X}_T\mathbf{X}_T^T)$, and $\boldsymbol{\rho}$ denotes the diagonal matrix of eigenvalues.

From Eq. 11.13, it is clear that $\mathbf{P}$ denotes the eigenvectors. Due to that the model Eq. 11.10 is a maximization problem, therefore, the optimal subspace $\mathbf{P}^*$ denotes the eigenvectors with respect to the first d largest eigenvalues $[\rho_1, \ldots, \rho_d]$, represented by

$$\mathbf{P}^* = [\mathbf{p}_1, \mathbf{p}_2, \ldots, \mathbf{p}_d] \tag{11.14}$$

For easy implementation, the proposed DRSA algorithm is summarized in Algorithm 11.1.

Algorithm 11.1 The proposed DRCA
Input: Heterogeneous data including source data $\mathbf{X}_S \in \Re^{D \times N_S}$, and target data $\mathbf{X}_T \in \Re^{D \times N_T}$, λ, d;
Procedure:

1. Compute the centroid of source data $\boldsymbol{\mu}_S = \frac{1}{N_S}\sum_{i=1}^{N_S} \mathbf{x}_S^i$;
2. Compute the centroid of target data $\boldsymbol{\mu}_T = \frac{1}{N_T}\sum_{i=1}^{N_T} \mathbf{x}_T^i$;
3. Compute the matrix $\mathbf{A}$ as

$$\mathbf{A} = \left((\boldsymbol{\mu}_S - \boldsymbol{\mu}_T)(\boldsymbol{\mu}_S - \boldsymbol{\mu}_T)^T\right)^{-1}(\mathbf{X}_S\mathbf{X}_S^T + \lambda \cdot \mathbf{X}_T\mathbf{X}_T^T)$$

4. Perform the eigenvalue decomposition of $\mathbf{A}$ by using Eq. 11.13;
5. Compute the optimal subspace $\mathbf{P}^* = [\mathbf{p}_1, \mathbf{p}_2, \ldots, \mathbf{p}_d]$ by Eq. 11.14;
6. New subspace projection with drift adaptation by $\mathbf{X}_S' = (\mathbf{P}^*)^T\mathbf{X}_S$ and $\mathbf{X}_T' = (\mathbf{P}^*)^T\mathbf{X}_T$;

Output: The basis transformation $\mathbf{P}$, the drift adapted (distribution aligned) data $\mathbf{X}_S'$ and $\mathbf{X}_T'$.

11.3.4 Remarks on DRCA

The proposed DRCA method is an unsupervised algorithm, in which the labels of the source and target data are not involved. In optimization of the subspace projection P, DRCA holds similar computation with the popular principal component analysis (PCA) and linear discriminant analysis (LDA) method by using eigenvalue decomposition, and the eigenvectors with respect to the first d largest eigenvalues are selected as a group of projection basis. Therefore, the computational complexity of DRCA is O(D3). Extensively, the proposed DRCA can also be further improved by considering the label information in source data, as LDA does. Therefore, a supervised or semi-supervised version can also be easily derived in the future work, if the full or partial label (class) information of the source data is available. Then, a discriminative subspace can be learned by maximizing the between-class scatter matrix and minimizing the within-class scatter matrix. We focus on the unsupervised algorithm for distribution alignment. Additionally, the parameters such as dimensionality d and the trade-off coefficient λ are tuned in the search space $d = \{2k, k = 0, 1, 2, 3, 4, 5, 6, 7\}$ and $\lambda = \{10k, k = -3, -2, -1, 0, 1, 2, 3, 4\}$.

11.4 Experiments

In this section, we conduct two experiments on different datasets with knowledge bias/shift and drift for testing the effectiveness of the proposed DRCA method. Specifically, we first test the proposed method on a public benchmark sensor drift dataset from an electronic nose. Then, we validate the proposed method on our own dataset with long-term drift by an E-nose.

11.4.1 Experiment on Benchmark Sensor Drift Data

For verification of the proposed DRCA method, the real sensor drift benchmark dataset of three years collected by an E-nose from Vergara et al. in UCSD [4] is used in our experiment. The dataset was gathered during the period of 36 months from January 2008 to February 2011 based on a gas delivery platform. The E-nose is with 16 gas sensors and exposed to six kinds of pure gaseous substances, such as acetone, acetaldehyde, ethanol, ethylene, ammonia, and toluene at different concentration levels. Totally, this dataset contains 13910 measurements (samples), which are divided into 10 batches of time series. The number of samples for each class with respect to each batch is summarized in Table 10.1. In feature extraction, eight features were extracted for each sensor, and consequently, a 128-dimensional feature vector (16×8) is formulated for each sample. We refer interested readers to as [4] for more technical details on feature extraction.

To reduce the scale variants among dimensions, the whitening processing (centralization) is conducted on the data. For visually observing the drift of the heterogeneous E-nose data, we have plotted the PCA scatter points in Fig. 11.2, in which the principal component projection (coefficients) is calculated on the raw clean data of batch 1, and the obtained PCA coefficients are used to compute the projected subspaces of other drifted batches. From Fig. 11.2, it is clear that the low-dimensional subspace distribution between batch 1 and other batches is significantly biased (inconsistent) due to drift impact over time.

Therefore, we can say that the drifted E-nose data is heterogeneous. Due to the distribution inconsistency between batch 1 (training data) and other batches (testing data), the recognition performance of the trained pattern classifier would be degraded, because it violates the basic assumption of machine learning that the training data and testing data should be with the same or similar probability distribution (i.e., independent identical distribution, *i.i.d.*).

To demonstrate the effectiveness of the proposed DRCA method for subspace adaptation and distribution alignment, we present the qualitative and quantitative experiments, respectively.

- **Qualitative result**

Motivated by the PCA scatter points in Fig. 11.2, we provide the domain regularized principal components (PC1 vs. PC2) by using the proposed DRCA method on Fig. 11.3. Totally, nine tasks with pairwise batches (batch 1 vs. batch i) ($i = 2, \ldots, 10$) are implemented, respectively. For each task, the projection coefficients is nominated as $\mathbf{P}_{1,i}$ $(i = 2, \ldots, 10)$, then the projected subspaces of batch 1 and batch i based on $\mathbf{P}_{1,i}$ are plotted. As a result, nine pairwise principal component scatter points plots highlighted by using dot rectangles are shown in Fig. 11.3. From the low-dimensional subspace by using the proposed DRCA, the distribution consistency between the drifted batch i and the clean batch 1 is improved.

It is worth noting that it may be impossible to directly calibrate the drifted sensor response due to the nonlinear dynamic behavior of sensor drift and complex sensing mechanism. Therefore, it is reasonable and necessary to handle drift compensation from the perspective of probability distribution alignment and subspace adaptation.

- **Quantitative result**

The essential task for the proposed method is to improve the classification performance. Therefore, the classification accuracy of six classes on each batch is reported as evaluation metric. In comparisons, two experimental settings by following [4] are given as follows.

(1) ***Setting 1***: Take batch 1 as source domain for model training, and test on batch K, $K = 2, \ldots, 10$ (target domains). The classification accuracy on batch K is reported.
(2) ***Setting 2***: Take batch $K - 1$ as source domain for model training, and test on batch K, $K = 2, \ldots, 10$ (target domains). The classification accuracy on batch K is reported.

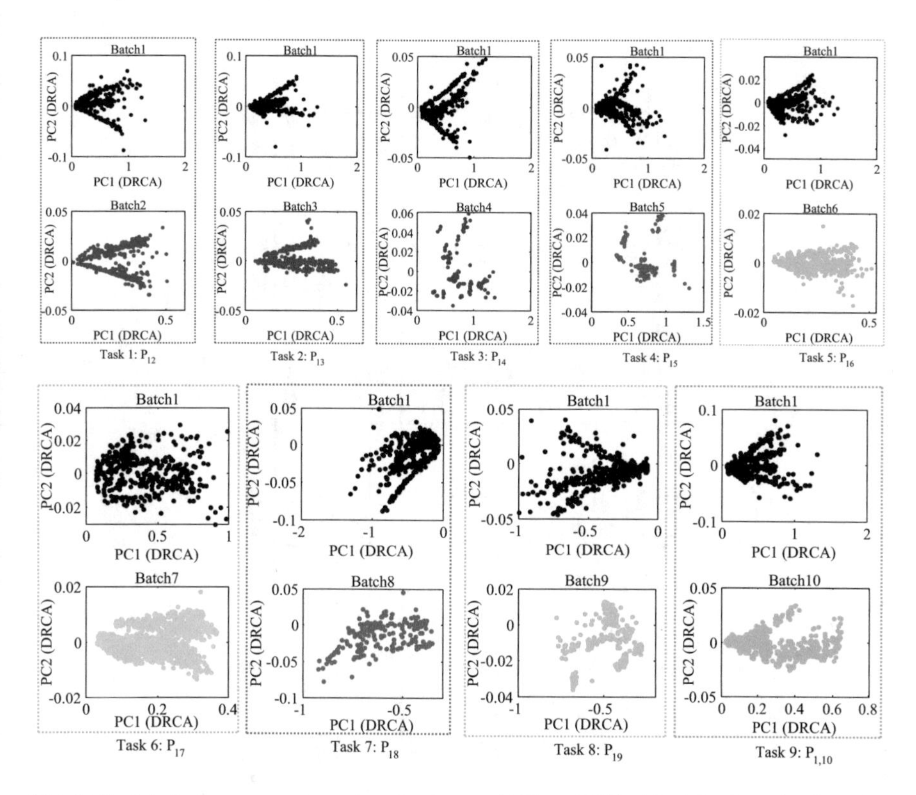

Fig. 11.3 Principal components projected subspace (PC1 vs. PC2) with drift adaptation by using the proposed DRCA method. Specifically, nine tasks are performed. For each task, a new subspace projection $\mathbf{P}_{1,i}$ ($i = 2, \ldots, 10$) is obtained by implementing DRCA on batch 1 and batch i ($i = 2, \ldots, 10$)

For classification, a multi-class SVM with RBF kernel (SVM-rbf) is used as classifier. In comparisons, we compare the proposed DRCA with two baseline subspace methods such as PCA and LDA trained on source data, and eight state-of-the-art results on this benchmark dataset such as CCPCA, SVM ensemble classifier, SVM-gfk, SVM-comgfk, ML-gfk, ML-comgfk, ELM-rbf, and OSC, and two representative calibration transfer methods in E-nose, such as generalized least squares weighting (GLSW) [18] and direct standardization (DS) [19], are compared.

We conducted the experiments on *Setting* 1 and *Setting* 2, respectively. The recognition results for different methods under Experimental *Setting* 1 are reported in Table 11.1, from which we observe that the proposed DRCA achieves the best classification performance. The average classification accuracy is 77.63%, which is 10% higher than the second best learning method, i.e., ML-comgfk. Moreover, for each batch, we also give the best parameters that the proposed method achieves the highest accuracy in Table 11.2.

Table 11.1 Recognition accuracy (%) under experimental **Setting 1**

Batch ID	Batch 2	Batch 3	Batch 4	Batch 5	Batch 6	Batch 7	Batch 8	Batch 9	Batch 10	Average
PCA_{SVM}	82.40	**84.80**	80.12	75.13	73.57	56.16	48.64	67.45	49.14	**68.60**
LDA_{SVM}	47.27	57.76	50.93	62.44	41.48	37.42	**68.37**	52.34	31.17	49.91
CC-PCA	67.00	48.50	41.00	35.50	55.00	31.00	56.50	46.50	30.50	45.72
SVM-rbf	74.36	61.03	50.93	18.27	28.26	28.81	20.07	34.26	34.47	38.94
SVM-gfk	72.75	70.08	60.75	75.08	73.82	54.53	55.44	69.62	41.78	63.76
SVM-comgfk	74.47	70.15	59.78	**75.09**	73.99	54.59	55.88	70.23	41.85	64.00
ML-rbf	42.25	73.69	75.53	66.75	77.51	54.43	33.50	23.57	34.92	53.57
ML-comgfk	80.25	74.99	78.79	67.41	**77.82**	**71.68**	49.96	50.79	**53.79**	67.28
ELM-rbf	70.63	66.44	66.83	63.45	69.73	51.23	49.76	49.83	33.50	57.93
OSC	**88.10**	66.71	54.66	53.81	65.13	**63.71**	36.05	40.21	40.08	56.50
GLSW	78.38	69.36	**80.75**	74.62	69.43	44.28	48.64	67.87	46.58	64.43
DS	69.37	46.28	41.61	58.88	48.83	32.83	23.47	**72.55**	29.03	46.98
DRCA	**89.15**	**92.69**	**87.58**	**95.94**	**86.52**	60.25	**62.24**	**72.34**	**52.00**	**77.63**

The boldface type denotes the best performance.

Table 11.2 Parameters' values of the DRCA under experimental **Setting 1**

Batch ID	Batch 2	Batch 3	Batch 4	Batch 5	Batch 6	Batch 7	Batch 8	Batch 9	Batch 10
λ	0.008	0.9	1000	0.03	0.06	8	0.01	0.001	0.08
d	30	40	39	12	12	17	8	11	10

Further, by following the experimental *Setting* 2, that is, model training on batch $K - 1$ and test on batch K, and the results are reported in Table 11.3. From Table 11.3, we can see that the proposed DRCA performs the second best performance (74.2% in average). From the quantitative classification accuracy, the effectiveness and competitiveness of the proposed DRCA model have been clearly demonstrated.

11.4.2 Experiment on E-Nose Data with Drift and Shift

In this section, a more complex E-nose dataset from our lab is used. This dataset includes three parts: master data (collected 5 years ago), slave 1 data (collected now), and slave 2 data (collected now). The "complex" is represented by multiple E-nose instruments of the same type. More specifically, not only the sensor drift is implied in the data, the sensor shift caused by bad reproducibility also happens. In data acquisition, the master and two slavery E-nose systems were developed. Four TGS series sensors and two variables for ambient temperature and humidity are integrated. Therefore, the dimension of features (DoF) is six. In this dataset, six kinds of gaseous contaminants such as formaldehyde, benzene, toluene, carbon monoxide, nitrogen dioxide, and ammonia are included. The detailed description of the dataset is shown in Table 11.4.

The PCA scatter points on the master data, slave 1 data, and slave 2 data, are shown in Fig. 11.4, respectively, in which we can see that the points from different classes are crossed. By implementing the proposed DRCA method on master $\rightarrow$ slave 1 and master $\rightarrow$ slave 2, respectively, the results are shown in Fig. 11.6, from which we can observe that the cross-problem of different classes is eased. Figure 11.5 qualitatively demonstrates the positive drift adaptation effect of the DRCA.

- **Settings**

The *master* data collected 5 years ago is used as source domain data (no drift). The slave 1 and slave 2 data collected after 5 years are used as target domain data (with drift and shift). Then, we conduct the experiments with two settings as follows.

Setting 1 (Drift): Due to that there is also sensor discreteness between slaves and master, the sensor calibration between slaves and master is made by using linear regression according to [20]. Therefore, only drift exists between the source and target data.

Table 11.3 Recognition accuracy (%) under experimental **Setting 2**

Batch ID	1 → 2	2 → 3	3 → 4	4 → 5	5 → 6	6 → 7	7 → 8	8 → 9	9 → 10	Average
PCA_{SVM}	82.40	**98.87**	83.23	72.59	36.70	74.98	58.16	84.04	30.61	69.06
LDA_{SVM}	47.27	46.72	70.81	85.28	48.87	75.15	77.21	62.77	30.25	60.48
SVM-rbf	74.36	87.83	90.06	56.35	42.52	83.53	**91.84**	62.98	22.64	68.01
SVM-gfk	72.75	74.02	77.83	63.91	70.31	77.59	78.57	86.23	15.76	68.56
SVM-comgfk	74.47	73.75	78.51	64.26	69.97	77.69	82.69	85.53	17.76	69.40
ML-rbf	42.25	58.51	75.78	29.10	53.22	69.17	55.10	37.94	12.44	48.17
ML-comgfk	80.25	**98.55**	84.89	**89.85**	**75.53**	**91.17**	61.22	**95.53**	**39.56**	**79.62**
ELM-rbf	70.63	40.44	64.16	64.37	72.70	80.75	88.20	67.00	22.00	63.36
GLSW	78.38	97.04	81.99	73.60	36.57	74.48	60.54	81.91	26.31	67.87
DS	69.37	53.59	67.08	37.56	36.30	26.57	49.66	42.55	25.78	45.38
DRCA	**89.15**	98.11	**95.03**	69.54	50.87	78.94	65.99	84.04	36.31	**74.22**

The boldface type denotes the best performance.

Table 11.4 Data description of the complex E-nose data

E-nose system	DoF	HCHO	Benzene	Toluene	CO	NO2	Ammonia	Total
Master (no drift)	6	126	72	66	58	38	60	420
Slave 1 (drift + shift)	6	108	108	106	98	107	81	608
Slave 2 (drift + shift)	6	108	87	94	95	108	84	576

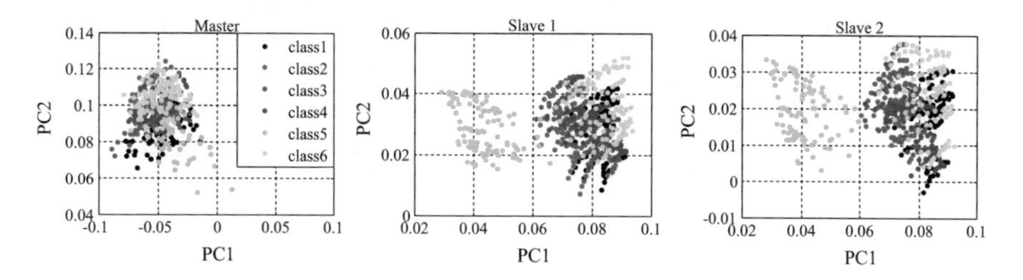

Fig. 11.4 PCA scatter points of the master, slave 1 and slave 2 data, respectively

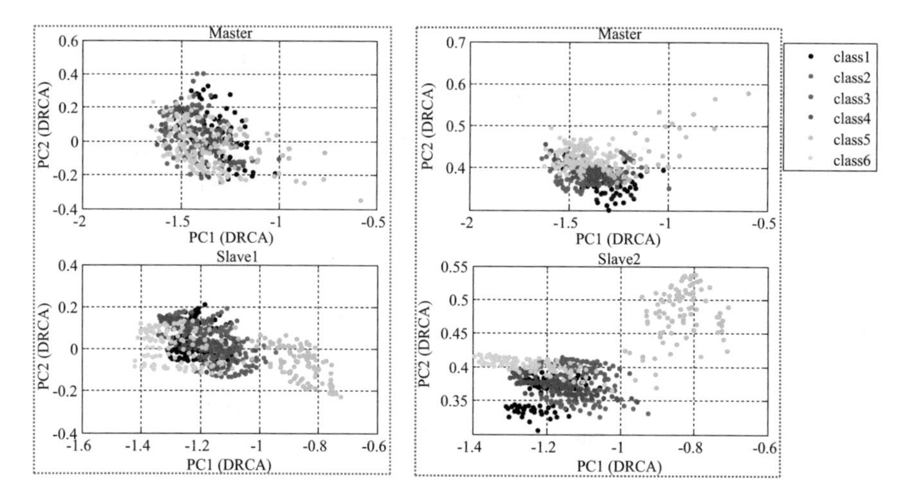

Fig. 11.5 Scatter points by using DRCA from master → slave 1 (red dot) and master → slave 2 (blue dot). Note that the analytes from class 1 to class 6 are toluene, benzene, ammonia, carbon monoxide, nitrogen dioxide, and formaldehyde, respectively

Setting 2 (Drift + Shift): The sensor calibration step is omitted, which implies that both the sensor drift and shift exist between the source and target data.

- **Results**

In this section, the average recognition accuracy of the six kinds of gases is reported. For each setting, two tasks including master → slave 1 and master → slave 2 are conducted. By following setting 1 (with shift calibration), the experimental results of the proposed DRCA method are presented in Table 11.5, in which SVM, PCA, and LDA and two representative calibration transfer methods (generalized least squares weighting (GLSW) [11] and direct standardization (DS) [18]) are compared. From the results, we can observe that the proposed method achieves the best recognition accuracies on two slaves, with 10% improvements. This demonstrates that the proposed DRCA can effectively address the long-term drift issue (5 years).

Further, to verify the proposed method for drift plus shift adaptation, the experimental results by following setting 2 are presented in Table 11.6. We can observe that the proposed DRCA still outperforms other methods. Additionally, by comparing the results between Tables 11.5 and 11.6, we get that the proposed method can well issue the sensor shift. Therefore, we confirm that the proposed method is effective in handling heterogeneous E-nose data caused by both drift and shift.

11.4.3 Parameter Sensitivity Analysis

In the proposed DRCA model, there are two parameters: the dimension d and regularization coefficient λ. To observe the performance variations in tuning the two parameters, we tune from the parameter set $d = \{2^k, k = 0, 1, 2, 3, 4, 5, 6, 7\}$ and $\lambda = \{10^k, k = -3, -2, -1, 0, 1, 2, 3, 4\}$. We select the benchmark sensor drift dataset for experiments. When tuning the parameter λ, the value of d is fixed and vice versa. The tuning results are shown in Fig. 11.6, which can help us quickly determine the range of the optimal parameters.

11.4.4 Discussion

The proposed DRCA method, from the viewpoint of machine learning, is a cross-domain subspace learning technique. The so-called cross-domain denotes the

Table 11.5 Recognition accuracy (%) with sensor shift calibration (Setting 1)

Task	SVM	PCA	LDA	GLSW	DS	DRCA
Master → slave 1	45.89	46.22	42.11	41.45	40.30	**57.07**
Master → slave 2	31.08	41.84	41.32	48.09	39.76	**52.95**

The boldface type denotes the best performance.

Table 11.6 Recognition accuracy (%) without sensor shift calibration (Setting 2)

Task	SVM	PCA	LDA	GLSW	DS	DRCA
Master → slave 1	51.97	51.97	51.97	47.53	40.46	**58.55**
Master → slave 2	60.59	60.59	56.77	59.38	40.63	**61.63**

The boldface type denotes the best performance.

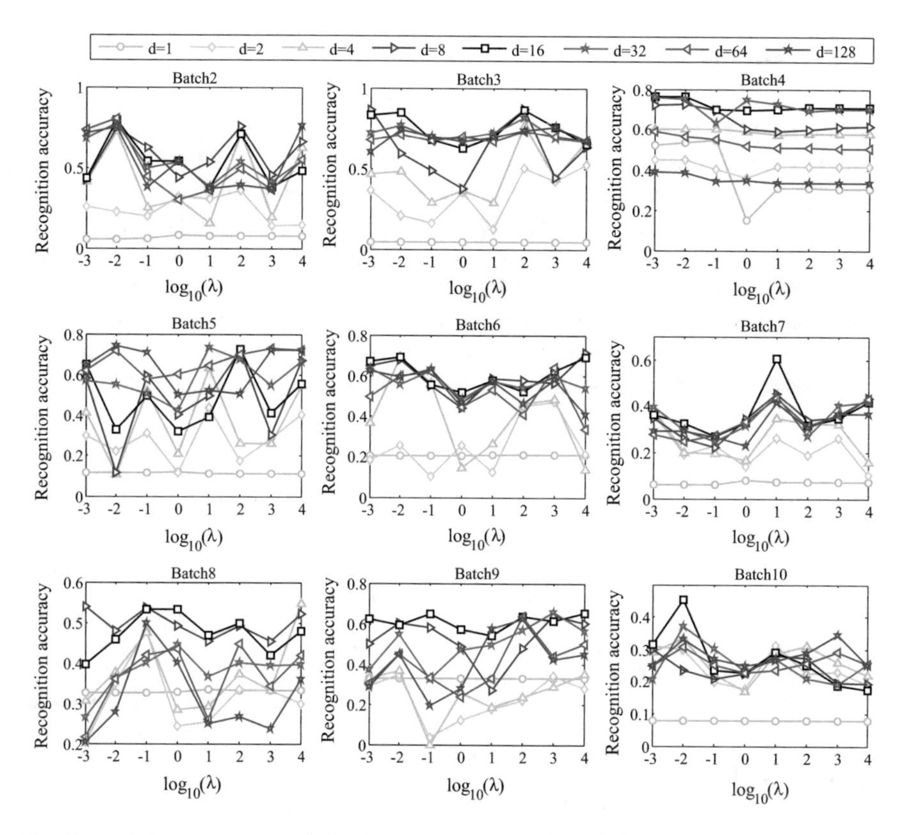

Fig. 11.6 Performance curves under different dimension d and λ

cross-system from master to slave in E-nose topic. The advantage of DRCA is that a common subspace can be learned for source domain (master) and target domain (slave), such that the drift information implied in target domain is eliminated in this common subspace. That is, in this common subspace, the master and slave share similar class information, which greatly helps the classification without drift blur. There is also a limitation that the proposed DRCA is essentially a linear technique, by transformation via a projection $\mathbf{P}$, which may not seriously guarantee the "drift-less" property. Therefore, our future improvement may be the nonlinear subspace learning by considering the kernel technique. Additionally, the proposed DRCA can be easily implemented in on-line manner. Specifically, the focus of DRCA is the learning of $\mathbf{P}$ based on source (master) and target (slave) dataset. In on-line application, the target dataset may be time-varying; therefore, the time interval may be defined by users, for updating the projection $\mathbf{P}$.

11.5 Summary

In this chapter, a domain regularized component analysis (DRCA) model for heterogeneous E-nose data with drift is presented. The proposed DRCA method is motivated by transfer learning, and the difference of probability distribution between source data and target data incurs the failure of machine learning in data mining. For learning, a common subspace for heterogeneous data, an intuitive idea by regularizing the target domain and minimizing the mean distribution discrepancy (MDD)-based domain distance is formulated. Experiments on synthetic dataset, benchmark sensor drift dataset, and sensor drift plus shift dataset demonstrate that the proposed DRCA method outperforms several state-of-the-art methods.

References

1. L. Zhang, F. Tian, S. Liu, L. Dang, X. Peng, X. Yin, Chaotic time series prediction of e-nose sensor drift in embedded phase space. Sens. Actuators B Chem. **182**, 71–79 (2013)
2. F. Hossein-Babaei, V. Ghafarinia, Compensation for the drift-like terms caused by environmental fluctuations in the response of chemoresistive gas sensors. Sens. Actuators B Chem. **143**, 641–648 (2010)
3. E. Martinelli, G. Magna, S. De Vito, R. Di Fuccio, G. Di Francia, A. Vergara, C. Di Natale, An adaptive classification model based on the artificial immune system for chemical sensor drift mitigation. Sens. Actuators B Chem. **177**, 1017–1026 (2013)
4. A. Vergara, S. Vembu, T. Ayhan, M.A. Ryan, M.L. Homer, R. Huerta, Chemical gas sensor drift compensation using classifier ensembles. Sens. Actuators B Chem. **167**, 320–329 (2012)
5. H. Liu, Z. Tang, Metal oxide gas sensor drift compensation using a dynamic classifier ensemble based on fitting. Sensors **13**, 9160–9173 (2013)
6. L. Zhang, D. Zhang, Domain adaptation extreme learning machines for drift compensation in E-nose systems. IEEE Trans. Instrum. Meas. **64**(7), 1790–1801 (2015)
7. Q. Liu, M. Ye, S.S. Ge, X. Du, Drift compensation for electronic nose by semi-supervised domain adaption. IEEE Sens. J. **14**(3), 657–665 (2014)
8. T. Artursson, T. Eklov, I. Lundstrom, P. Martensson, M. Sjostrom, M. Holmberg, Drift correction for gas sensors using multivariate methods. J. Chemometr. **14**(5–6), 711–723 (2000)
9. M. Padilla, A. Perera, I. Montoliu, A. Chaudry, K. Persaud, S. Marco, Drift compensation of gas sensor array data by orthogonal signal correction. Chemometr. Intell. Lab. Syst. **100**, 28–35 (2010)
10. S. Di Carlo, M. Falasconi, E. Sanchez, A. Scionti, G. Squillero, A. Tonda, Increasing pattern recognition accuracy for chemical sensing by evolutionary based drift compensation. Pattern Recogn. Lett. **32**, 1594–1603 (2011)
11. A. Ziyatdinov, S. Marco, A. Chaudry, K. Persaud, P. Caminal, A. Perera, Drift compensation of gas sensor array data by common principal component analysis. Sens. Actuators B Chem. **146**(2), 460–465 (2010)
12. M. Turk, A. Pentland, Eigenfaces for recognition. J. Cogn. Neurosci. **3**(1), 71–86 (1991)
13. P.N. Belhumeur, J. Hespanha, D.J. Kriegman, Eigenfaces vs. fisherfaces: recognition using class specific linear projection. IEEE Trans. Pattern Anal. Mach. Intell. **19**(7), 711–720 (1997)
14. X. He, S. Yan, Y. Hu, P. Niyogi, H.J. Zhang, Face recognition using laplacianfaces. IEEE Trans. Pattern Anal. Mach. Intell. **27**(3), 328–340 (2005)

15. S. Yan, D. Xu, B. Zhang, H. Zhang, Q. Yang, S. Lin, Graph embedding and extensions: a general framework for dimensionality reduction. IEEE Trans. Pattern Analysis Mach. Intell. **29**(1), 40–51 (2007)
16. S. Yan, J. Liu, X. Tang, T. Huang, A parameter-free framework for general supervised subspace learning. IEEE Trans. Inf. Forensics Secur. **2**(1), 69–76 (2007)
17. S. Yan, D. Xu, Q. Yang, L. Zhang, X. Tang, H. Zhang, Multilinear discriminant analysis for face recognition. IEEE Trans. Image Proc. **16**(1), 212–220 (2007)
18. L. Fernandez, S. Guney, A. Gutiérrez-Gálvez, S. Marco, Calibration transfer in temperature modulated gas sensor arrays. Sens. Actuators B Chem. **231**, 276–284 (2016)
19. J. Fonollosa, L. Fernández, A. Gutiérrez, R. Huerta, S. Marco, Calibration transfer and drift counteraction in chemical sensor arrays using direct standardization. Sens. Actuators B Chem. **236**, 1044–1053 (2016)
20. L. Zhang, F. Tian, C. Kadri, B. Xiao, H. Li, L. Pan, H. Zhou, On-line sensor calibration transfer among electronic nose instruments for monitoring volatile organic chemicals in indoor air quality. Sens. Actua. B **160**(1), 899–909 (2011)

Chapter 12
Cross-Domain Subspace Learning Approach

Abstract Extreme learning machine (ELM) was proposed for solving a single-layer feed-forward network (SLFN) with fast learning speed and has been confirmed to be effective and efficient for pattern classification and regression in different fields. ELM originally focuses on the supervised, semi-supervised, and unsupervised learning problems, but just in the single domain. To our best knowledge, ELM with cross-domain learning capability in subspace learning has not been exploited very well. Inspired by a cognitive-based extreme learning machine, this chapter presents a unified subspace transfer framework called cross-domain extreme learning machine (CdELM), which aims at learning a common (shared) subspace across domains. Experiments on drifted E-nose datasets demonstrate that the proposed CdELM method significantly outperforms other compared methods.

Keywords Extreme learning machine (ELM) · Subspace learning
Domain adaptation · Electronic nose

12.1 Introduction

Many machine learning models (e.g., neural networks and support vector machines) have been proposed to solve classification and regression problems. Recently, extreme learning machines (ELM) [1–5], as a cognitive-based technique, are proposed for "generalized" single-layer feed-forward networks (SLFNs) [1, 2, 4, 6, 7]. ELM can analytically determine the output layer using Moor–Penrose generalized inverse by adopting the square loss of prediction error. Huang et al. [4, 6, 8, 9] have rigorously proved that, in theory, ELM can approximate any continuous functions. Moreover, different from traditional learning algorithms [10], ELM tends to achieve not only the smallest training error but also the smallest norm of output weights for better generalization performance [3, 7]. Its variants [11–19] also focus on the regression, classification, and pattern recognition applications.

The data distribution obtained in different stages with different sampling conditions would be different, which is identified as cross-domain problem.

L. Zhang et al., *Electronic Nose: Algorithmic Challenges*,
https://doi.org/10.1007/978-981-13-2167-2_12

Traditional ELM assumes that the data distribution between training and testing data should be similar and therefore cannot address this issue. To handle this problem, domain adaptation has been proposed for heterogeneous data (i.e., cross-domain problem), by leveraging a few labeled instances from another domain with similar semantic [20, 21]. Inspired by domain adaptation, Zhang and Zhang [22] proposed a domain adaptation ELM (DAELM) for classification across tasks (source domain vs. target domain), which is the first chapter to study the cross-domain learning mechanism in ELM. However, DAELM was proposed as a cross-domain classifier, and how to learn a shared (common) subspace with source and target domains, to our knowledge, has never been studied in ELM. Therefore, in this chapter, we extend ELMs to handle cross-domain problems by transfer learning and subspace learning and explore its capability in the multi-domain application for ELMs. Inspired by DAELM, we propose a cross-domain extreme learning machine (CdELM) for common subspace learning, and the basic idea of the proposed CdELM method is illustrated in Fig. 12.1, in which we aim at learning

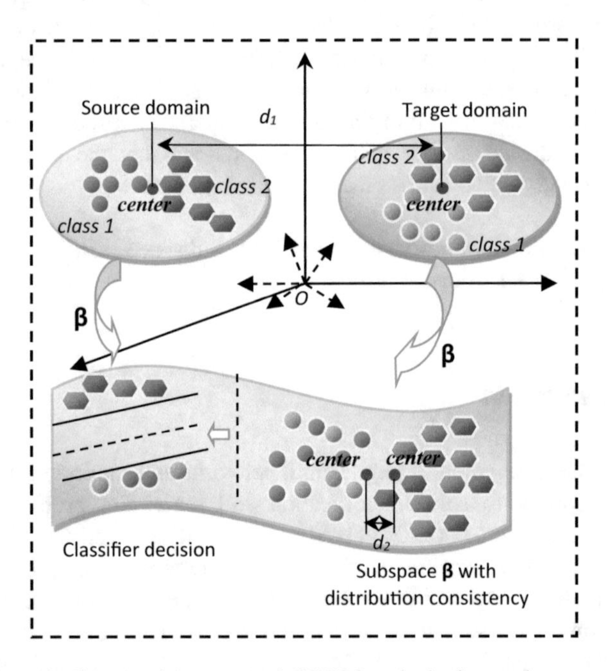

Fig. 12.1 Schematic diagram of the proposed CdELM method; after a subspace projection $\boldsymbol{\beta}$, the source domain and target domain of different space distribution lie in a latent subspace with good distribution consistency (the centers of both domains become very close, and drift is removed). Formally, the upper coordinate system denotes the raw data points of source domain and target domain in three dimensions. We use the word "center" to represent the mean of each domain data. From the upper figure, we can see that the difference between the mean of source domain and the mean of target domain is large. After a subspace projection $\boldsymbol{\beta}$ in the below figure, we can see that the values of d_2 become smaller, which demonstrate that the distribution difference becomes small, and both domains of different space distribution lie in a latent common subspace with good distribution consistency

a shared subspace $\boldsymbol{\beta}$. Notably, the proposed CdELM is different from DAELM that we would like to learn a shared subspace for source and target domains, instead of a shared classifier.

12.2 Related Work

12.2.1 Review of ELM

Briefly, the principle of ELM is described as follows. Given the training data $\mathbf{X} = [\mathbf{x}_1, \mathbf{x}_2, \ldots, \mathbf{x}_N] \in \Re^{N\times n}$, where n is the dimensionality, N is the number of training samples, and $\mathbf{T} = [\mathbf{t}_1, \mathbf{t}_2, \ldots, \mathbf{t}_N] \in \Re^{N\times m}$ denotes the labels with respect to the data $\mathbf{X}$, where m is the number of classes. The output of the hidden layer is denoted as $\mathcal{H}(\mathbf{x}_i) \in \Re^{1\times L}$, where L is the number of hidden nodes and $\mathcal{H}(\cdot)$ is the activation function. The output weights between the hidden layer and the output layer being learned are denoted as $\boldsymbol{\beta} \in \Re^{L\times m}$. The regularized ELM aims at minimizing the training error and the norm of the output weights for better generalization performance, formulated as follows:

$$\begin{aligned} &\min_{\boldsymbol{\beta}} \mathcal{L} = \frac{1}{2}\|\boldsymbol{\beta}\|^2 + \frac{C}{2}\sum_{i=1}^{N}\|\xi_i\|^2 \\ &\text{s.t.} \mathcal{H}(\mathbf{x}_i)\boldsymbol{\beta} = \mathbf{t}_i - \xi_i, \quad i = 1, \ldots, N \end{aligned} \tag{12.1}$$

where ξ_i denotes the prediction error with respect to the ith training pattern $\mathbf{x}_i$, and C is a penalty constant on the training errors.

By substituting the constraint term in Eq. 12.1 into the objective function, an equivalent unconstrained optimization problem can be obtained as follows:

$$\min_{\boldsymbol{\beta}} \mathcal{L} = \frac{1}{2}\|\boldsymbol{\beta}\|^2 + \frac{C}{2}\sum_{i=1}^{N}\|\mathbf{T} - \mathbf{H}\boldsymbol{\beta}\|^2 \tag{12.2}$$

where $\mathbf{H} = [\mathcal{H}(\mathbf{x}_1); \mathcal{H}(\mathbf{x}_2); \ldots; \mathcal{H}(\mathbf{x}_N)] \in \Re^{N\times L}$.

The minimization problem Eq. 12.2 is a regularized least square problem. By setting the gradient of $\mathcal{L}$ with respect to $\boldsymbol{\beta}$ to zero, we can get the closed-form solution of $\boldsymbol{\beta}$. There are two cases while solving $\boldsymbol{\beta}$. If N is larger than L, the gradient equation is overdetermined, and the closed-form solution can be obtained as

$$\boldsymbol{\beta}^* = \left(\mathbf{H}^{\mathrm{T}}\mathbf{T} + \frac{\mathbf{I}}{C}\right)^{-1}\mathbf{H}^{\mathrm{T}}\mathbf{T} \tag{12.3}$$

where $\mathbf{I}$ denotes the identity matrix of size L.

Second, if the number N of training samples is smaller than L, an underdetermined least square problem would be handled. In this case, the solution of Eq. 12.1 can be obtained as

$$\boldsymbol{\beta}^* = \mathbf{H}^{\mathrm{T}}\left(\mathbf{H}\mathbf{H}^{\mathrm{T}} + \frac{\mathbf{I}}{C}\right)^{-1}\mathbf{T} \tag{12.4}$$

where $\mathbf{I}$ denotes the identity matrix of size N.

Therefore, for classification problem, the solution of $\boldsymbol{\beta}$ can be computed by using Eq. 12.3 or Eq. 12.4. We direct the interested readers to [3] for more details on ELM theory and the algorithms.

12.2.2 Subspace Learning

Subspace learning aims at learning a low-dimensional subspace. There are several common methods, such as principal component analysis (PCA) [23], linear discriminant analysis (LDA) [24], and manifold learning-based locality preserving projections (LPPs) [25]. All of these methods suppose that the data distribution is consistent; namely, they are only applicable to single domain. However, this assumption is often violated in many real-world applications. So, for heterogeneous data, we proposed a new cross-domain learning method to learn a shared subspace.

12.3 The Proposed CdELM Method

12.3.1 Notations

In this chapter, source domain and target domain are defined by subscript "S" and "T", respectively. The training data of source and target domain is denoted as $\mathbf{X}_{\mathrm{S}} = [\mathbf{x}_{\mathrm{S}}^1, \ldots, \mathbf{x}_{\mathrm{S}}^{N_{\mathrm{S}}}] \in \Re^{D \times N_{\mathrm{S}}}$ and $\mathbf{X}_{\mathrm{T}} = [\mathbf{x}_{\mathrm{T}}^1, \ldots, \mathbf{x}_{\mathrm{T}}^{N_{\mathrm{T}}}] \in \Re^{D \times N_{\mathrm{T}}}$, respectively, where D is the number of dimensions, and N_{S} and N_{T} are the number of training samples in both domains. Let $\boldsymbol{\beta} \in \Re^{L \times d}$ represent the basis transformation that maps the ELM space of source and target data to some subspace with dimension of d. $\|\cdot\|_{\mathrm{F}}$ and $\|\cdot\|_2$ denote the Frobenius norm and l_2-norm. $\mathrm{Tr}(\cdot)$ denotes the trace operator, and $(\cdot)^{\mathrm{T}}$ denotes the transpose operator. Throughout this chapter, matrix is written in capital boldface, vector is presented in lower boldface, and variable is in italics.

12.3.2 Model Formulation

As illustrated in Fig. 12.1, the distribution between the source domain and target domain is different. Therefore, the performance of the learned classifier by the source domain will be dramatically degraded. Inspired by subspace learning and ELM, the main idea of the proposed CdELM is to learn a shared subspace $\boldsymbol{\beta}$ in ELM space rather than a classifier $\boldsymbol{\beta}$. Therefore, the source domain and target domain share the similar feature distribution in the latent projection $\boldsymbol{\beta}$.

Firstly, by mapping the source data and target data into the ELM space, and then we could obtain $\mathbf{H}_\mathrm{S} = [\mathbf{h}_\mathrm{S}^1, \ldots, \mathbf{h}_\mathrm{S}^{N_\mathrm{S}}] \in \Re^{L \times N_\mathrm{S}}$ and $\mathbf{H}_\mathrm{T} = [\mathbf{h}_\mathrm{T}^1, \ldots, \mathbf{h}_\mathrm{T}^{N_\mathrm{T}}] \in \Re^{L \times N_\mathrm{T}}$, where $\mathbf{h}_\mathrm{S}^i = g(\mathbf{W}^\mathrm{T}\mathbf{x}_\mathrm{S}^i + \mathbf{b}^\mathrm{T})$ and $\mathbf{h}_\mathrm{T}^j = g(\mathbf{W}^\mathrm{T}\mathbf{x}_\mathrm{T}^j + \mathbf{b}^\mathrm{T})$ are the output (column) vector of the hidden layer with respect to the input $\mathbf{x}_\mathrm{S}^i$ and $\mathbf{x}_\mathrm{T}^j$, respectively, $i = 1, 2, \ldots, N_\mathrm{S}$, $j = 1, 2, \ldots, N_\mathrm{T}$, $g(\cdot)$ is a activation function, L is the number of randomly generated hidden nodes, $\mathbf{W} \in \Re^{D \times L}$ and $\mathbf{b} \in \Re^{1 \times L}$ are randomly generated weights.

In the learned subspace $\boldsymbol{\beta}$, we expect that the distribution between source domain and target domain should be consistent, and the discrimination can also be improved for recognition. Inspired by linear discriminant analysis (LDA), we aim at minimizing the intra-class scatter matrix $\mathbf{S}_W^S$ and simultaneously maximizing the inter-class scatter matrix $\mathbf{S}_B^S$ of the source data, such that the separability can be promised in the learned linear subspace. Therefore, for source domain, it is rational to maximize the following term

$$\max_{\boldsymbol{\beta}} \frac{\mathrm{Tr}(\boldsymbol{\beta}^\mathrm{T}\mathbf{S}_B^S\boldsymbol{\beta})}{\mathrm{Tr}(\boldsymbol{\beta}^\mathrm{T}\mathbf{S}_W^S\boldsymbol{\beta})} \tag{12.5}$$

where the inter-class scatter matrix and intra-class scatter matrix can be computed as $\mathbf{S}_B^S = \sum_{c=1}^{C} (\boldsymbol{\mu}_\mathrm{S}^c - \boldsymbol{\mu}_\mathrm{S})(\boldsymbol{\mu}_\mathrm{S}^c - \boldsymbol{\mu}_\mathrm{S})^\mathrm{T}$ and $\mathbf{S}_W^S = \sum_{c=1}^{C} \sum_{k=1,\mathbf{h}_\mathrm{S}^k \in G_c}^{n_c} (\mathbf{h}_\mathrm{S}^k - \boldsymbol{\mu}_\mathrm{S}^c)(\mathbf{h}_\mathrm{S}^k - \boldsymbol{\mu}_\mathrm{S}^c)^\mathrm{T}$, where $\boldsymbol{\mu}_\mathrm{S}$ represents the center of source data, $\boldsymbol{\mu}_\mathrm{S}^c$ represents the center of class c of source data in the raw space, C represents the number of categories, G_c represents a collection belonging to class c, and n_c represents the number of class c.

For learning such a subspace $\boldsymbol{\beta}$ that maximizes the formulation Eq. 12.5, we should also ensure that the projection does not distort the data from target domain, such that much more available information can be kept in the new subspace representation. Therefore, it is rational to maximize the following term,

$$\max_{\boldsymbol{\beta}} \mathrm{Tr}((\boldsymbol{\beta}^\mathrm{T}\mathbf{H}_\mathrm{T})(\boldsymbol{\beta}^\mathrm{T}\mathbf{H}_\mathrm{T})^\mathrm{T}) = \max_{\boldsymbol{\beta}} \mathrm{Tr}(\boldsymbol{\beta}^\mathrm{T}\mathbf{H}_\mathrm{T}\mathbf{H}_\mathrm{T}^\mathrm{T}\boldsymbol{\beta}) \tag{12.6}$$

Naturally, after projected by $\boldsymbol{\beta}$, the feature distributions between the mapped source domain $\mathbf{H}_\mathrm{S} = [\mathbf{h}_\mathrm{S}^1, \ldots, \mathbf{h}_\mathrm{S}^{N_\mathrm{S}}] \in \Re^{L \times N_\mathrm{S}}$ and target domain $\mathbf{H}_\mathrm{T} = [\mathbf{h}_\mathrm{T}^1, \ldots, \mathbf{h}_\mathrm{T}^{N_\mathrm{T}}] \in \Re^{L \times N_\mathrm{T}}$ can become similar. Therefore, it is rational to have an idea that the mean distribution discrepancy (MDD) between $\mathbf{H}_\mathrm{S}$ and $\mathbf{H}_\mathrm{T}$ can be minimized. That is, the distance between the centers of the two domains should be minimized. Therefore, the MDD minimization is formulated as

$$\min\left\|\frac{1}{N_{\mathrm{S}}}\sum_{i=1}^{N_{\mathrm{S}}}\boldsymbol{\beta}^{\mathrm{T}}\mathbf{h}_{\mathrm{S}}^{i}-\frac{1}{N_{\mathrm{T}}}\sum_{j=1}^{N_{\mathrm{T}}}\boldsymbol{\beta}^{\mathrm{T}}\mathbf{h}_{\mathrm{T}}^{j}\right\|_{\mathrm{F}}^{2} \tag{12.7}$$

With the merits of ELM, we expect that the norm of $\boldsymbol{\beta}$ is minimized,

$$\min_{\boldsymbol{\beta}}\|\boldsymbol{\beta}\|_{\mathrm{F}}^{2} \tag{12.8}$$

After a detailed description of the four specific parts in the proposed CdELM model, by incorporating the Eq. 12.5 into Eq. 12.8, a complete CdELM model is formulated as follows

$$\min_{\boldsymbol{\beta}}\frac{\mathrm{Tr}(\boldsymbol{\beta}^{\mathrm{T}}\mathbf{S}_{W}^{S}\boldsymbol{\beta})+\lambda_{0}\|\boldsymbol{\beta}\|_{\mathrm{F}}^{2}+\lambda_{1}\left\|\frac{1}{N_{\mathrm{S}}}\sum_{i=1}^{N_{\mathrm{S}}}\boldsymbol{\beta}^{\mathrm{T}}\mathbf{h}_{\mathrm{S}}^{i}-\frac{1}{N_{\mathrm{T}}}\sum_{j=1}^{N_{\mathrm{T}}}\boldsymbol{\beta}^{\mathrm{T}}\mathbf{h}_{\mathrm{T}}^{j}\right\|_{\mathrm{F}}^{2}}{\mathrm{Tr}(\boldsymbol{\beta}^{\mathrm{T}}\mathbf{S}_{B}^{S}\boldsymbol{\beta})+\lambda_{2}\mathrm{Tr}(\boldsymbol{\beta}^{\mathrm{T}}\mathbf{H}_{\mathrm{T}}\mathbf{H}_{\mathrm{T}}^{\mathrm{T}}\boldsymbol{\beta})} \tag{12.9}$$

where λ_0, λ_1, and λ_2 denote the trade-off parameters.

Let $\boldsymbol{\mu}_{\mathrm{S}}=\frac{1}{N_{\mathrm{S}}}\sum_{i=1}^{N_{\mathrm{S}}}\mathbf{h}_{\mathrm{S}}^{i}$ and $\boldsymbol{\mu}_{\mathrm{T}}=\frac{1}{N_{\mathrm{T}}}\sum_{j=1}^{N_{\mathrm{T}}}\mathbf{h}_{\mathrm{T}}^{j}$ be the centers of source domain and target domain in ELM space, then the minimization problem in Eq. 12.9 can be finally written as

$$\begin{aligned}
&\min_{\boldsymbol{\beta}}\frac{\mathrm{Tr}(\boldsymbol{\beta}^{\mathrm{T}}\mathbf{S}_{W}^{S}\boldsymbol{\beta})+\lambda_{0}\|\boldsymbol{\beta}\|_{\mathrm{F}}^{2}+\lambda_{1}\left\|\boldsymbol{\beta}^{\mathrm{T}}\left(\frac{1}{N_{\mathrm{S}}}\sum_{i=1}^{N_{\mathrm{S}}}\mathbf{h}_{\mathrm{S}}^{i}\right)-\boldsymbol{\beta}^{\mathrm{T}}\left(\frac{1}{N_{\mathrm{T}}}\sum_{j=1}^{N_{\mathrm{T}}}\mathbf{h}_{\mathrm{T}}^{j}\right)\right\|_{\mathrm{F}}^{2}}{\mathrm{Tr}(\boldsymbol{\beta}^{\mathrm{T}}\mathbf{S}_{B}^{S}\boldsymbol{\beta})+\lambda_{2}\mathrm{Tr}(\boldsymbol{\beta}^{\mathrm{T}}\mathbf{H}_{\mathrm{T}}\mathbf{H}_{\mathrm{T}}^{\mathrm{T}}\boldsymbol{\beta})}\\
&=\min_{\boldsymbol{\beta}}\frac{\mathrm{Tr}(\boldsymbol{\beta}^{\mathrm{T}}\mathbf{S}_{W}^{S}\boldsymbol{\beta})+\lambda_{0}\|\boldsymbol{\beta}\|_{\mathrm{F}}^{2}+\lambda_{1}\|\boldsymbol{\beta}^{\mathrm{T}}\boldsymbol{\mu}_{\mathrm{S}}-\boldsymbol{\beta}^{\mathrm{T}}\boldsymbol{\mu}_{\mathrm{T}}\|_{\mathrm{F}}^{2}}{\mathrm{Tr}(\boldsymbol{\beta}^{\mathrm{T}}\mathbf{S}_{B}^{S}\boldsymbol{\beta})+\lambda_{2}\mathrm{Tr}(\boldsymbol{\beta}^{\mathrm{T}}\mathbf{H}_{\mathrm{T}}\mathbf{H}_{\mathrm{T}}^{\mathrm{T}}\boldsymbol{\beta})}\\
&=\min_{\boldsymbol{\beta}}\frac{\mathrm{Tr}(\boldsymbol{\beta}^{\mathrm{T}}\mathbf{S}_{W}^{S}\boldsymbol{\beta})+\lambda_{0}\mathrm{Tr}(\boldsymbol{\beta}^{\mathrm{T}}\boldsymbol{\beta})+\lambda_{1}\mathrm{Tr}((\boldsymbol{\beta}^{\mathrm{T}}\boldsymbol{\mu}_{\mathrm{S}}-\boldsymbol{\beta}^{\mathrm{T}}\boldsymbol{\mu}_{\mathrm{T}})(\boldsymbol{\beta}^{\mathrm{T}}\boldsymbol{\mu}_{\mathrm{S}}-\boldsymbol{\beta}^{\mathrm{T}}\boldsymbol{\mu}_{\mathrm{T}})^{\mathrm{T}})}{\mathrm{Tr}(\boldsymbol{\beta}^{\mathrm{T}}\mathbf{S}_{B}^{S}\boldsymbol{\beta}+\lambda_{2}\boldsymbol{\beta}^{\mathrm{T}}\mathbf{H}_{\mathrm{T}}\mathbf{H}_{\mathrm{T}}^{\mathrm{T}}\boldsymbol{\beta})}\\
&=\min_{\boldsymbol{\beta}}\frac{\mathrm{Tr}(\boldsymbol{\beta}^{\mathrm{T}}\mathbf{S}_{W}^{S}\boldsymbol{\beta}+\lambda_{0}\boldsymbol{\beta}^{\mathrm{T}}\boldsymbol{\beta}+\lambda_{1}(\boldsymbol{\beta}^{\mathrm{T}}\boldsymbol{\mu}_{\mathrm{S}}-\boldsymbol{\beta}^{\mathrm{T}}\boldsymbol{\mu}_{\mathrm{T}})(\boldsymbol{\beta}^{\mathrm{T}}\boldsymbol{\mu}_{\mathrm{S}}-\boldsymbol{\beta}^{\mathrm{T}}\boldsymbol{\mu}_{\mathrm{T}})^{\mathrm{T}})}{\mathrm{Tr}(\boldsymbol{\beta}^{\mathrm{T}}\mathbf{S}_{B}^{S}\boldsymbol{\beta}+\lambda_{2}\boldsymbol{\beta}^{\mathrm{T}}\mathbf{H}_{\mathrm{T}}\mathbf{H}_{\mathrm{T}}^{\mathrm{T}}\boldsymbol{\beta})}\\
&=\min_{\boldsymbol{\beta}}\frac{\mathrm{Tr}(\boldsymbol{\beta}^{\mathrm{T}}(\mathbf{S}_{W}^{S}+\lambda_{0}\mathbf{I}+\lambda_{1}(\boldsymbol{\mu}_{\mathrm{S}}-\boldsymbol{\mu}_{\mathrm{T}})(\boldsymbol{\mu}_{\mathrm{S}}-\boldsymbol{\mu}_{\mathrm{T}})^{\mathrm{T}})\boldsymbol{\beta})}{\mathrm{Tr}(\boldsymbol{\beta}^{\mathrm{T}}(\mathbf{S}_{B}^{S}+\lambda_{2}\mathbf{H}_{\mathrm{T}}\mathbf{H}_{\mathrm{T}}^{\mathrm{T}})\boldsymbol{\beta})}
\end{aligned} \tag{12.10}$$

where $\mathbf{I}$ is an identity matrix of size L.

12.3.3 Model Optimization

In the minimization problem Eq. 12.10, there are many possible solutions of $\boldsymbol{\beta}$ (i.e., nonunique solutions). To guarantee the unique property of solution, we impose an equality constraint on the optimization problem, and then, Eq. 12.10 can be written as

$$\begin{aligned}&\min_{\boldsymbol{\beta}} \mathrm{Tr}\left(\boldsymbol{\beta}^{\mathrm{T}}\left(\mathbf{S}_W^S + \lambda_0\mathbf{I} + \lambda_1(\boldsymbol{\mu}_\mathrm{S} - \boldsymbol{\mu}_\mathrm{T})(\boldsymbol{\mu}_\mathrm{S} - \boldsymbol{\mu}_\mathrm{T})^{\mathrm{T}}\right)\boldsymbol{\beta}\right)\\ &\text{s.t.}\ \mathrm{Tr}\left(\boldsymbol{\beta}^{\mathrm{T}}\left(\mathbf{S}_B^S + \lambda_2\mathbf{H}_\mathrm{T}\mathbf{H}_\mathrm{T}^{\mathrm{T}}\right)\boldsymbol{\beta}\right) = \eta\end{aligned} \tag{12.11}$$

where η is a positive constant.

To solve Eq. 12.11, the Lagrange multiplier function is written as

$$\begin{aligned}\mathcal{L}(\boldsymbol{\beta},\rho) = {} & \boldsymbol{\beta}^{\mathrm{T}}\left(\mathbf{S}_W^S + \lambda_0\mathbf{I} + \lambda_1(\boldsymbol{\mu}_\mathrm{S} - \boldsymbol{\mu}_\mathrm{T})(\boldsymbol{\mu}_\mathrm{S} - \boldsymbol{\mu}_\mathrm{T})^{\mathrm{T}}\right)\boldsymbol{\beta}\\ & - \rho\left(\boldsymbol{\beta}^{\mathrm{T}}\left(\mathbf{S}_B^S + \lambda_2\mathbf{H}_\mathrm{T}\mathbf{H}_\mathrm{T}^{\mathrm{T}}\right)\boldsymbol{\beta} - \eta\right)\end{aligned} \tag{12.12}$$

where ρ denotes the Lagrange multiplier coefficient.

By setting the partial derivation of $\mathcal{L}(\boldsymbol{\beta},\rho)$ with respect to $\boldsymbol{\beta}$ to be 0, we have

$$\begin{aligned}&\frac{\partial\mathcal{L}(\boldsymbol{\beta},\rho)}{\partial\boldsymbol{\beta}} = 0 \rightarrow\\ &\left(\mathbf{S}_B^S + \lambda_2\mathbf{H}_\mathrm{T}\mathbf{H}_\mathrm{T}^{\mathrm{T}}\right)^{-1}\left(\mathbf{S}_W^S + \lambda_0\mathbf{I} + \lambda_1(\boldsymbol{\mu}_\mathrm{S} - \boldsymbol{\mu}_\mathrm{T})(\boldsymbol{\mu}_\mathrm{S} - \boldsymbol{\mu}_\mathrm{T})^{\mathrm{T}}\right)\boldsymbol{\beta} = \rho\boldsymbol{\beta}\end{aligned} \tag{12.13}$$

From Eq. 12.13, we can observe that $\boldsymbol{\beta}$ can be obtained by solving the following eigenvalue decomposition problem,

$$\mathbf{A}\boldsymbol{\beta} = \rho\boldsymbol{\beta} \tag{12.14}$$

where $\mathbf{A} = \left(\mathbf{S}_B^S + \lambda_2\mathbf{H}_\mathrm{T}\mathbf{H}_\mathrm{T}^{\mathrm{T}}\right)^{-1}\left(\mathbf{S}_W^S + \lambda_0\mathbf{I} + \lambda_1(\boldsymbol{\mu}_\mathrm{S} - \boldsymbol{\mu}_\mathrm{T})(\boldsymbol{\mu}_\mathrm{S} - \boldsymbol{\mu}_\mathrm{T})^{\mathrm{T}}\right)$, and ρ denotes the eigenvalues.

From Eq. 12.14, it is clear that $\boldsymbol{\beta}$ denotes the eigenvectors. Due to that the model Eq. 12.11 is a minimization problem, therefore, the optimal subspace denotes the eigenvectors with respect to the first d smallest eigenvalues $[\rho_1, \ldots, \rho_d]$ represented by

$$\boldsymbol{\beta}^* = [\boldsymbol{\beta}_1, \boldsymbol{\beta}_2, \ldots, \boldsymbol{\beta}_d] \tag{12.15}$$

For easy implementation, the proposed algorithm is summarized in Algorithm 12.1.

Algorithm 12.1
Input: Heterogeneous data including data $\mathbf{X}_S \in \Re^{D \times N_S}$ and data $\mathbf{X}_T \in \Re^{D \times N_T}$, hidden nodes L, λ_0, λ_1, λ_2, and d
Procedure:

1. Randomly generated weights $\mathbf{W} \in \Re^{D \times L}$, $\mathbf{b} \in \Re^{1 \times L}$, and choose the activation function $g(\cdot)$.
   Compute $\mathbf{H}_S = [\mathbf{h}_S^1, \ldots, \mathbf{h}_S^{N_S}]$ and $\mathbf{H}_T = [\mathbf{h}_T^1, \ldots, \mathbf{h}_T^{N_T}]$, where $\mathbf{h}_S^i = g(\mathbf{W}^T\mathbf{x}_S^i + \mathbf{b}^T)$ and $\mathbf{h}_T^j = g(\mathbf{W}^T\mathbf{x}_T^j + \mathbf{b}^T)$
2. Compute the inter-class scatter matrix $\mathbf{S}_B^S$ and the intra-class scatter matrix $\mathbf{S}_W^S$
3. Compute the matrix $\mathbf{A}$ as

$$(\mathbf{S}_B^S + \lambda_2 \mathbf{H}_T \mathbf{H}_T^T)^{-1} (\mathbf{S}_W^S + \lambda_0 \mathbf{I} + \lambda_1 (\boldsymbol{\mu}_S - \boldsymbol{\mu}_T)(\boldsymbol{\mu}_S - \boldsymbol{\mu}_T)^T)$$

4. Perform the eigenvalue decomposition of $\mathbf{A}$ by using Eq. 12.14;
5. Compute the optimal subspace $\boldsymbol{\beta}^* = [\boldsymbol{\beta}_1, \boldsymbol{\beta}_2, \ldots, \boldsymbol{\beta}_d]$ by Eq. 12.15;

Output: The basis transformation $\boldsymbol{\beta}$

12.4 Experiments

12.4.1 Data Description

We validate the proposed method on our own datasets. This dataset includes three subsets: master data (collected 5 years ago), slave 1 data (collected now), and slave 2 data (collected now). In data acquisition, the master and two slavery E-nose systems were developed in [26]. Each system consists of four TGS series sensors and an extra temperature and humidity module. Therefore, the dimension of each sample is 6. This dataset includes six kinds of gaseous contaminants (i.e., six classes), such as formaldehyde, benzene, toluene, carbon monoxide, nitrogen dioxide, and ammonia. The detailed description of the dataset is shown in Table 12.1. For visually observing the heterogeneous E-nose data, the PCA scatter

Table 12.1 Data description of the E-nose data

E-nose system	HCHO	Benzene	Toluene	CO	NO_2	Ammonia	Total
Master	126	72	66	58	38	60	420
Slave1	108	108	106	98	107	81	608
Slave2	108	108	94	95	108	84	576

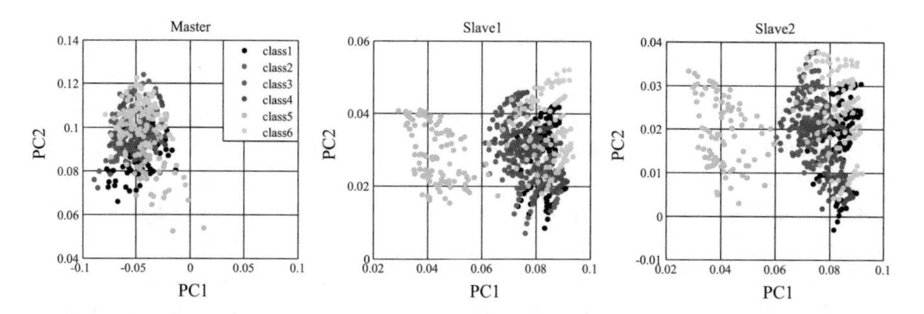

Fig. 12.2 PCA scatter points of the master, slave 1, and slave 2 data, respectively

points on the master data, slave 1 data, and slave 2 data are shown in Fig. 12.2, in which we can see that the points from different classes are overlapped.

12.4.2 Experimental Settings

The master data collected 5 years ago is used as source domain data (no drift). The slave 1 and slave 2 data collected now are used as target domain data (with drift). Then, we conduct the experiments with two settings as follows.

Setting 1: During CdELM training and classifier learning, the labels of the target domain data are unavailable, and only the source labels are used. The classification accuracy on slave1 data (or slave2 data) is reported.
Setting 2: The only difference between *Setting* 1 and *Setting* 2 is that, in classifier training, partial labeled data of target domain can be used. Specifically, for each class in the target domain, k labeled samples can be used for classifier learning, where the values k = 1, 3, 5, 7, 9 are discussed in this chapter.

12.4.3 Single-Domain Subspace Projection Methods

To show the effectiveness of the proposed method, we have chosen 12 machine learning methods. First, three baseline methods such as support vector machine (SVM), principal component analysis (PCA), and linear discriminant analysis (LDA) are compared. Second, five semi-supervised learning method-based manifold learning, including locality preservation projection (LPP) [25], multidimensional scaling (MDS) [27], neighborhood component analysis (NCA) [28], neighborhood preserving embedding (NPE) [29], and local Fisher discriminant analysis (LFDA) [30], are explored and compared. Finally, the popular subspace transfer learning method, sampling geodesic flow (SGF) [31], is also compared.

12.4.4 Classification Results

In this section, the experimental results on each setting are reported to validate the performance of the proposed CdELM method. Under each setting, two tasks including master → slave 1 and master → slave 2 are conducted.

Under the setting 1, we first observe the qualitative result by implementing the proposed CdELM method on master → slave 1 and master → slave 2, respectively. The result is shown in Fig. 12.3, in which the separability among data points from different classes (represented as different symbols) is much improved in the learned common subspace compared to Fig. 12.2. Further, the odor classification accuracy of the target domain data has been presented in Table 12.2. From the results, we can observe that the proposed CdELM achieves the highest accuracies on two tasks. While the activation function in hidden layer is Gaussian (RBF function), the best performance of the CdELM is achieved. This demonstrates that the proposed CdELM has a good performance for cross-domain pattern recognition scenarios.

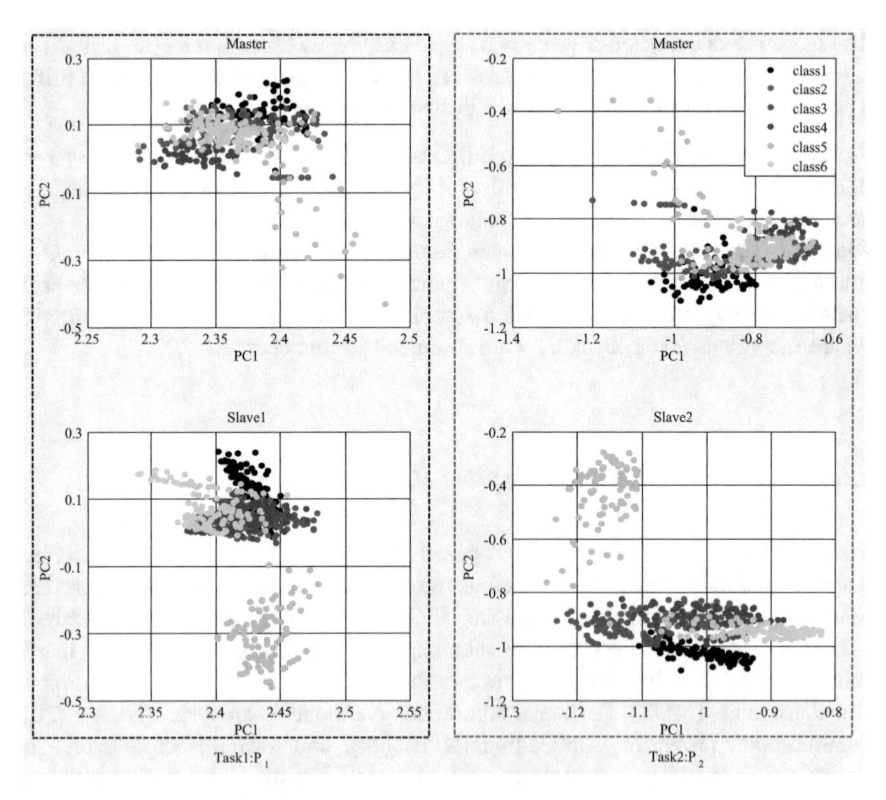

Fig. 12.3 Scatter points by using CdELM from master → slave 1 (black dot) and master → slave 2 (red dot)

Table 12.2 Recognition accuracy (%) with sensor calibration under **Setting 1**

Cross-domain tasks	SVM	ELM (sigmoid)	ELM (rbf)	LPP	NPE	NCA	MDS	LFDA	SGF	CdELM (sigmoid)	CdELM (rbf)
Master → slave1	51.97	55	54.59	53.95	53.62	41.28	51.15	61.84	55.10	64.90	**66.39**
Master → slave2	60.59	59.83	61.16	57.81	54.69	33.85	58.51	61.63	57.49	68.11	**68.45**

The boldface type denotes the best performance.

Table 12.3 Recognition accuracy (%) with sensor calibration under **Setting 2** (task 1)

Methods	n_t					Average
	1	3	5	7	9	
SVM	59.14	63.22	62.80	70.49	70.76	65.28
ELM (sigmoid)	59.05	65.10	69.52	71.20	72.18	67.41
ELM (rbf)	62.46	65.83	67.61	69.36	69.87	67.03
LPP	65.46	69.83	71.45	72.08	71.48	70.06
NPE	64.78	64.07	63.49	71.55	71.84	67.15
NCA	52.49	50.85	53.81	50.00	63.00	54.03
MDS	61.13	64.75	65.57	70.32	72.92	66.94
LFDA	62.13	67.12	71.63	76.86	74.91	70.53
SGF	66.61	66.95	67.13	70.14	72.74	68.71
CdELM (sigmoid)	67.88	71.53	73.50	**75.34**	76.36	72.92
CdELM (rbf)	**68.34**	**73.32**	**74.42**	75.07	**77.35**	**73.70**

The boldface type denotes the best performance.

Under the setting 2, k labeled data for each class in the target domain is leveraged for classifier learning. The recognition accuracy of the first task (i.e., master $\rightarrow$ slave 1) is reported in Table 12.3, and the second task (i.e., master $\rightarrow$ slave 2) is reported in Table 12.4. Notably, all the compared methods follow the same setting conditions. From Table 12.3 and Table 12.4, we can observe that the proposed CdELM still outperforms other methods. Therefore, we confirm that the proposed method is effective in handling heterogeneous measurement data.

12.4.5 Parameter Sensitivity

In the proposed CdELM model, there are three parameters: λ_0, λ_1, and d. We focus on observing the performance variations in tuning λ_0 and λ_1 according to 10^t, where $t = \{-6, -4, -2, 0, 2, 4, 6\}$. To show the performance with respect to each parameter, one is tuned by frozen the other one. The parameter λ_1 tuning results by fixing λ_0 are shown in Fig. 12.4, and the parameter λ_0 tuning results by fixing λ_1 are shown in Fig. 12.5, from which the best parameters λ_0 and λ_1 can be witnessed. Further, we tune the subspace dimensionality d from the parameter set $d = \{1, 2, 3, 4, 5, 6, 7, 8, 9, 10\}$, and the result is shown Fig. 12.6 by fixing other model parameters.

Table 12.4 Recognition accuracy (%) with sensor calibration under **Setting 2** (task 2)

Methods	n_t					Average
	1	3	5	7	9	
SVM	69.65	72.76	73.63	74.16	74.90	73.02
ELM (sigmoid)	64.18	67.35	67.91	68.09	69.29	67.36
ELM (rbf)	65.35	68.41	68.63	68.71	69.60	68.14
LPP	69.82	74.19	73.63	72.85	76.82	73.46
NPE	71.05	71.15	71.43	72.28	74.14	72.01
NCA	56.32	47.49	52.01	55.62	58.03	53.90
MDS	72.11	73.48	73.81	75.28	75.29	74.00
LFDA	65.26	70.61	73.08	75.47	77.01	72.29
SGF	68.07	71.51	73.08	73.03	73.56	71.85
CdELM (sigmoid)	**74.47**	75.35	75.78	77.56	78.09	76.25
CdELM (rbf)	74.13	**76.06**	**77.69**	**78.95**	**79.18**	**77.21**

The boldface type denotes the best performance.

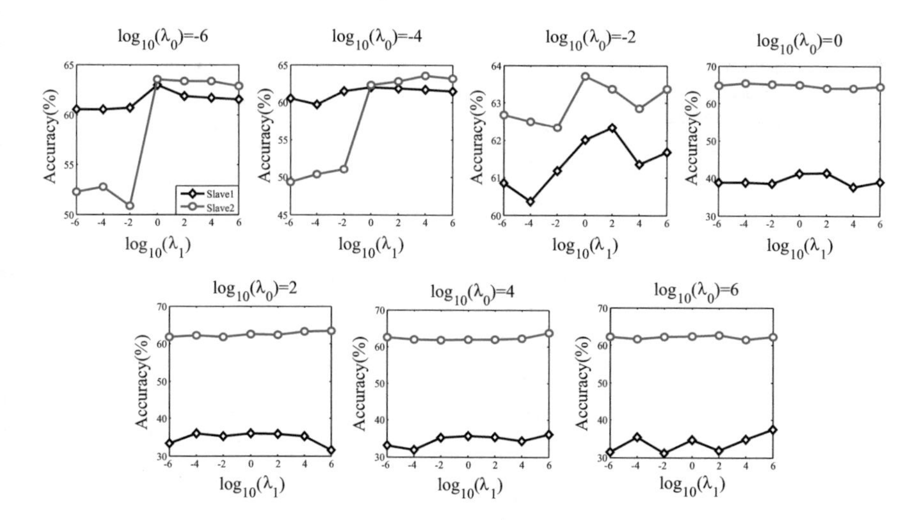

Fig. 12.4 Performance curves with respect to λ_1 under different values of λ_0

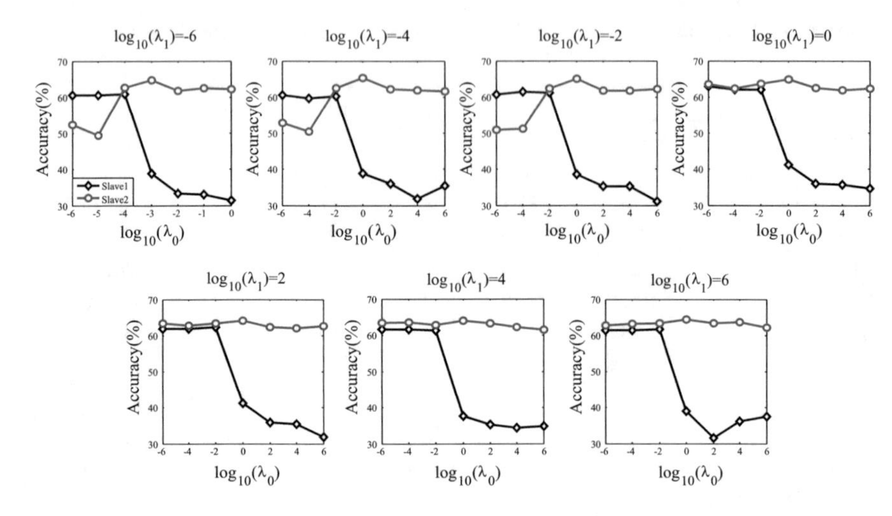

Fig. 12.5 Performance curves with respect to λ_0 under different values of λ_1

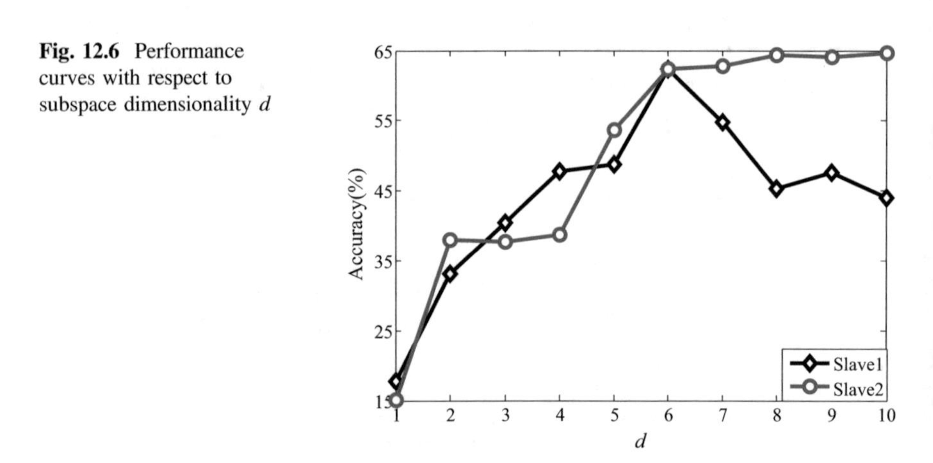

Fig. 12.6 Performance curves with respect to subspace dimensionality d

12.5 Summary

In this chapter, we present a cross-domain common subspace learning approach for heterogeneous data classification, which is called cross-domain extreme learning machine (CdELM). The method is motivated by subspace learning, domain adaptation, and cognitive-based extreme learning machine. Extensive experiments on drifted E-nose datasets demonstrate that the proposed method outperforms other compared methods.

References

1. G.B. Huang, An insight into extreme learning machines: random neurons, random features and kernels. Cognit. Comput. **6**(3), 376–390 (2014)
2. G.B. Huang, What are extreme learning machines? Filling the gap between frank Rosenblatt's dream and John von Neumann's puzzle. Cognit. Comput. **7**(3), 263–278 (2015)
3. G.B. Huang, Q.Y. Zhu, C.K. Siew, Extreme learning machine: theory and applications. Neurocomputing **70**(1–3), 489–501 (2006)
4. G.B. Huang, H. Zhou, X. Ding, R. Zhang, Extreme learning machine for regression and multiclass classification. IEEE Trans. Syst. Man Cybern. B Cybern. **42**(2), 513–529 (2012)
5. E. Cambria, Extreme learning machines [trends & controversies]. IEEE Intell. Syst. **28**(6), 30–59 (2013)
6. G.B. Huang, L. Chen, C.K. Siew, Universal approximation using incremental constructive feedforward networks with random hidden nodes. IEEE Trans. Neural Netw. **17**(4), 879–892 (2006)
7. G.B. Huang, Q.Y. Zhu, C.K. Siew, Extreme learning machine: a new learning scheme of feedforward neural networks, in *Proceedings of IEEE International Joint Conference on Neural Networks*, vol. 2. July 2004, pp. 985–990
8. G.B. Huang, L. Chen, Convex incremental extreme learning machine. Neurocomputing **70** (16–18), 3056–3062 (2007)
9. G.B. Huang, L. Chen, Enhanced random search based incremental extreme learning machine. Neurocomputing **71**(16–18), 3460–3468 (2008)
10. D.E. Rumelhart, G.E. Hinton, R.J. Williams, Learning representations by back-propagation errors. Nature **323**, 533–536 (1986)
11. G.B. Huang, Q.Y. Zhu, K.Z. Mao, C.K. Siew, P. Saratchandran, Can threshold networks be trained directly? IEEE Trans. Circuits Syst. II Exp. Briefs **53**(3), 187–191 (2006)
12. N.Y. Liang, G.B. Huang, P. Saratchandran, N. Sundararajan, A fast and accurate on-line sequential learning algorithm for feedforward networks. IEEE Trans. Neural Netw. **17**(6), 1411–1423 (2006)
13. M.B. Li, G.B. Huang, P. Saratchandran, N. Sundararajan, Fully complex extreme learning machine. Neurocomputing **68**, 306–314 (2005)
14. G. Feng, G.B. Huang, Q. Lin, R. Gay, Error minimized extreme learning machine with growth of hidden nodes and incremental learning. IEEE Trans. Neural Netw. **20**(8), 1352–1357 (2009)
15. H.J. Rong, G.B. Huang, N. Sundararajan, P. Saratchandran, Online sequential fuzzy extreme learning machine for function approximation and classification problems. IEEE Trans. Syst. Man Cybern. B Cybern. **39**(4), 1067–1072 (2009)
16. Z. Huang, Y. Yu, J. Gu, H. Liu, An efficient method for traffic sign recognition based on extreme learning machine. IEEE Trans. Cybern. (2016) (appear online)
17. Y. Yu, Z. Sun, Sparse coding extreme learning machine for classification. Neurocomputing (2016) (accepted)
18. L. Zhang, D. Zhang, Evolutionary cost-sensitive extreme learning machine. IEEE Trans. Neural Networks Learn. Syst. https://doi.org/10.1109/tnnls.2016.2607757 (2016)
19. L. Zhang, D. Zhang, Robust visual knowledge transfer via extreme learning machine based domain. IEEE Trans. Image Process. **25**(10), 4959–4973 (2016)
20. S.J. Pan, I.W. Tsang, J.T. Kwok, Q. Yang, Domain adaptation via transfer component analysis. IEEE Trans. Neural Netw. **22**(2), 199–210 (2011)
21. L. Zhang, W. Zuo, D. Zhang, LSDT: latent sparse domain transfer learning for visual adaptation. IEEE Trans. Image Process. **25**(3), 1177–1191 (2016)
22. L. Zhang, D. Zhang, Domain adaptation extreme learning machines for drift compensation in e-nose systems. IEEE Trans. Instrum. Meas. **64**(7), 1790–1801 (2015)
23. M. Turk, A. Pentland, Eigenfaces for recognition. J. Cogn. Neurosci. **3**(1), 71–86 (1991)

24. P.N. Belhumeur, J. Hespanha, D.J. Kriegman, Eigenfaces vs. fisherfaces: recognition using class specific linear projection. IEEE Trans. Pattern Anal. Mach. Intell. **19**(7), 711–720 (1997)
25. X. He, P. Niyogi, Locality preserving projections, in *NIPS* (2004)
26. L. Zhang, F. Tian, Performance study of multilayer perceptrons in a low-cost electronic nose. IEEE Trans. Instrum. Meas. **63**(7), 1670–1679 (2014)
27. W.S. Torgerson, Multidimensional scaling: I. Theory and method. Psychometrika **17**(4), 401–419 (1952)
28. J. Goldberger, S. Roweis, G. Hinton, R. Salakhutdinov, Neighborhood component analysis, in *NIPS* (2004)
29. X. He, D. Cai, S. Yan, H.J. Zhang, Neighborhood preserving embedding, in *ICCV* (2005)
30. M. Sugiyama, Local fisher discriminant analysis for supervised dimensionality reduction, in *ICML*, pp. 905–912 (2006)
31. R. Gopalan, R. Li, R. Chellappa, Domain adaptation for object recognition: an unsupervised approach, in *ICCV* (2011)

Chapter 13
Domain Correction-Based Adaptive Extreme Learning Machine

Abstract This chapter presents a novel domain correction and adaptive extreme learning machines framework (DC-AELM) with transferring capability to solve the drift and interference problem of E-nose. The framework consists of two parts: (1) domain correction (DC) that makes the distributions of two domains close; (2) adaptive extreme learning machine (AELM) that learns a transferable classifier at decision level. This method is motivated by the idea of transfer learning, especially from the perspective of domain correction and decision making, to realize the knowledge transfer for interference suppression and drift compensation. Experiments on a background interference dataset and a public benchmark sensor drift dataset via E-nose verify the effectiveness of the proposed DC-AELM method.

Keywords Interference · Electronic nose · Transfer learning · Domain correction · Machine learning

13.1 Introduction

The background interference and drift caused by the cross-sensitivity and ambient factors will significantly degrade performance of E-nose [1]. Generally, sensor drift can also be recognized as a kind of serious interference to E-nose [2–4]. Some researchers tried to suppress these interferences by designing appropriate filters. However, the interferences cannot be effectively separated from target signal by traditional filters, due to that the interference signals are almost entirely mixed with target signals under similar frequencies. Methods for interference suppression can be divided into four categories: (1) component correction methods, such as principal component analysis (PCA) [5], independent component analysis (ICA) [6–8], orthogonal signal correction (OSC) [9, 10]; (2) adaptive methods, such as the adaptive self-organizing map (SOM) [11], domain adaptation methods [12]; (3) machine learning methods; (4) classier ensemble, it was used to suppress sensor drift, which shows that the recognition accuracy of the drift data was improved [13–15]. Other current methods used to suppress the interference are given in [16–18].

L. Zhang et al., *Electronic Nose: Algorithmic Challenges*,
https://doi.org/10.1007/978-981-13-2167-2_13

The above methods can suppress the special interferences to some extent, but the effects are limited by their weak generalization to a new kind of interference. The interference problem is still a challenging issue in machine olfaction community and sensory field, because of data distribution mismatch. Transfer learning [19], which tries to transfer the knowledge from source task to a target task with few labeled data and different distribution, is an appropriate method for interference suppression.

Motivated by the fact that transfer learning can be used to transfer the knowledge between two domains, a novel domain correction and adaptive extreme learning machines framework (DC-AELM) with transferring capability is proposed to solve the interference problem in E-nose. The framework consists of two parts: (1) domain correction (DC) which makes the distributions of two domains close; (2) adaptive extreme learning machine (AELM) which realizes the knowledge transfer at the decision level. In this chapter, the source domain data is regular without drift and interference, and the target domain data is noised and drifted.

13.2 Related Work

13.2.1 Transfer Learning

Transfer learning is a machine learning method which can apply the knowledge from different but related tasks to target tasks. It is different from traditional machine learning that always has two basic assumptions: (1) The training data and future data must be in the same feature space and have the same distribution; (2) in order to obtain a robust prediction model, there must be enough training samples to learn. The difference between traditional machine learning and transfer learning is shown in Fig. 13.1. The aim of transfer learning is to solve the problem that there are few or even not any labeled data in target tasks, while the data in source tasks is sufficient and related with target tasks but outdated (i.e., the data distribution in source is different from that in target domain). That is, transfer learning has the ability that a model can recognize or apply the knowledge learned in source tasks to new target tasks. The unified definition of transfer learning is as follows [19].

Given a source domain D_S and learning task T_S, a target domain D_T and learning task T_T, transfer learning aims to help improve the learning of the target predictive function $f_T(\cdot)$ in D_T using the knowledge in D_S and T_S, where $D_S \neq D_T$, or $T_S \neq T_T$.

Traditional machine learning tries to train each learning system by all the three sample sets of apples, pears, and oranges, respectively. Transfer learning tries to transfer the knowledge from apple and pear samples to train the recognition system for orange when there are only a small number of orange samples. Herein, the sample sets of apples, pears, and oranges are metaphors, representing different kinds of samples for different tasks.

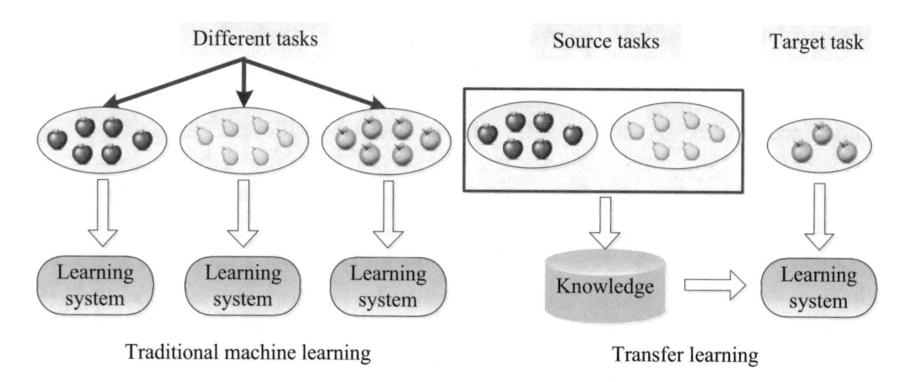

Fig. 13.1 Learning process difference between traditional machine learning and transfer learning

Similarly, the distribution of interference E-nose data is different from that of regular E-nose data. Therefore, it is worthwhile to use transfer learning theory to solve this problem. Many researchers have paid much attention to compensate drift or calibrate transfer among multi-instruments using the idea of transfer learning. A drift correction autoencoder (DCAE) was proposed to address the drift issue in E-nose [20]. DCAE learned to model and correct the influential factors explicitly with the help of transfer samples. Windowed piecewise direct standardization (WPDS) and standardization error-based model improvement (SEMI) were used in combination to solve the calibration transfer problem among many E-nose systems [21]. So the transfer ability of prediction models was improved by these two steps (WPDS and SEMI). Both of these two methods were from the perspective of sensor response correction to transfer knowledge. The cross-domain discriminative subspace learning (CDSL) was proposed to address the transfer calibration among many E-nose systems [22]. The recognition model was trained by source domain dataset from a master E-nose, but tested on another target domain dataset from a slave system. This method was from the perspective of feature representation to transfer knowledge. A domain adaptation extreme learning machine framework (DAELM) was proposed to compensate drift in E-nose [23]. It learned a robust classifier by leveraging a limited number of labeled data from target domain for drift compensation in E-nose. A transfer sample-based coupled task learning (TCTL) framework was proposed to solve the problems of instrumental variation and sensor drift [24]. Both of these two methods transferred the knowledge from the perspective of decision-making level.

13.2.2 Extreme Learning Machine

Extreme learning machine (ELM) was proposed by Huang et al. to solve a single-layer feed-forward network [25–27]. It is an effective and efficient algorithm for pattern classification and regression in many fields. The weights of input layer to the hidden layer are randomly selected. ELM analytically determines the output weights between the hidden layer and the output layer using Moore–Penrose generalized inverse by adopting the square loss of prediction error, which then involves in solving a regularized least square problem efficiently in closed form [23]. In recent years, many improved versions about ELM have been developed [28–31]. The definition of ELM is as follows [25].

Given N samples $[\mathbf{x}_1, \mathbf{x}_2, \ldots, \mathbf{x}_N]$ and the corresponding truth $[\mathbf{t}_1, \mathbf{t}_2, \ldots, \mathbf{t}_N]$, where $\mathbf{x}_i = [x_{i1}, x_{i2}, \ldots, x_{in}]^{\mathrm{T}} \in \Re^n$ and $\mathbf{t}_i = [t_{i1}, t_{i2}, \ldots, t_{im}]^{\mathrm{T}} \in \Re^m$, n and m is the number of input and output neurons, respectively. The output of the hidden layer is denoted as $H(\mathbf{x}_i) \in \Re^L$ where L is the number of hidden nodes and $H(\cdot)$ is the activation function. $\boldsymbol{\beta} \in \Re^{L \times m}$ is the output weight between the hidden layer and the output layer. ELM aims to compute the output weights by minimizing the squared loss summation of prediction errors and the norm of the output weights to prevent the over-fitting, the formulas are as follows:

$$\begin{aligned} &\min_{\boldsymbol{\beta}} \mathcal{L}_{\mathrm{ELM}} = \frac{1}{2}||\boldsymbol{\beta}||^2 + C \cdot \frac{1}{2} \cdot \sum_{i=1}^{N} ||\boldsymbol{\xi}_i||^2 \\ &\text{s.t.} \quad \mathcal{H}(\mathbf{x}_i)\boldsymbol{\beta} = \mathbf{t}_i - \boldsymbol{\xi}_i, \quad i = 1, \ldots, N \end{aligned} \tag{13.1}$$

where $\boldsymbol{\xi}_i$ is the prediction error to the ith training sample, and C is a penalty constant.

The equivalent unconstrained optimization problem of Eq. 13.1 can be represented as follows:

$$\min_{\boldsymbol{\beta} \in \Re^{L \times M}} \mathcal{L}_{\mathrm{ELM}} = \frac{1}{2}||\boldsymbol{\beta}||^2 + C \cdot \frac{1}{2} \cdot ||\mathbf{T} - \mathbf{H}\beta||^2 \tag{13.2}$$

where $\mathbf{H} = [\mathcal{H}(\mathbf{x}_1); \mathcal{H}(\mathbf{x}_2); \ldots, \mathcal{H}(\mathbf{x}_N)] \in \Re^{N \times L}$, $\mathbf{T} = [\mathbf{t}_1, \mathbf{t}_2, \cdots, \mathbf{t}_N]^{\mathrm{T}}$. The solution of $\boldsymbol{\beta}$ can be obtained by setting the gradient of the objective function Eq. 13.2 with respect to $\boldsymbol{\beta}$ to zero.

$$\boldsymbol{\beta} = \left(\mathbf{H}^{\mathrm{T}}\mathbf{H} + \frac{\mathbf{I}}{C}\right)^{-1} \mathbf{H}^{\mathrm{T}}\mathbf{T} \tag{13.3}$$

where $\mathbf{I}$ is the identity matrix.

In ELM, the weights of input layer to the hidden layer are randomly selected and there is no need to solve it iteratively. It only aims to compute the weights from the hidden layer to the output layer, and there is no iteration process. Compared with the back-propagation neural network (BP) algorithm, the training speed is greatly improved. Considering the real-time requirement of E-nose, ELM is the best choice due to its high efficiency in the training process.

Although ELM has provided better prediction accuracy when there are a number of label data from source domain to train, the transferring capability of ELM is poor with a limited number of labeled data from target domain. We aim to propose a method to solve this problem. By enhancing the adaptive performance of classier and improving the robustness of prediction model, the method can provide good prediction accuracy when there are interferences, so that it can realize the purpose of suppressing the interference in E-nose. It is very meaningful to train a classifier using very few labeled new samples (target domains) without dropping up the old data (source domain) and realize effectively knowledge transfer (i.e., interference suppression) from source domain to multiple target domains.

13.3 The Proposed Method

13.3.1 Notations

Herein, the training data of source are denoted as $\mathbf{X}_\mathrm{S} = [\mathbf{x}_\mathrm{S}^1, \ldots, \mathbf{x}_\mathrm{S}^{N_\mathrm{S}}] \in \Re^{d \times N_\mathrm{S}}$ and the corresponding true values are $\mathbf{T}_\mathrm{S} = [\mathbf{t}_\mathrm{S}^1, \ldots, \mathbf{t}_\mathrm{S}^{N_\mathrm{S}}] \in \Re^{m \times N_\mathrm{S}}$, where d and m is the number of input and output neurons, respectively, N_S is the number of training samples in source domains. The data of target domain is denoted as $\mathbf{X}_\mathrm{T}$, and most samples are unlabeled, with only a small number of labeled samples (i.e., transfer samples). The transfer samples of target domain are denoted as $\mathbf{M}_\mathrm{T} = [\mathbf{m}_\mathrm{T}^1, \ldots, \mathbf{m}_\mathrm{T}^{N_\mathrm{T}}] \in \mathbf{X}_\mathrm{T}$, and the corresponding true values are $\mathbf{T}_\mathrm{T} = [\mathbf{t}_\mathrm{T}^1, \ldots, \mathbf{t}_\mathrm{T}^{N_\mathrm{T}}] \in \Re^{m \times N_\mathrm{T}}$, and N_T is the number of transfer samples. And the transfer samples of source domain are $\mathbf{M}_\mathrm{S} = [\mathbf{m}_\mathrm{S}^1, \ldots, \mathbf{m}_\mathrm{S}^{N_\mathrm{T}}] \in X_\mathrm{S}$. $(\cdot)^\mathrm{T}$ denotes the transpose operator.

13.3.2 Domain Correction and Adaptive Extreme Learning Machine

A novel domain correction and adaptive extreme learning machines framework (DC-AELM) is proposed to transfer knowledge. The framework consists of two parts.

(1) The domain correction (DC) is used to solve the problem of different distribution between two domains. It aims to learn a basis transformation $\mathbf{P}$ that maps the original space of target data to the space of source data, which makes the feature distributions of target data close to source data.
(2) An adaptive extreme learning machine (AELM) is used to learn a robust classier. It aims to realize the knowledge transfer at the decision level and makes the robustness of prediction model improved.

Figure 13.2 is the schematic diagram of the proposed DC-AELM method. In the figure, the circles and triangles represent samples from different classes, respectively. And the samples with same shape are from the same class while different colors represent different measurement datasets (with or without interference). After a basic transform **P**, the feature distribution of target domain is close to that of source domain ($a \rightarrow b$); then, a classifier $\boldsymbol{\beta}$ is trained using all labeled samples from the source domain by leveraging a limited number of labeled samples from target domain ($b \rightarrow c$). This method transfers the knowledge from two perspectives of domain correction directly and decision making.

(1) Domain Adaptation

As illustrated in Fig. 13.2, the aim of domain correction is to learn a basis transformation **P** which makes the feature distributions of target data close to that of source data and makes the prediction model trained by the source data be used to predict the target data. In order to obtain the basis transformation **P**, some transfer samples are needed. The transfer samples in target data $\mathbf{M}_{\mathrm{T}}$ are selected randomly from each class, and the transfer samples in source data $\mathbf{M}_{\mathrm{S}}$ are selected by Kennard–Stone sequential (KSS) method [32], which can represent the characteristics of each class. So in this step, the **P** should be computed by Eq. 13.4.

$$\min_{\mathbf{P}} ||\mathbf{M}_{\mathrm{S}} - \mathbf{M}_{\mathrm{T}}\mathbf{P}||_2^2 + \lambda||\mathbf{P}||_2^2 \tag{13.4}$$

where λ is the regularization parameter. And the solution of **P** is,

$$\mathbf{P} = \left((\mathbf{M}_{\mathrm{T}})^{\mathrm{T}} \cdot (\mathbf{M}_{\mathrm{T}}) + \lambda\mathbf{I}\right)^{-1} \cdot (\mathbf{M}_{\mathrm{T}})^{\mathrm{T}} \cdot (\mathbf{M}_{\mathrm{S}}) \tag{13.5}$$

Based on Eqs. 13.4 and 13.5, the basis transformation **P** can be obtained, so that the target domain samples are close to the source domain. Equation 13.4 can be used to suppress the interference in E-nose, which can approximately adjust the distribution of target domain (i.e., interference samples).

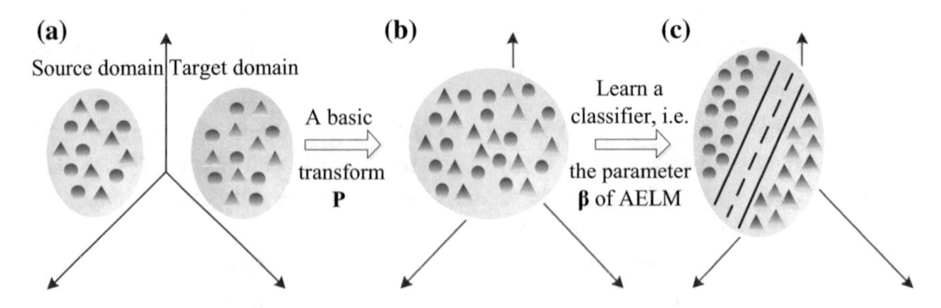

Fig. 13.2 Schematic diagram of the proposed DC-AELM method

(2) Adaptive extreme learning machine (AELM)

In this chapter, we assume that all samples in the source domain are labeled data, and the transfer samples in target domain are also labeled samples. The proposed AELM aims to learn a classifier $\boldsymbol{\beta}$ using all labeled samples from the source domain $\mathbf{X}_S = [\mathbf{x}_S^1, \ldots, \mathbf{x}_S^{N_S}] \in \Re^{d \times N_S}$ by leveraging a limited number of labeled samples from target domain ($\mathbf{M}'_T$ obtained by transforming $\mathbf{M}_T = [\mathbf{m}_T^1, \ldots, \mathbf{m}_T^{N_T}]$ by $\mathbf{P}$). The AELM can be formulated as,

$$\begin{aligned} &\min_{\boldsymbol{\beta}} \tfrac{1}{2}||\boldsymbol{\beta}||^2 + C_S \cdot \tfrac{1}{2} \cdot \sum_{i=1}^{N_S} ||\xi_S^i||^2 + C_T \cdot \tfrac{1}{2} \cdot \sum_{j=1}^{N_T} ||\xi_T^j||^2 \\ &\text{s.t.} \begin{cases} \mathbf{H}_S^i \boldsymbol{\beta} = \mathbf{t}_S^i - \xi_S^i, & i = 1, \ldots, N_S \\ \mathbf{H}_T^j \boldsymbol{\beta} = \mathbf{t}_T^j - \xi_T^j, & j = 1, \ldots, N_T \end{cases} \end{aligned} \quad (13.6)$$

where $\mathbf{H}_S^i$, ξ_S^i, and $\mathbf{t}_S^i$ are the output of hidden layer, the prediction error, and the label of the ith training sample $\mathbf{x}_S^i$ from the source domain; $\mathbf{H}_T^j$, ξ_T^j, and $\mathbf{t}_T^j$ are the output of hidden layer, the prediction error, and the label of the jth transfer sample $\mathbf{m}_T^j$ from the target domain. C_S and C_T are tradeoff parameters. From Eq. 13.6, it can be found that the learned classifier is robust due to the introduction of very few labeled transfer samples from target domain. It realizes the knowledge transfer between source domain and target domain. For interference suppression, C_T can be made large so that the classifier $\boldsymbol{\beta}$ is robust for data of target domain.

By substituting the constraint term into the objective function, an equivalent unconstrained optimization problem can be obtained as follows:

$$\min_{\boldsymbol{\beta}} \mathcal{L} = \frac{1}{2}||\boldsymbol{\beta}||^2 + C_S \frac{1}{2}||\mathbf{T}_S - \mathbf{H}_S \boldsymbol{\beta}||^2 + C_T \frac{1}{2}||\mathbf{T}_T - \mathbf{H}_T \boldsymbol{\beta}||^2 \quad (13.7)$$

where $\mathbf{T}_S = [\mathbf{t}_S^1, \ldots, \mathbf{t}_S^{N_S}]$, $\mathbf{H}_S = [\mathbf{H}_S^1, \ldots, \mathbf{H}_S^{N_S}]$, $\mathbf{T}_T = [\mathbf{t}_T^1, \ldots, \mathbf{t}_T^{N_T}]$, $\mathbf{H}_T = [\mathbf{H}_T^1, \ldots, \mathbf{H}_T^{N_T}]$.

To solve the optimization of Eq. 13.7, set the gradient of Eq. 13.7 with respect to $\boldsymbol{\beta}$ to be zero. That is,

$$\frac{\partial \mathcal{L}}{\partial \boldsymbol{\beta}} = \boldsymbol{\beta} - C_S \cdot \mathbf{H}_S^T \cdot (\mathbf{T}_S - \mathbf{H}_S \boldsymbol{\beta}) - C_T \cdot \mathbf{H}_T^T \cdot (\mathbf{T}_T - \mathbf{H}_T \boldsymbol{\beta}) = 0 \quad (13.8)$$

$$\boldsymbol{\beta} = (\mathbf{I} + C_S \mathbf{H}_S^T \mathbf{H}_S + C_T \mathbf{H}_T^T \mathbf{H}_T)^{-1} (C_S \mathbf{H}_S^T \mathbf{T}_S + C_T \mathbf{H}_T^T \mathbf{T}_T) \quad (13.9)$$

where $\mathbf{I}$ is the identity matrix.

(3) Domain correction and adaptive extreme learning machine method (DC-AELM)

In Eq. 13.5, a basis transformation $\mathbf{P}$ is calculated to make the feature distributions of target data close to source data. In Eq. 13.9, a classifier $\boldsymbol{\beta}$ is a robust prediction model which is trained using all labeled samples from the source domain by leveraging a limited number of labeled samples from target domain. The knowledge transfer is realized by different perspectives, one is from the perspective of variable correction, and the other is from the perspective of decision-making level.

If the difference of two distributions is very large, the distribution of target domain can be adjusted roughly by domain correction so that the target domain is close to the source domain. This process is called initial adjustment. However, this step cannot guarantee that the previously trained classifier is applicable for the samples from target domain. The robustness of classifier can be improved by using Eq. 13.9. This process is called meticulous adjustment. For the interference with a certain degree of instability, both initial adjustment (transferring knowledge by domain correction) and meticulous adjustment (transferring knowledge by decision-making level) are necessary. Merge Eqs. 13.4 and. 13.6 to improve the knowledge transfer capability.

$$\begin{aligned} \min_{\mathbf{P},\boldsymbol{\beta}} & ||\mathbf{M}_\mathrm{S} - \mathbf{M}_\mathrm{T}\mathbf{P}||_2^2 + \lambda||\mathbf{P}||_2^2 + \frac{1}{2}||\boldsymbol{\beta}||^2 \\ & + C_\mathrm{S} \cdot \frac{1}{2} \cdot \sum_{i=1}^{N_s} ||\boldsymbol{\xi}_\mathrm{S}^i||^2 + C_\mathrm{T} \cdot \frac{1}{2} \cdot \sum_{j=1}^{N_\mathrm{T}} ||\boldsymbol{\xi}_\mathrm{T}^j||^2 \\ \text{s.t.} & \begin{cases} \mathbf{H}_\mathrm{S}^i\boldsymbol{\beta} = \mathbf{t}_\mathrm{S}^i - \boldsymbol{\xi}_\mathrm{S}^i, & i = 1, \ldots, N_\mathrm{S} \\ \mathbf{H}_\mathrm{T}^j\boldsymbol{\beta} = \mathbf{t}_\mathrm{T}^j - \boldsymbol{\xi}_\mathrm{T}^j, & j = 1, \ldots, N_\mathrm{T} \end{cases} \end{aligned} \tag{13.10}$$

The framework of the proposed DC-AELM method is shown in Fig. 13.3, and the algorithm of the proposed DC-AELM method is listed in Algorithm 13.1.

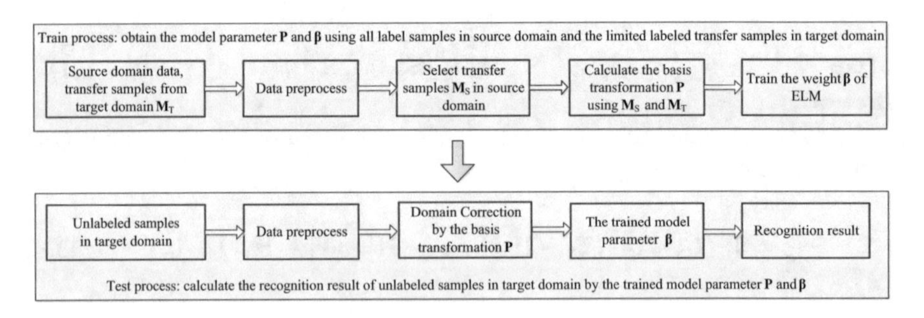

Fig. 13.3 Framework of the proposed DC-AELM

Algorithm 13.1 DC-AELM

Input:

Samples $\{\mathbf{X}_S, \mathbf{T}_S\}$ of the source domain.
Transfer samples $\{\mathbf{M}_T, \mathbf{T}_T\}$ of target domain.
The tradeoff parameter C_S and C_T for source and target domain.

Output:

Output weight $\boldsymbol{\beta}$ of ELM.
Recognition accuracy (%) of unlabeled data in target domain.

Procedure:

1. Data preprocessing.
2. Select transfer samples $\mathbf{M}_S$ in source domain.
3. Domain correction: calculate the basis transformation $\mathbf{P}$ in Eq. 13.5.
4. Train the weight $\boldsymbol{\beta}$ of ELM with all labeled samples $\{\mathbf{X}_S, \mathbf{T}_S\}$ in source domain and the transformed transfer samples $\{\mathbf{M}_T^{'}, \mathbf{T}_T\}$ in target domain; ($\mathbf{M}_T^{'}$ obtained by transforming $\mathbf{M}_T = [\mathbf{m}_T^1, \ldots, \mathbf{m}_T^{N_T}]$ by $\mathbf{P}$).
5. Calculate the predicted output of the unlabeled samples of in target domain.
6. Calculate the recognition accuracy (%) of unlabeled data.

The merits of the proposed DC-AELM method are as follows:

(1) Most of the existing conventional methods are from the perspective of component correction (such as PCA, ICA, OSC) for interference suppression. The proposed DC-AELM method is from the perspective of transfer learning, especially from the perspective of domain correction and decision making, to realize the knowledge transfer for interference suppression.
(2) There is a certain degree of uncertainty of the interference, and the proposed method is applicable to both weaker interference (background interference) and strong interference (sensor drift with a long time).

13.4 Experiments

Two experiments on different interference datasets are used to test the effectiveness of the proposed DC-AELM method. Specifically, a background interference dataset obtained by our own designed E-nose used for detection of bacteria in wound infection is used to validate the proposed method. Then, the proposed method is tested on a public benchmark sensor drift dataset from an E-nose.

13.4.1 Experiment on Background Interference Data

In order to make the E-nose system portable and miniaturized, in the case of satisfying the detection requirement, the carrier gas is clean air which takes the place of zero air produced by specific instrument. The clean air is obtained by filtering the air using common filter, but not all of the interference can be cleared by it in some special application scenarios, such as in a hospital. Due to the presence of background interference, the distribution of testing data is different from that of training data. Thus, it is urgent to solve the problem of background interference.

In this section, a complex dataset obtained by designed E-nose is used for verification of the proposed DC-AELM method. The E-nose is used for detection of bacteria in wound infection. It is composed of 30 gas sensors according to the sensitivities to typical metabolites of common bacteria in wound infection and a module with three sensors for temperature, humidity, and atmospheric pressure. The information of gas sensors is shown in Table 13.1. For the feature extraction, the maximum value point in each sensor responses sequence is used as the feature point, i.e., a 34-dimensional feature vector for each sample listed in Table 13.2.

Three kinds of bacteria and their mixture (i.e., eight kinds of samples totally, S1: culture medium, S2: *Escherichia coli*, S3: *Staphylococcus aureus*, S4: *Pseudomonas aeruginosa*, S5: mixture of *E. coli* and *S. aureus*, S6: mixture of *S. aureus* and

Table 13.1 Gas sensor information

Sensor	Type	Producer (country)
TGS813, TGS816, TGS822, TGS826, TGS2600, TGS2602, TGS2610C, TGS2610D, TGS2611C, TGS2611D, TGS2620	MOS	Figaro (Japan)
MP135A, MP4, MP503, MP901, MQ135, MQ136, MQ137, MQ138, MQ3B, WSP2110	MOS	Winsen (China)
GSBT-11, MS1100	MOS	Ogam (Korea)
SP3-AQ2-01	MOS	FIS (Japan)
TGS4161	Electrochemistry	Figaro (Japan)
NH3-3E100SE, 4OXV, 4S, 4HS	Electrochemistry	CITY (UK)
CH2O/M-10	Electrochemistry	Membrapor (Switzerland)

Table 13.2 Number of samples in each dataset

	S1	S2	S3	S4	S5	S6	S7	S8	Total
Train	35	36	37	40	33	34	32	36	283
1	15	16	16	17	15	15	14	16	124
2	16	16	16	16	16	16	16	16	128
3	8	8	8	8	8	8	8	8	64

Table 13.3 Recognition accuracy (%) on the background interference dataset

Method	Air with alcohol volatile (number of transfer samples: 1, 2, 3, 4)				Gas collected in the ward of hospital (number of transfer samples: 1, 2, 3, 4)			
ELM	74.4				39.0625			
ICA-ELM	75.2				53.125			
Filter-ELM	83.36				61.25			
DC-ELM	55.2	55.2	57.6	59.2	53.125	59.38	64.06	65.63
AELM	**82.4**	**83.2**	88	92.8	81.25	**89.06**	93.75	95.31
DC-AELM	77.6	78.4	**88.8**	**95.2**	**82.8125**	**89.06**	**95.31**	**98.44**

The boldface type denotes the best performance.

P. aeruginosa, S7: mixture of *E. coli* and *P. aeruginosa*, S8: mixture of *S. aureus*, *P. aeruginosa* and *E. coli*) are detected. In this dataset, there are three sub-datasets. Sub-dataset 1 shown in Table 13.3 is a multi-class sample set composed of eight kinds of bacteria samples with clean air as carrier gas. The samples in this dataset are the standard samples without interference in source domain. Sub-dataset 2 shown in Table 13.2 is a multi-class sample set composed of eight kinds of bacteria samples which are obtained by using the air mixed with medical alcohol volatiles as the carrier gas. In the ward of trauma department, medical alcohol (C_2H_5OH, 70–75%) is often used for sterilization, and it is volatile. Therefore, the content of alcohol in hospital air is higher than that in a normal environment. The medical alcohol has an impact on most sensors in the E-nose, so it is a kind of interference. Therefore, for validation of the proposed DC-AELM, the air mixed with medical alcohol volatiles as the carrier gas is used to simulate the hospital environment. Sub-dataset 3 shown in Table 13.2 is a dataset for further verifying the performance of the proposed method in practical application. The carrier gas is the actual environmental air collected in the ward of trauma department in hospital. The experiment is a practical application of E-nose in real-world scenario. In order to reduce the scale variants among dimensions, the centralization process is conducted on the data.

To observe the space distribution variation with background interference, the PCA algorithm is applied to the dataset. The first principal component (PC1) is the horizontal axis, and the second principal component (PC2) is the vertical axis. The projected 2-D subspace for the same kind of bacteria in each sub-dataset is shown in Fig. 13.4, from which the significant changes of data space distribution caused by background interference can be observed.

Sub-dataset 1 is taken as source domain for model training, while Sub-dataset 2 and 3 (target domains) are used as test datasets. The classification accuracy on sub-dataset 2 and 3 is presented.

1. Sub-dataset 1, it is a multi-class sample set composed of eight kinds of bacteria samples with clean air as carrier gas. It is used to train a prediction model by dividing into training and test data, and the proportion of training and testing data is approximately 2:1 for generalization.

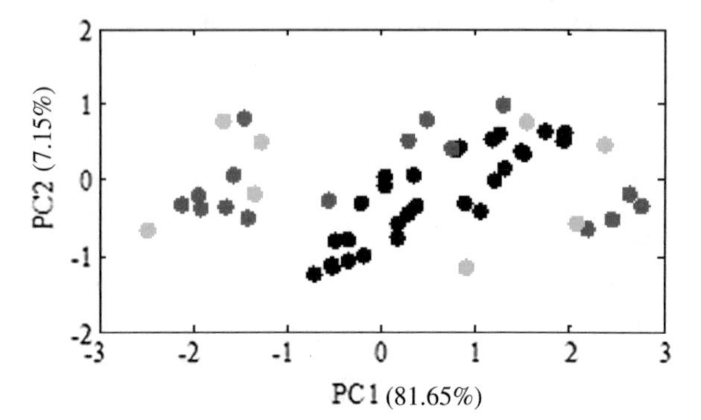

Fig. 13.4 Principal component subspace of the interference data by using PCA (black circle: standard bacterial samples, red circle: medical alcohol volatile as background interference; green circle: hospital gas as background interference)

2. Sub-dataset 2, it is a multi-class sample set composed of eight kinds of bacteria samples which are obtained by using the air mixed with medical alcohol volatiles as the carrier gas.
3. Sub-dataset 3 is obtained by using the actual environmental air collected in the ward of trauma department in hospital as carrier gas.

13.4.2 Experiment on Sensor Drift Data

The long-term sensor drift is one kind of interference. The reasons for drift are unknown dynamic processes in the sensor system, e.g., poisoning, aging of sensors, or changing of environmental parameters. The sensor response after drift cannot be calibrated directly due to the nonlinear dynamic behavior of sensor drift. Therefore, drift compensation by data distribution adaptation and machine learning is an interesting method.

(1) Dataset description

To verify the proposed DC-AELM method, the real sensor drift benchmark dataset of three years collected by an E-nose from Vergara et al. in UCSD [11] was used in this chapter. The dataset was gathered during the period from January 2008 to February 2011 with 36 months in a gas delivery platform. This dataset with 13,910 measurements, which were divided into 10 batches of time series, was from an E-nose with 16 gas sensors. For each sensor, eight features were extracted, and a 128-dimensional feature vector (8 features $\times$ 16 sensors) for each observation was formulated as a result. The detection substances were acetone, acetaldehyde, ethanol, ethylene, ammonia, and toluene at different concentration levels. The details of the dataset are presented in Table 11.1. To observe the space distribution variation with drift, the PCA algorithm is applied to the dataset. The projected 2-D subspace for all data in each batch is shown in Fig. 11.4, from which the significant changes of data space distribution caused by drift can be observed. In order to reduce the scale variants among dimensions, the centralization process is conducted on the data.

In this chapter, take batch 1 as source domain for model training, and test on batch K, $K = 2,\ldots,10$ (target domains). The classification accuracy on batch K is reported.

(2) Discriminant results and Comparisons

The principle component analysis algorithm (PCA) and orthogonal signal correction algorithm (OSC) are classical methods for drift suppression in E-nose. For comparison, the recognition accuracies of PCA algorithm, OSC algorithm, domain correction (DC, the first step of DC-AELM) with the standard ELM as classification algorithm, adaptive ELM (AELM, the second step of DC-AELM) which transfers the knowledge in decision level and the proposed DC-AELM method are shown in Tables 13.3 and 13.4. The trade-off parameters C_S and C_T are set as 0.01 and 0.1, respectively. For latter three methods, the recognition accuracies of different number of transfer samples are given. Different columns correspond to the results of different batches. The observations are as follows.

(1) From Table 13.3, it can be found that the proposed DC-AELM method with five transfer samples has the highest recognition accuracy, except for batch 2 and 10.
(2) The two parts of DC-AELM (i.e., DC with standard ELM and AELM) are used to suppress the interference independently. The effect of DC-ELM is very poor. However, the AELM has a significant effect on drift suppression. Therefore, the method from the perspective of decision level is effective for interference suppression.

13.5 Summary

In this chapter, a novel domain correction and adaptive extreme learning machine (DC-AELM) is proposed to suppress the interference and drift in E-nose. The method is motivated by the probability distribution difference between source domain (without interference/drift) and target domain (with interference/drift). The first step aims to solve the distribution mismatch between two domains by using domain correction. The second step aims to learn a cross-domain classifier at decision level by using an adaptive extreme learning machine (AELM). Experiments on an interference dataset and a benchmark drift dataset demonstrate the effectiveness of the presented method.

Table 13.4 Recognition accuracy (%) on the sensor drift dataset

Method	Batch 2	Batch 3	Batch 4	Batch 5	Batch 6	Batch 7	Batch 8	Batch 9	Batch 10	Average
ELM	77.12	72.18	67.95	64.67	72.13	57.08	54.01	56.55	31.37	61.45
DC-ELM(1)	81.05	82.43	56.77	32.69	56.57	52.94	30.75	64.64	46.53	56.04
DC-ELM(2)	84.65	82.57	60.25	89.34	62.82	61.26	34.49	76.77	47.47	66.62
DC-ELM(3)	77.88	80.73	71.80	73.91	29.50	41.85	28.16	84.47	49.03	59.70
DC-ELM(4)	70.35	87.40	83.73	75.03	47.23	45.20	34.83	80.85	43.90	63.17
DC-ELM(5)	73.71	87.74	87.63	75.13	53.07	46.62	37.21	87.36	42.48	65.66
AELM(1)	77.03	81.53	60.01	68.33	83.91	67.85	67.79	70.87	40.69	68.67
AELM(2)	77.54	82.49	75.28	96.26	78.94	87.78	43.79	95.75	45.82	75.96
AELM(3)	77.88	95.97	78.88	89.09	76.33	85.70	52.06	**100**	60.25	79.57
AELM(4)	82.51	98.48	81.12	92.39	73.42	90.19	59.42	**100**	63.45	82.33
AELM(5)	87.28	98.48	78.95	97.44	84.12	90.40	69.63	**100**	**65.62**	85.77
DC-AELM(1)	81.38	89.16	64.41	36.70	62.57	59.77	33.37	71.45	47.54	60.71
DC-AELM(2)	**87.75**	86.31	71.31	96.85	83.62	79.97	44.08	99.58	49.31	77.64
DC-AELM(3)	78.11	96.26	77.39	97.36	54.72	65.71	52.45	99.79	53.95	75.08
DC-AELM(4)	74.62	99.16	74.72	93.86	88.90	92.82	62.48	98.96	50.62	81.79
DC-AELM(5)	75.53	**99.36**	**95.65**	**97.41**	**96.80**	**94.87**	**79.966**	**100**	59.23	**88.76**

The boldface type denotes the best performance.

References

1. J. Feng, F. Tian, J. Yan, Q. He, Y. Shen, L. Pan, A background elimination method based on wavelet transform in wound infection detection by electronic nose. Sens. Actuators, B Chem. **157**(2), 395–400 (2011)
2. S. Marco, A. Gutierrez-Galvez, Signal and data processing for machine olfaction and chemical sensing: a review. IEEE Sens. J. **12**(11), 3189–3214 (2012)
3. G. Korotcenkov, B.K. Cho, Instability of metal oxide-based conductometric gas sensors and approaches to stability improvement (short survey). Sens. Actuators, B Chem. **156**(2), 527–538 (2011)
4. S.D. Carlo, M. Falasconi, Drift correction methods for gas chemical sensors in artificial olfaction systems: techniques and challenges. Adv. Chem. Sens. (2012). https://doi.org/10.5772/33411
5. T. Artursson, T. Eklöv, I. Lundström et al., Drift correction for gas sensors using multivariate methods. J. Chemometr. **14**, 711–723 (2012)
6. F. Tian, J. Yan, S. Xu, J. Feng, Background interference elimination in wound infection detection by electronic nose based on reference vector-based independent component analysis. Inf. Technol. J. **11**(7), 850–858 (2012)
7. O. Tomic, H. Ulmer, J.E. Haugen, Standardization methods for handling instrument related signal shift in gas-sensor array measurement data. Anal. Chim. Acta **472**, 99–111 (2002)
8. C.D. Natale, E. Martinelli, A.D. Amico, Counteraction of environmental disturbances of electronic nose data by independent component analysis. Sens. Actuators, B Chem. **82**, 158–165 (2002)
9. J. Feng, F. Tian, P. Jia, Q. He, Y. Shen, S. Fan, Improving the performance of electronic nose for wound infection detection using orthogonal signal correction and particle swarm optimization. Sens. Rev. **34**(4), 389–395 (2014)
10. M. Padilla, A. Perera, I. Montoliu, A. Chaudry, K. Persaud, S. Marco, Drift compensation of gas sensor array data by orthogonal signal correction. Chemometr. Intell. Lab. Syst. **100**(1), 28–35 (2010)
11. M. Zuppa, C. Distante, P. Siciliano, K.C. Persaud, Drift counteraction with multiple self-organising maps for an electronic nose. Sens. Actuators, B Chem. **98**(2), 305–317 (2004)
12. S.D. Vito, G. Fattoruso, M. Pardo et al., Semi-supervised learning techniques in artificial olfaction: a novel approach to classification problems and drift counteraction. IEEE Sens. J. **12**(11), 3215–3224 (2012)
13. A. Vergara, S. Vembu, T. Ayhan, M.A. Ryan, M.L. Homer, R. Huerta, Chemical gas sensor drift compensation using classifier ensembles. Sens. Actuators, B Chem. **167**, 320–329 (2012)
14. E. Martinelli, G. Magna, A. Vergara, C.D. Natale, Cooperative classifiers for reconfigurable sensor arrays. Sens. Actuators, B Chem. **199**, 83–92 (2014)
15. E. Martinelli, G. Magna, S.D. Vito, R.D. Fuccio, G.D. Francia, A. Vergara, C.D. Natale, An adaptive classification model based on the artificial immune system for chemical sensor drift mitigation. Sens. Actuators, B Chem. **177**, 1017–1026 (2013)
16. A.C. Romain, J. Nicolas, Long term stability of metal oxide-based gas sensors for e-nose environmental applications: an overview. Sens. Actuators, B Chem. **146**(9), 502–506 (2010)
17. L. Zhang, F. Tian, S. Liu, L. Dang, X. Peng, X. Yin, Chaotic time series prediction of E-nose sensor drift in embedded phase space. Sens. Actuators, B Chem. **182**(1), 71–79 (2013)
18. D.A.P. Daniel, K. Thangavel, R. Manavalan, R.S.C. Boss, *ELM-Based Ensemble Classifier for Gas Sensor Array Drift Dataset* (Springer, India, 2014). https://doi.org/10.1007/978-81-322-1680-3_10
19. S.J. Pan, Q. Yang, A survey on transfer learning. IEEE Educ. Activities Dept. **22**(10), 1345–1359 (2010)
20. K. Yan, D. Zhang, Correcting instrumental variation and time-varying drift: a transfer learning approach with autoencoders. IEEE Trans. Instrum. Meas. **65**(9), 2012–2022 (2016)

21. K. Yan, D. Zhang, Improving the transfer ability of prediction models for electronic noses. Sens. Actuators, B Chem. **220**, 115–124 (2015)
22. L. Zhang, Y. Liu, P. Deng, Odor recognition in multiple e-nose systems with cross-domain discriminative subspace learning. IEEE Trans. Instrum. Meas. (2017). https://doi.org/10.1109/TIM.2017.2669818
23. L. Zhang, D. Zhang, Domain adaptation extreme learning machines for drift compensation in e-nose systems. IEEE Trans. Instrum. Meas. **64**(7), 1790–1801 (2015)
24. K. Yan, D. Zhang, Calibration transfer and drift compensation of e-noses via coupled task learning. Sens. Actuators, B Chem. **225**, 288–297 (2016)
25. G. Huang, Q. Zhu, C. Siew, Extreme learning machine: theory and applications. Neurocomputing **70**(1), 489–501 (2006)
26. G. Feng, G.B. Huang, Q. Lin, R. Gay, Error minimized extreme learning machine with growth of hidden nodes and incremental learning. IEEE Trans. Neural Netw. **20**(8), 1352–1357 (2009)
27. G.B. Huang, H. Zhou, X. Ding, R. Zhang, Extreme learning machine for regression and multiclass classification. IEEE Trans. Syst. Man Cybern. Part B **42**(2), 513–529 (2012)
28. W. Zong, G.B. Huang, Y. Chen, Weighted extreme learning machine for imbalance learning. Neurocomputing **101**(3), 229–242 (2013)
29. Z. Bai, G.B. Huang, D. Wang, H. Wang, M.B. Westover, Sparse extreme learning machine for classification. IEEE Trans. Cybern. **44**(10), 1858–1870 (2014)
30. X. Li, W. Mao, W. Jiang, Fast sparse approximation of extreme learning machine. Neurocomputing **128**(5), 96–103 (2014)
31. G. Huang, S. Song, J.N. Gupta, C. Wu, Semi-supervised and unsupervised extreme learning machines. IEEE Trans. Cybern. **44**(12), 2405–2417 (2014)
32. L. Zhang, F. Tian, C. Kadri et al., On-line sensor calibration transfer among electronic nose instruments for monitoring volatile organic chemicals in indoor air quality. Sens. Actuators, B Chem. **160**(1), 899–909 (2011)

Chapter 14
Multi-feature Semi-supervised Learning Approach

Abstract The concerns of this chapter are threefold: first, due to that each sensor feature can be exploited in different modality, multiple feature modalities for each sensor may be extracted; second, consider that the manual labeling of artificial olfaction data in real-time detection is difficult and hardly impossible, semi-supervised learning strategy is expected to be a breakthrough and overcome the problem of insufficient labeled data in artificial olfactory system; third, in E-Nose community, classifier learning is generally independent from feature extraction, such that the recognition capability of an E-Nose is limited due to the achieved suboptimal performance. Motivated by these concerns, in this chapter, from a new machine learning perspective, we aim at proposing a multi-feature kernel semi-supervised learning framework nominated as MFKS, whose merits can be composed of three points. (1) A multi-feature joint learning with low-rank constraint is developed for exploiting the multiple feature modalities from each sensor. The relatedness of all sub-classifiers learned on multiple feature modalities is preserved by imposing a low-rank constraint on the group classifier as regularization. (2) With a manifold assumption, a Laplacian graph manifold regularization is incorporated for semi-supervised learning and overcomes the flaw of insufficient labeled data in E-Nose. (3) The feature level and classifier level in artificial olfactory system are learned simultaneously in a complete framework, such that the recognition performance of an E-Nose can be optimally achieved. Experiments on two olfaction datasets including a large-scale 16-sensor data with 36-month drift and a small-scale temperature-modulated sensor data demonstrate that the proposed approach outperforms other algorithms.

Keywords Artificial olfactory system · Electronic nose · Multi-feature learning
Semi-supervised learning

L. Zhang et al., *Electronic Nose: Algorithmic Challenges*,
https://doi.org/10.1007/978-981-13-2167-2_14

14.1 Introduction

14.1.1 Problem Statement

From the existing work of artificial olfactory system and E-Nose, we can observe that different methods in feature level and classifier level (model-level) have been proposed by many researchers. Though the performance of E-Nose in discriminative capability has been gradually improved, the data processing methods proposed in feature level are independent from the pattern recognition unit in classifier level, such that the extracted sensor features cannot be adaptive to the classification model if the data structure becomes more complex (e.g., high-dimensional data and sensor drift data) [1]. Briefly, sensor drift in artificial olfaction is commonly caused by unknown dynamic process such as poisoning, aging, environmental variation (ambient temperature, humidity, pressure, etc.), which has deteriorated the performance of an E-Nose in both feature level and classifier level [2, 3]. One salient feature of drift is that the conditional probability distribution $P(\mathbf{X}_{\text{nodrift}}|\mathbf{Y}) \neq P(\mathbf{X}_{\text{drift}}|\mathbf{Y})$, where $\mathbf{X}_{\text{nodrift}} \in \Re^{d \times N}$ and $\mathbf{X}_{\text{drift}} \in \Re^{d \times N}$ denote the sensor dataset before and after drift, and $\mathbf{Y} \in \Re^{N}$ denotes the labels. The distribution differences also imply that the statistical characteristics such as mean and covariance change.

Therefore, some independent feature-level data processing methods or classifier-level prediction models would not adapt to the drifted and noisy E-Nose data in real applications due to the temporal data structure variation.

14.1.2 Motivation

From the perspective of machine learning, both feature and classifier learning in E-Nose are supervised and enforced separately. However, with the change of data structure caused by complex and long-term E-Nose drift in time series, it is a huge burden for manual labeling of each observation. Consequently, the model learning in E-Nose would face with a new dilemma that the labeled data is insufficient.

On the one hand, in classifier-level learning, for addressing the issue of insufficient labeled data, a semi-supervised learning idea motivated by machine learning theory has been introduced in artificial olfaction [4, 5]. Semi-supervised learning [6] was initially proposed for handling the problem of insufficient labeled data, which includes two important assumptions: manifold assumption and cluster assumption [7]. Manifold assumption implies that the data spans a low-dimensional manifold space in which the structure information of unlabeled data can be preserved and the nearby data points are more likely to have the same label. The cluster assumption implies that the data points on the same cluster are more likely to have the same label. The differences between the two assumptions lie in that cluster assumption is local while manifold assumption is global. Therefore, manifold regularization has become a mainstream of semi-supervised learning in computer vision community for image classification and multimedia application [8, 9]. The idea of Laplacian graph manifold was generally proposed for dimension reduction and graph embedding, which implies the manifold structure preservation from high dimension to low dimension [10–14]. However, manifold learning-based dimension reduction suffered from a fact that there is no explicit mapping matrix for low-dimensional projection. Instead of dimension reduction, in this chapter, we tend to propose a semi-supervised learning framework with Laplacian graph manifold regularization for handling noisy E-Nose data.

On the other hand, for feature-level data processing, we aim at proposing a multi-feature joint learning mechanism by integrating the classifier learning together, which is different from the existing feature selection and feature fusion algorithms that are independent from classifier learning. In other words, the motivation of this chapter is to jointly learn the multi-feature and semi-supervised classifier simultaneously for pursuit of the optimal performance in system-level.

14.2 Related Work

In E-Nose system, machine learning-based methods including representation learning, ensemble learning, transfer learning, and deep learning have been proposed for drift compensation in very recent years. Specifically, the literature reviews are divided into four categories.

- In representation learning-based methods, Ziyatdinov et al. [15] proposed a principal component analysis (PCA) method for learning a transformation such that the drift direction can be captured. Carlo et al. [16] proposed to learn a shift-transformation matrix by using evolutionary algorithm in raw data space with an objective of improving the classification accuracy of E-Nose.
- In ensemble learning-based methods, Vergara et al. [1], Dang et al. [17], and Liu et al. [18] proposed classifier ensemble method based on multiple SVM models that aim at improving the robustness of gases recognition with drift data. However, the weight of each SVM is obtained independent from the model learning, which would result in a local solution. Besides, how to guarantee the diversity of multiple SVM classifiers is still an open problem.
- In transfer learning-based methods, Liu et al. [19] proposed a novel idea of semi-supervised domain adaptation method for drift compensation in electronic nose. Zhang et al. [20, 21] proposed a transfer extreme learning machine-based domain adaptation for fast classifier learning and drift compensation. Domain adaptation-based transfer learning has been proved to be effective in improving the generalization of classifier with drift-knowledge adaptation.
- In deep learning-based methods, Martin et al. [22] proposed to learn deep feature in unsupervised manner by using deep restricted Boltzmann machines for bacteria identification in blood by an E-Nose. Similarly, Liu et al. [23] proposed for gas recognition with concept drift by using deep learning

techniques such as deep Boltzmann machine and sparse auto-encoder. A good advantage of deep learning is the feature representation ability, but depends on a large-scale training data which is still comparatively deficient in E-Nose.

14.3 Multi-feature Kernel Semi-supervised Joint Learning Model

14.3.1 Notations

Let $\mathbf{X}_i = \left[\mathbf{X}_i^l, \mathbf{X}_i^u\right] = \left[\mathbf{x}_i^1, \mathbf{x}_i^2, \cdots, \mathbf{x}_i^N\right] \in \Re^{d \times N}, i = 1, \ldots, m$ denotes the training data of the ith feature modality, where $\mathbf{X}_i^l \in \Re^{d \times N_l}$ and $\mathbf{X}_i^u \in \Re^{d \times N_u}$ represent the labeled data and unlabeled data, respectively, $\mathbf{x}_i^j \in \Re^d$ denotes the jth observation sample of the ith feature modality, $N = N_l + N_u$ denotes the number of samples, d denotes the dimension (i.e., the number of sensors) of each observation, and m denotes the number of feature types. Correspondingly, let $\mathbf{Y} = [\mathbf{Y}_l, \mathbf{Y}_u] \in \Re^{C \times N}$ denotes the label matrix, where $\mathbf{Y}_j \in \Re^C, j = 1, \ldots, N$. C denotes the number of classes. In this chapter, $\|\cdot\|_F$ denotes the Frobenius norm of a matrix, $\|\cdot\|_2$ represents the ℓ_2-norm, $\|\mathbf{P}\|_*$ denotes nuclear norm, $Tr(\cdot)$ denotes the trace operator, rank$(\cdot)$ denotes the rank function, $\langle\cdot\rangle$ denotes inner product operator, and $(\cdot)^{\mathrm{T}}$ denotes the transpose operator. The labels are defined as follows.

- For labeled data, $\mathbf{Y}_l \in \Re^{C \times N_l}$, where $Y_{l,j}^c = 1$ if the jth sample belongs to class c (c = 1,...,C), and -1 otherwise.
- For unlabeled data, $\mathbf{Y}_u = 0 \in \Re^{C \times N_u}$.

Considering the nonlinear property of multi-sensor system, the raw data $\mathbf{X}_i$ can be mapped into a high-dimensional space (e.g., reproduced kernel Hilbert space $\mathcal{H}$, RKHS) by using a nonlinear transformation φ. Therefore, we introduce kernel matrix $\mathbf{K}_i \in \Re^{N \times N}, i = 1, \ldots, m$ for representing the training data $\mathbf{X}_i$ in RKHS $\mathcal{H}$ as follows.

$$\mathbf{K}_i = \langle\varphi(\mathbf{X}_i), \varphi(\mathbf{X}_i)\rangle = \varphi(\mathbf{X}_i)^{\mathrm{T}}\varphi(\mathbf{X}_i) = \kappa(\mathbf{X}_i, \mathbf{X}_i) \tag{14.1}$$

where $\kappa(\cdot)$ denotes the kernel function (e.g., Gaussian RBF, sigmoid, polynomial). In this chapter, the Gaussian RBF function represented as $\kappa\left(\mathbf{x}_i, \mathbf{x}_j\right) = \exp\left(-\|\mathbf{x}_i - \mathbf{x}_j\|_2^2 \big/ 2\sigma^2\right)$ is used as kernel function. The kernel parameter σ can be tuned for pursuit of the best performance.

14.3.2 Model

For realizing multi-feature joint learning with m kernel feature matrix in semi-supervised manner, the proposed MFKS is to solve the following general minimization problem.

$$\min_{\mathbf{F},\mathbf{P}_i,\mathbf{B}_i} \sum_{i=1}^{m} \|\mathbf{F} - \mathbf{K}_i\mathbf{P}_i - \mathbf{1}_N\mathbf{B}_i\|_F^2 + \gamma \cdot \text{rank}(\mathbf{P}) + \text{Loss}(\mathbf{F},\mathbf{Y}) + \lambda \cdot \psi(\mathbf{F}) \tag{14.2}$$

where $\mathbf{P}_i \in \Re^{N\times C}$ denotes the classifier parameter matrix, $\mathbf{B}_i \in \Re^{1\times C}$ denotes the classifier bias, $\mathbf{P} = [\mathbf{P}_1, \mathbf{P}_2, \ldots, \mathbf{P}_m] \in \Re^{N\times mC}$ denotes the group classifier consisting of m sub-classifiers, $1_N \in \Re^N$ is a full-one vector, $\mathbf{F} \in \Re^{N\times C}$ represents the predicted label matrix, $\text{Loss}(\cdot)$ is the least square-alike loss function, and γ and λ are regularization parameters.

Specifically, the rational explanations behind of the four terms in the proposed model Eq. 14.2 are presented as follows.

- The first term represents joint label prediction based on m feature modalities by simultaneously learning classifier $\mathbf{P}_i(i = 1,\ldots,m)$ for each type of feature.
- The second term is the regularization of the learned group classifier $\mathbf{P}$ which is restricted to be low rank, behind the rational is that we expect the relatedness among the learned $\mathbf{P}_1, \mathbf{P}_2, \ldots, \mathbf{P}_m$ to be well preserved by using minimized low-rank constraint, such that the structural information of multiple features can be shared in the proposed model. A good work of low-rank representation has been introduced for subspace clustering and segmentation in [24].
- The third term denotes the semi-supervised loss function, which is formulated based on least square-alike loss as follows.

$$\text{Loss}(\mathbf{F},\mathbf{Y}) = Tr(\mathbf{F} - \mathbf{Y})^{\mathrm{T}}\mathbf{U}(\mathbf{F} - \mathbf{Y}) \tag{14.3}$$

where $\mathbf{U}$ is diagonal selection matrix defined as follows

$$\mathbf{U}_{jj} = \begin{cases} \infty, & \text{if } \mathbf{x}_i^j \text{ is labeled} \\ 0, & \text{otherwise} \end{cases}; \quad j = 1,\ldots,N;\ i = 1,\ldots,m \tag{14.4}$$

- The last term denotes the manifold regularization on the predicted label $\mathbf{F}$ for exploring label consistency in semi-supervised learning. With the manifold assumption, we expect that nearby points are more likely to have the same labels, which can be mathematically formulated as the following minimization problem,

$$\begin{aligned}
&\min_f \sum_{p,q} A_{p,q} \| f(\mathbf{x}_p) - f(\mathbf{x}_q) \|_2^2 \\
&= \min_f \sum_{p,q} \left(\| f(\mathbf{x}_p) \|_2^2 + \| f(\mathbf{x}_q) \|_2^2 - 2 f(\mathbf{x}_p) f(\mathbf{x}_q)^{\mathrm{T}} \right) A_{p,q} \\
&= \min_f \sum_p \left(\| f(\mathbf{x}_p) \|_2^2 \sum_q A_{p,q} \right) + \sum_q \left(\| f(\mathbf{x}_q) \|_2^2 \sum_p A_{p,q} \right) \\
&\quad - 2 \sum_{p,q} f(\mathbf{x}_p) f(\mathbf{x}_q)^{\mathrm{T}} A_{p,q} \\
&= \min_{\mathbf{F}} \mathrm{Tr}\left(2\mathbf{F}^{\mathrm{T}}\mathbf{D}\mathbf{F}\right) - \mathrm{Tr}\left(2\mathbf{F}^{\mathrm{T}}\mathbf{A}\mathbf{F}\right) \\
&= \min_{\mathbf{F}} \mathrm{Tr}\left(2\mathbf{F}^{\mathrm{T}}\mathbf{L}\mathbf{F}\right)
\end{aligned} \tag{14.5}$$

where $f(\cdot)$ denotes the label predictor, $\mathbf{D}$ is a diagonal matrix with entries $D_{pp} = \sum_q A_{pq}$, $\mathbf{L} = \mathbf{D} - \mathbf{A}$ is the Laplacian graph matrix, and $\mathbf{A}$ is the affinity matrix computed as,

$$A_{p,q} = \begin{cases} 1, & \text{if } \mathbf{x}_p \in \mathrm{N}_k(\mathbf{x}_q) \text{ or } \mathbf{x}_q \in \mathrm{N}_k(\mathbf{x}_p) \\ 0, & \text{otherwise} \end{cases} \tag{14.6}$$

where $\mathrm{N}_k(\mathbf{x})$ represents the k nearest neighbors of sample $\mathbf{x}$.

Therefore, in terms of Eq. 14.5, the last term $\psi(\mathbf{F})$ in the proposed model Eq. 14.2 can be written as follows.

$$\psi(\mathbf{F}) = Tr\left(\mathbf{F}^{\mathrm{T}} \left(\sum_{i=1}^{m} \alpha_i^r \mathbf{L}_i \right) \mathbf{F} \right) \tag{14.7}$$

where the Laplacian matrix $\mathbf{L} = \sum_{i=1}^{m} \alpha_i^r \mathbf{L}_i$ is represented a weighted summation of $\mathbf{L}_i$ for multi-feature learning, $0 < \alpha_i^r < 1$ represents the coefficient of the ith feature, $\sum_{i=1}^{m} \alpha_i = 1$, and $r > 1$. Note that the setting of $r > 1$ is to better exploit the complementary information of multiple features and avoid that trivial solution with only the best feature used (e.g., $\alpha_i = 1$).

In summary, by combining Eqs. 14.2, 14.3, and 14.7, the proposed MFKS model can be completely written as follows.

$$\begin{aligned}
&\min_{\mathbf{F},\mathbf{P}_i,\mathbf{B}_i,\alpha_i} \sum_{i=1}^{m} \| \mathbf{F} - \mathbf{K}_i \mathbf{P}_i - \mathbf{1}_N \mathbf{B}_i \|_{\mathrm{F}}^2 + \gamma \cdot \mathrm{rank}(\mathbf{P}) \\
&\quad + Tr(\mathbf{F} - \mathbf{Y})^{\mathrm{T}} \mathbf{U} (\mathbf{F} - \mathbf{Y}) + \lambda \cdot Tr\left(\mathbf{F}^{\mathrm{T}} \left(\sum_{i=1}^{m} \alpha_i^r \mathbf{L}_i \right) \mathbf{F} \right) \\
&\text{s.t.} \sum_{i=1}^{m} \alpha_i = 1, 0 < \alpha_i < 1
\end{aligned} \tag{14.8}$$

Note that the rank of $\mathbf{P}$ is a non-convex operator, and it is generally addressed by using its convex surrogate of the rank [24], that is, the nuclear norm or trace norm defined as $\|\mathbf{P}_*\|$ (i.e., the sum of singular values of $\mathbf{P}$). It can be represented as,

$$\|\mathbf{P}\|_* = Tr\left(\mathbf{P}^{\mathrm{T}}\left(\mathbf{P}\mathbf{P}^{\mathrm{T}}\right)^{-\frac{1}{2}}\mathbf{P}\right) \tag{14.9}$$

Finally, with the low-rank constraint of Eq. 14.9, the proposed MFKS model Eq. 14.8 is specifically formulated as follows.

$$\begin{aligned} &\min_{\mathbf{F},\mathbf{P}_i,\mathbf{B}_i,\alpha_i} \sum_{i=1}^{m} \|\mathbf{F} - \mathbf{K}_i\mathbf{P}_i - \mathbf{1}_N\mathbf{B}_i\|_{\mathrm{F}}^2 + \frac{\gamma}{2} \cdot \|\mathbf{P}\|_* \\ &+ Tr(\mathbf{F} - \mathbf{Y})^{\mathrm{T}}\mathbf{U}(\mathbf{F} - \mathbf{Y}) + \lambda \cdot Tr\left(\mathbf{F}^{\mathrm{T}}\left(\sum_{i=1}^{m} \alpha_i^r \mathbf{L}_i\right)\mathbf{F}\right) \\ &\text{s.t.} \sum_{i=1}^{m} \alpha_i = 1, 0 < \alpha_i < 1 \end{aligned} \tag{14.10}$$

14.3.3 Optimization Algorithm

From the structure of the proposed MFKS model Eq. 14.10, we observe that the objective function is non-convex *w.r.t.* four variables $\mathbf{F}$, $\mathbf{P}_i$, $\mathbf{B}_i$, and α_i. However, when fix other three variables (e.g., $\mathbf{P}_i$, $\mathbf{B}_i$, and α_i), the objective function is convex *w.r.t.* another variable (i.e., $\mathbf{F}$). Therefore, the solutions $\mathbf{F}$, $\mathbf{P}_i$, $\mathbf{B}_i$, and α_i can be efficiently solved by a variable-alternative optimization approach. Specifically, the optimization process of the proposed model is presented as follows.

- *Initialize* $\mathbf{F}^{(0)}$, $\alpha_i^{(0)}$, $\mathbf{P}_i^{(0)}$, *and* $\mathbf{B}_i^{(0)}$

First, we initialize $\mathbf{P}_i^{(0)} = 0$ and $\mathbf{B}_i^{(0)} = 0$. According to the equality constraint $\sum_{i=1}^{m} \alpha_i = 1$, we initialize $\alpha_i^{(0)} = 1/m, \forall i$, the initialized $\mathbf{F}$ can be solved analytically by setting the derivative of following objective function *w.r.t.* $\mathbf{F}$ as 0,

$$\min_{\mathbf{F}} Tr(\mathbf{F} - \mathbf{Y})^{\mathrm{T}}\mathbf{U}(\mathbf{F} - \mathbf{Y}) + \lambda \cdot Tr\left(\mathbf{F}^{\mathrm{T}}\left(\sum_{i=1}^{m} \alpha_i^r \mathbf{L}_i\right)\mathbf{F}\right) \tag{14.11}$$

Then, $\mathbf{F}$ can be initialized as,

$$\mathbf{F}^{(0)} = \left(\mathbf{U} + \lambda \cdot \sum_{i=1}^{m} \left(\alpha_i^{(0)}\right)^r \mathbf{L}_i\right)^{-1} \mathbf{U}\mathbf{Y} \tag{14.12}$$

- *Update* $\mathbf{P}_i^{(t)}$ *and* $\mathbf{B}_i^{(t)}$

Second, after fixing $\mathbf{F}$ and α_i, the optimization problem of model Eq. 14.10 becomes,

$$\begin{aligned} &\min_{\mathbf{P}_i,\mathbf{B}_i} \sum_{i=1}^{m} \|\mathbf{F}-\mathbf{K}_i\mathbf{P}_i-\mathbf{1}_N\mathbf{B}_i\|_\mathrm{F}^2 + \frac{\gamma}{2}\cdot\|\mathbf{P}\|_* \\ &= \min_{\mathbf{P}_i,\mathbf{B}_i} \sum_{i=1}^{m} \|\mathbf{F}-\mathbf{K}_i\mathbf{P}_i-\mathbf{1}_N\mathbf{B}_i\|_\mathrm{F}^2 + \frac{\gamma}{2}\cdot Tr\left(\mathbf{P}^\mathrm{T}\left(\mathbf{P}\mathbf{P}^\mathrm{T}\right)^{-\frac{1}{2}}\mathbf{P}\right) \end{aligned} \tag{14.13}$$

By setting the derivative of the objective function $J(\mathbf{P}_i, \mathbf{B}_i)$ in Eq. 14.13 *w.r.t.* $\mathbf{P}_i$ and $\mathbf{B}_i$ to be 0, respectively, one can obtain,

$$\begin{aligned} \frac{\partial J(\mathbf{P}_i,\mathbf{B}_i)}{\partial \mathbf{P}_i} &= -\mathbf{K}_i^\mathrm{T}(\mathbf{F}-\mathbf{K}_i\mathbf{P}_i-\mathbf{1}_N\mathbf{B}_i) + \frac{\gamma}{2}\left(\mathbf{P}\mathbf{P}^\mathrm{T}\right)^{-\frac{1}{2}}\mathbf{P}_i = 0 \\ \frac{\partial J(\mathbf{P}_i,\mathbf{B}_i)}{\partial \mathbf{B}_i} &= -\mathbf{1}_N^\mathrm{T}(\mathbf{F}-\mathbf{K}_i\mathbf{P}_i-\mathbf{1}_N\mathbf{B}_i) = 0 \end{aligned} \tag{14.14}$$

Therefore, for iteration t, $\mathbf{P}_i^{(t)}$ and $\mathbf{B}_i^{(t)}$ *w.r.t.* the ith feature can be solved in closed form as follows.

$$\begin{aligned} \mathbf{P}_i^{(t)} &= \left(\mathbf{K}_i^\mathrm{T}\mathbf{K}_i + \gamma\cdot\mathbf{D}_p^{(t-1)}\right)^{-1}\left(\mathbf{K}_i^\mathrm{T}\mathbf{F}^{(t-1)} - \mathbf{K}_i^\mathrm{T}\mathbf{1}_N\mathbf{B}_i^{(t-1)}\right) \\ \mathbf{B}_i^{(t)} &= \frac{\mathbf{1}_N^\mathrm{T}\mathbf{F} - \mathbf{1}_N^\mathrm{T}\mathbf{K}_i\mathbf{P}_i^{(t)}}{\mathbf{1}_N^\mathrm{T}\mathbf{1}_N} \end{aligned} \tag{14.15}$$

where $\mathbf{D}_p^{(t-1)} = 0.5\left(\mathbf{P}^{(t-1)}\left(\mathbf{P}^{(t-1)}\right)^\mathrm{T}\right)^{-0.5}$ is a diagonal matrix and $\mathbf{P}^{(t-1)} = \left[\mathbf{P}_1^{(t-1)}, \ldots, \mathbf{P}_m^{(t-1)}\right]$ is the group classifier.

- *Update* $\mathbf{F}^{(t)}$

Third, after fixing α_i, $\mathbf{P}_i$ and $\mathbf{B}_i$, the optimization problem Eq. 14.10 becomes,

$$\begin{aligned} &\min_{\mathbf{F}} \sum_{i=1}^{m} \|\mathbf{F}-\mathbf{K}_i\mathbf{P}_i-\mathbf{1}_N\mathbf{B}_i\|_\mathrm{F}^2 + Tr(\mathbf{F}-\mathbf{Y})^\mathrm{T}\mathbf{U}(\mathbf{F}-\mathbf{Y}) \\ &+\lambda\cdot Tr\left(\mathbf{F}^\mathrm{T}\left(\sum_{i=1}^{m}\alpha_i^r\mathbf{L}_i\right)\mathbf{F}\right) \end{aligned} \tag{14.16}$$

From Eq. 14.16, it is easy to solve $\mathbf{F}$ by setting the derivative *w.r.t.* $\mathbf{F}$ to be 0, as follows.

$$\frac{\partial J(\mathbf{F})}{\partial \mathbf{F}} = \left(m\mathbf{I} + \mathbf{U} + \lambda \cdot \sum_{i=1}^{m} \alpha_i^r \mathbf{L}_i\right)\mathbf{F} - \mathbf{U}\mathbf{Y} - \sum_{i=1}^{m} (\mathbf{K}_i \mathbf{P}_i + \mathbf{1}_N \mathbf{B}_i) = 0 \tag{14.17}$$

where $\mathbf{I} \in \Re^{N \times N}$ is an identity matrix. Then, from Eq. 14.17, $\mathbf{F}^{(t)}$ in iteration t can be updated as,

$$\mathbf{F}^{(t)} = \left(m\mathbf{I} + \mathbf{U} + \lambda \cdot \sum_{i=1}^{m} \left(\alpha_i^{(t-1)}\right)^r \mathbf{L}_i\right)^{-1} \left(\mathbf{U}\mathbf{Y} + \sum_{i=1}^{m} \left(\mathbf{K}_i \mathbf{P}_i^{(t)} + \mathbf{1}_N \mathbf{B}_i^{(t)}\right)\right) \tag{14.18}$$

- *Update* $\alpha_i^{(t)}$

Fourth, after obtaining $\mathbf{F}$, $\mathbf{P}_i$, and $\mathbf{B}_i$, the optimization problem Eq. 14.10 becomes,

$$\begin{aligned} &\min_{\alpha_i} \mathrm{Tr}\left(\mathbf{F}^{\mathrm{T}}\left(\sum_{i=1}^{m} \alpha_i^r \mathbf{L}_i\right)\mathbf{F}\right) \\ &\text{s.t.} \sum_{i=1}^{m} \alpha_i = 1, \quad 0 < \alpha_i < 1 \end{aligned} \tag{14.19}$$

The Lagrange equation of Eq. 14.19 can be written as,

$$J(\alpha, \mu) = Tr\left(\mathbf{F}^{\mathrm{T}}\left(\sum_{i=1}^{m} \alpha_i^r \mathbf{L}_i\right)\mathbf{F}\right) - \mu \cdot \left(\sum_{i=1}^{m} \alpha_i - 1\right) \tag{14.20}$$

By setting the derivative of Eq. 14.20 *w.r.t.* α_i and μ to 0, respectively, we have,

$$\begin{cases} r\alpha_i^{r-1} Tr\left(\mathbf{F}^{\mathrm{T}} \mathbf{L}_i \mathbf{F}\right) - \mu = 0 \\ \sum_{i=1}^{m} \alpha_i - 1 = 0 \end{cases} \tag{14.21}$$

By solving the equation group Eq. 14.21, $\alpha_i^{(t)}, i = 1, \ldots, m$ in iteration t can be easily updated as,

$$\alpha^{(t)} \left(\frac{1}{Tr\left(\left(\mathbf{F}^{(t)}\right)^{\mathrm{T}} \mathbf{L}_i \mathbf{F}^{(t)}\right)}\right)^{\frac{1}{r-1}} \Big/ \sum_{i=1}^{m} \left(\frac{1}{Tr\left(\left(\mathbf{F}^{(t)}\right)^{\mathrm{T}} \mathbf{L}_i \mathbf{F}^{(t)}\right)}\right)^{\frac{1}{r-1}} \tag{14.22}$$

Finally, the optimization of the proposed MFKS is illustrated as Algorithm 14.1.

Algorithm 14.1 MFKS

Input:

The training data of m feature matrix $\mathbf{X}_i \in \mathbb{R}^{d \times N}, i = 1, \ldots, m$;
The training labels $\mathbf{Y} \in \mathbb{R}^{C \times N}$;
Parameters λ, γ, and r;

Procedure:

1. Compute the kernel matrix $\mathbf{K}_i$ using Eq. 14.1;
2. Compute the graph Laplacian matrix $\mathbf{L}_i$ with Eqs. 14.5 and 14.6;
3. Compute the selection matrix $\mathbf{U}$ using Eq. 14.4;
4. Initialize $\alpha_i^{(0)} \leftarrow 1/m, \mathbf{P}_i^{(0)} \leftarrow \mathbf{0}, \mathbf{B}_i^{(0)} \leftarrow \mathbf{0}, i = 1, \ldots, m$;
5. Initialize $\mathbf{P}^{(0)} = \left[\mathbf{P}_1^{(0)}, \ldots, \mathbf{P}_m^{(0)}\right]$ and $\mathbf{D}_p^{(0)} = \frac{1}{2}\left(\mathbf{P}^{(0)}\left(\mathbf{P}^{(0)}\right)^{\mathrm{T}}\right)^{-\frac{1}{2}}$;
6. Initialize $\mathbf{F}^{(0)}$ according to Eq. 14.12;
7. **repeat**

   Compute $\mathbf{P}_i^{(t)}$ and $\mathbf{B}_i^{(t)}$ according to Eq. 14.15;
   Update $\mathbf{F}^{(t)}$ according to Eq. 14.18;
   Update $\alpha_i^{(t)}$ according to Eq. 14.22;
   Update $\mathbf{P}^{(t)} = \left[\mathbf{P}_1^{(t)}, \ldots, \mathbf{P}_m^{(t)}\right]$ and $\mathbf{D}_p^{(t)} = \frac{1}{2}\left(\mathbf{P}^{(t)}\left(\mathbf{P}^{(t)}\right)^{\mathrm{T}}\right)^{-\frac{1}{2}}$
   until convergence;
8. **Output**: $\mathbf{P}_i$ and $\mathbf{B}_i$, $i = 1,\ldots,m$;

14.3.4 Classification

The classification denotes the testing phase after MFKS training. The classification parameters $\{\mathbf{P}_i\}_{i=1}^m$ and $\{\mathbf{B}_i\}_{i=1}^m$ can be obtained by solving the proposed model Eq. 14.10 using the proposed Algorithm 14.1 on the provided training data $\mathbf{X} = \{\mathbf{X}_i\}_{i=1}^m$. For real-time application, we perform classification and predict the label of a new observation $\mathbf{z}$ represented by m features $\mathbf{z} = \{\mathbf{z}_i\}_{i=1}^m$ using the following decision function,

$$\mathrm{label}(\mathbf{z}) = \arg \max_{j \in \{1,\ldots,C\}} \sum_{i=1}^{m} (\kappa(\mathbf{z}_i, \mathbf{X}_i) \cdot \mathbf{P}_i + \mathbf{B}_i) \tag{14.23}$$

where $\kappa(\cdot)$ is the kernel function. The classification procedure of the proposed MFKS is summarized in Algorithm 14.2.

Algorithm 14.2 Classification of MFKS

Input:

Training set $\{\mathbf{X}_i\}_{i=1}^m$, training labels $\mathbf{Y}$, and one test sample $\{\mathbf{z}_i\}_{i=1}^m$ with m kinds of features;

Procedure:

1. Obtain the optimal $\{\mathbf{P}_i\}_{i=1}^m$ and $\{\mathbf{B}_i\}_{i=1}^m$ by solving the MFKS model Eq. 14.10 using Algorithm 14.1.
2. Compute the kernel mapping $\mathbf{K}_{\mathbf{z}_i} = \kappa(\mathbf{z}_i, \mathbf{X}_i)$ of $\{\mathbf{z}_i\}_{i=1}^m$.

Output:

$$\text{identity}(\mathbf{z}) \leftarrow \arg \max_{j\in\{1,\ldots,C\}} \left[\sum_{i=1}^{m}(\mathbf{K}_{\mathbf{z}_i} \cdot \mathbf{P}_i + \mathbf{B}_i)\right]_j$$

14.3.5 Convergence

To explore the convergence behavior of the proposed Algorithm 14.1, we first provide a lemma as follows.

Lemma 14.1 *For alternative optimization, when update one variable with other variables fixed, it will not increase the objective function value. Four claims with short proofs are given as follows*:

Claim 1. $J\left(\mathbf{P}_i^{(t)}, \mathbf{B}_i^{(t)}, \mathbf{F}^{(t)}, \alpha_i^{(t)}\right) \geq J\left(\mathbf{P}_i^{(t+1)}, \mathbf{B}_i^{(t)}, \mathbf{F}^{(t)}, \alpha_i^{(t)}\right)$

Proof When fix $\mathbf{B}_i^{(t)}$, $\mathbf{F}^{(t)}$, $\alpha_i^{(t)}$ and update $\mathbf{P}_i^{(t+1)}$, the objective function is convex *w.r.t.* $\mathbf{P}_i$. As shown in Eq. 14.14 by setting the derivative of the objective function *w.r.t.* $\mathbf{P}_i$ to be 0, then it is clear that the expression of *claim 1* holds.

Claim 2. $J\left(\mathbf{P}_i^{(t+1)}, \mathbf{B}_i^{(t)}, \mathbf{F}^{(t)}, \alpha_i^{(t)}\right) \geq J\left(\mathbf{P}_i^{(t+1)}, \mathbf{B}_i^{(t+1)}, \mathbf{F}^{(t)}, \alpha_i^{(t)}\right)$

Proof Similar to the proof of *claim 1*, the objective function becomes convex *w.r.t.* $\mathbf{B}_i$ when fix $\mathbf{P}_i^{(t+1)}$, $\mathbf{F}^{(t)}$, $\alpha_i^{(t)}$ and update $\mathbf{B}_i^{(t+1)}$. Then, *claim 2* is proven.

Claim 3. $J\left(\mathbf{P}_i^{(t+1)}, \mathbf{B}_i^{(t+1)}, \mathbf{F}^{(t)}, \alpha_i^{(t)}\right) \geq J\left(\mathbf{P}_i^{(t+1)}, \mathbf{B}_i^{(t+1)}, \mathbf{F}^{(t+1)}, \alpha_i^{(t)}\right)$

Proof When $\mathbf{P}_i^{(t+1)}$, $\mathbf{B}_i^{(t+1)}$, $\alpha_i^{(t)}$ are fixed, and update $\mathbf{F}^{(t)}$, the optimization problem becomes Eq. 14.16 which is convex *w.r.t.* $\mathbf{F}$. By setting the derivative of the objective function Eq. 14.16 *w.r.t.* **F** to be 0, its solution in Eq. 14.17 makes *claim 3* hold.

Claim 4. $J\left(\mathbf{P}_i^{(t+1)}, \mathbf{B}_i^{(t+1)}, \mathbf{F}^{(t+1)}, \alpha_i^{(t)}\right) \geq J\left(\mathbf{P}_i^{(t+1)}, \mathbf{B}_i^{(t+1)}, \mathbf{F}^{(t+1)}, \alpha_i^{(t+1)}\right)$

Proof As can be seen from Eq. 14.21, with $\mathbf{P}_i^{(t+1)}$, $\mathbf{B}_i^{(t+1)}$, $\mathbf{F}^{(t+1)}$ fixed, the update rule of $\alpha_i^{(t+1)}$ is obtained by setting the derivatives of objective function Eq. 14.20 *w.r.t.* α_i to be 0. Also, since the second-order derivative *w.r.t.* α_i is positive with the condition that $r > 1, \alpha_i > 0$, i.e., there is

$$\frac{\partial^2 J(\alpha_i, \mu)}{\partial \alpha_i^2} = r(r-1)\alpha_i^{r-2} Tr\left(\mathbf{F}^{\mathrm{T}} \mathbf{L}_i \mathbf{F}\right) > 0$$

Thus, the update rule Eq. 14.22 of α_i can make the objective function Eq. 14.20 decrease, and *claim 4* is proven. In summary, the convergence of the proposed Algorithm 14.1 is summarized as the following theorem.

Theorem 14.1 *The objective function Eq.* 14.10 *monotonically decreases until convergence after several iterations by using Algorithm* 14.1.

Proof Suppose the updated $\mathbf{P}_i^{(t)}$, $\mathbf{B}_i^{(t)}$, $\mathbf{F}^{(t)}$, and $\alpha_i^{(t)}$ are $\mathbf{P}_i^{(t+1)}$, $\mathbf{B}_i^{(t+1)}$, $\mathbf{F}^{(t+1)}$, and $\alpha_i^{(t+1)}$, respectively. In terms of the *claim 1*, *claim 2*, *claim 3,* and *claim 4* presented in Lemma 14.1, we have the following inequalities,

$$\begin{aligned} J\left(\mathbf{P}_i^{(t)}, \mathbf{B}_i^{(t)}, \mathbf{F}^{(t)}, \alpha_i^{(t)}\right) &\geq J\left(\mathbf{P}_i^{(t+1)}, \mathbf{B}_i^{(t)}, \mathbf{F}^{(t)}, \alpha_i^{(t)}\right) \\ &\geq J\left(\mathbf{P}_i^{(t+1)}, \mathbf{B}_i^{(t+1)}, \mathbf{F}^{(t)}, \alpha_i^{(t)}\right) \\ &\geq J\left(\mathbf{P}_i^{(t+1)}, \mathbf{B}_i^{(t+1)}, \mathbf{F}^{(t+1)}, \alpha_i^{(t)}\right) \\ &\geq J\left(\mathbf{P}_i^{(t+1)}, \mathbf{B}_i^{(t+1)}, \mathbf{F}^{(t+1)}, \alpha_i^{(t+1)}\right) \end{aligned}$$

Therefore, the Theorem 14.1 is proven.

14.3.6 Computational Complexity

We briefly analyze the computational complexity of the MFKS involving T iterations and m modalities. Before stepping into the learning phase, the complexity of computing the Laplacian matrices is $O(mN^3)$. In learning, each iteration involves three update steps, and the complexity in T iterations is $O(m^2N^2T)$. Hence, the computational complexity of our method is $O(mN^3) + O(m^2N^2T)$. Note that the Laplacian matrices are not involved in iterations and therefore the computational complexity of computing Laplacian matrix is $O(mN^3)$.

14.3.7 Remarks on Optimality Condition

The convergence of MFKS shown in Algorithm 14.1 is demonstrated in Theorem 14.1. For insight of the optimality condition, the objective function value of Eq. 14.10 over 20 iterations on E-nose dataset used in this chapter for six kinds of gases recognition is described in Fig. 14.1a. One can observe that after ten iterations, the algorithm can converge to a stable value. Furthermore, the classifier gap $\delta \mathbf{P}^{(t)}$ is introduced for optimality condition check. The classifier gap is defined as,

$$\delta \mathbf{P}^{(t)} = \sum_{i=1}^{m} \left\| \mathbf{P}_i^{(t)} - \mathbf{P}_i^{(t-1)} \right\|_{\mathrm{F}} \Big/ \left\| \mathbf{P}_i^{(t)} \right\|_{\mathrm{F}} \tag{14.24}$$

The convergence of $\delta \mathbf{P}^{(t)}$ over 20 iterations is described in Fig. 14.1b. It is clearly seen that model gap can quickly converge to a small and stable value after 10 iterations. Therefore, we can conclude that the optimality condition of the proposed MFKS model can be achieved during 10 iterations by using the proposed optimization Algorithm 14.1.

From the viewpoint of machine learning, the proposed semi-supervised MFKS model and algorithm can also be used for supervised learning by setting $\mathbf{F} = \mathbf{Y}$ in Eq. 14.10, such that the first two terms are kept for purely supervised learning, i.e.,

$$\min_{\mathbf{P}_i, \mathbf{B}_i} \sum_{i=1}^{m} \| \mathbf{Y} - \mathbf{K}_i \mathbf{P}_i - \mathbf{1}_N \mathbf{B}_i \|_{\mathrm{F}}^2 + \frac{\gamma}{2} \cdot \| \mathbf{P} \|_* \tag{14.25}$$

From Eq. 14.25, the supervised MFKS can be recognized as a multi-feature learning framework with low-rank constraint on the group classifier **P**.

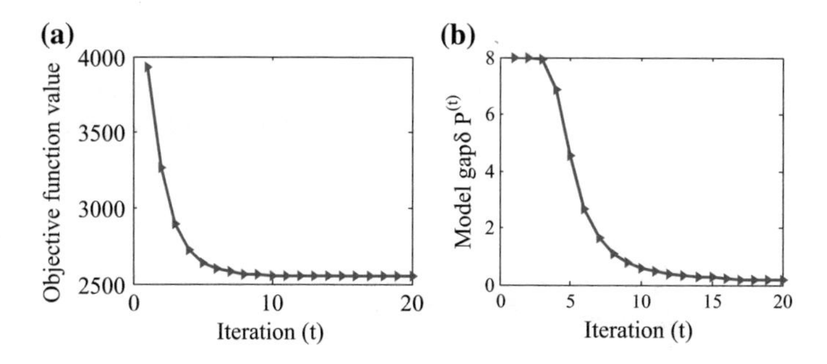

Fig. 14.1 Convergence analysis for optimality condition

14.4 Experiments on Drifted E-Nose Data

In this section, the recently released E-nose olfactory dataset with drift in UCI Machine Learning Repository [1, 25] is used to evaluate the proposed MFKS model and algorithm.

14.4.1 Description of Data

The artificial olfactory dataset contains 13,910 measurements (observation samples) of six kinds of gases including acetone, acetaldehyde, ethanol, ethylene, ammonia, and toluene at different concentration levels. The dataset is divided into ten batches versus time. The dataset was sampled by an E-Nose system with 16 gas sensors. For each sensor, eight kinds of features were extracted from each sensor (i.e., $m = 8$, in this chapter). We refer interested readers to as [1] for details on how to select the eight features. Figure 14.2 illustrates the normalized sensor response and the multi-feature formulation in this chapter. For each kind of feature, a 16-dimensional feature vector for each measurement is formulated as a result. The details of the dataset are presented in Table 11.1, which shows the number of samples for each gas.

14.4.2 Experimental Setup

For evaluating the proposed semi-supervised learning framework, we adopt the following experimental setting.

Semi-supervised setting: Take batch 1 as fixed training set and tested on batch K, $K = 2,\ldots,10$.

Description: In the semi-supervised setting, the batch 1 is used as labeled training data, while the data from batch K ($K = 2,\ldots,10$) *is used as unlabeled data for semi-supervised learning. Besides, we also randomly select* $L = 10, 20, 30, 40,$

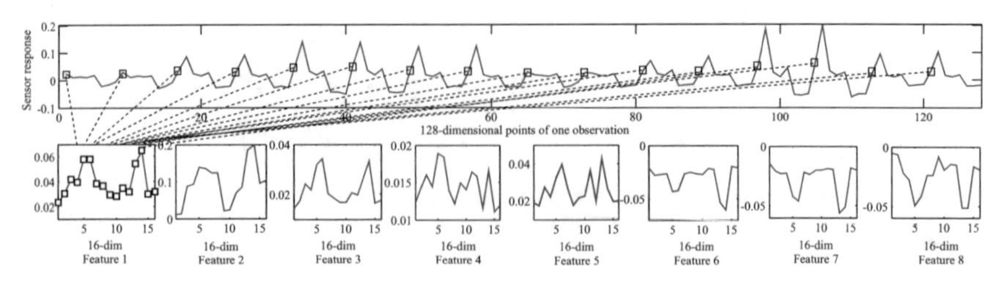

Fig. 14.2 Multi-feature representation of one observation in batch 1 with 16 sensors under Acetone. Each square denotes the first feature of each sensor

and 50 *labeled data from batch K* (K = 2,…,10), *respectively, and the remaining are still unlabeled for semi-supervised learning.*

14.4.3 Parameter Setting

In default, the RBF kernel function is used for kernel matrix computing. For finding out the best results, the kernel parameter σ is adjusted from the given set $\{2^{-4}, 2^{-3}, \ldots, 2^4\}$ and the regularization coefficient γ is adjusted from the given set $\{10^{-4}, 10^{-3}, \ldots, 10\}$. The regularization coefficient of graph manifold regularization λ is set as 1. The number of maximum iterations in MFKS model is set as 10 throughout the chapter. The raw sensor features are normalized appropriately into the interval (-1,1) for easier convergence in optimization.

14.4.4 Compared Methods

Following the semi-supervised experimental setting and parameter setting, we compare the proposed semi-supervised MFKS model with three SVM methods including RBF kernel-based SVM (SVM-rbf), geodesic flow kernel-based SVM (SVM-gfk), and combination kernel-based SVM (SVM-comgfk); two semi-supervised methods including RBF kernel-based manifold regularization (ML-rbf) and combination geodesic flow kernel-based manifold regularization (ML-comgfk) [19], and extreme learning machine (ELM) based on RBF activation function (ELM-rbf) [26].

14.4.5 Results and Analysis

– **Performance comparisons**

The recognition accuracies of all methods including SVM-rbf, SVM-gfk, SVM-comgfk, ML-rbf, ML-comgfk, ELM-rbf, and the proposed MFKS on the testing data from batch 2 to batch 10 have been reported in Table 14.1. From the results, we can see that the proposed MFKS outperforms other methods except for batch 7, batch 9, and batch 10, where SVM-comgfk and ML-comgfk show a comparative performance. The average recognition accuracies from batch 2 to batch 10 for MFKS (10) and MFKS (20) are 68.07% and 77.75%, which are superior to other methods. Note that for MFKS (L), the L denotes the number of labeled data leveraged from batch K (K = 2,…,10). From the comparisons, the effectiveness of the proposed MFKS is clearly demonstrated. This also implies that semi-supervised learning can be used to improve the drift-counteraction property of an E-Nose system.

– **Performance with increasing number *L***

In training process on batch 1, a few number L of labeled data from batch K ($K = 2,\ldots,10$) are leveraged for semi-supervised learning. Therefore, we discuss the performance with different number L of leveraged labeled data for each test batch, respectively. The recognition accuracies for each batch with different number L from 10 to 50 by using the proposed MFKS model are illustrated in Fig. 14.3. From the results, we observe that the performance is positively improved with increasing number L; however, when $L = 50$, the performance becomes weak. From the results, we can see the accuracy change with L. Specifically, we get that for $L = 10, 20, 30, 40$, and 50, the average recognition accuracy is 68.1, 77.8, 82.0, 87.0, and 84.4%, respectively. Note that the accuracies when $L = 10$ and 20 have been reported in Table 14.1.

14.5 Experiments on Modulated E-Nose Data

14.5.1 Description of Data

The temperature modulated E-Nose system is developed by our group. Generally, the system consists of three parts: voltage control (i.e., sensor temperature control), sensor array, and data acquisition. The printed circuit board (PCB) comprises of a gas sensor array (i.e., TGS2620 and TGS 2602), DC power supply interface, voltage control interface and data acquisition interface. We claim the data to be "modulated" due to that the sensors' heating voltage changes linearly from 3 to 5 V in period with a frequency of 20 MHz during the sampling process.

In sampling experiments, three gases including formaldehyde (HCHO), nitrogen dioxide (NO_2), and carbon monoxide (CO) are experimented by using our temperature-modulated E-Nose system, respectively. The modulated gas sensor response curves in steady state period for one observation of each gas are illustrated in Fig. 14.4. In feature extraction, 25 points are *uniformly sampled* from each sensor curve with 5000 points shown in Fig. 14.4. Therefore, for each observation, a 50-dimensional vector (two sensors) is obtained. Motivated by the proposed multi-feature learning, the 50-dimensional vector is represented by five feature modalities, for each the dimension is ten according to Fig. 14.5.

Totally, the number of samples for HCHO, NO_2, and CO is 100, 113, and 96, respectively. Note that for each feature type, the input dimension is 12 consisting of ten points for gas sensors in Fig. 14.5 and two extra values for the ambient temperature and humidity in each observation. The temperature range changes from 10 to 35 °C, and the humidity range is from 40 to 70%.

Table 14.1 Comparison of recognition accuracy with semi-supervised setting

Batch ID	Batch 2	Batch 3	Batch 4	Batch 5	Batch 6	Batch 7	Batch 8	Batch 9	Batch 10	Average
SVM-rbf	74.36	61.03	50.93	18.27	28.26	28.81	20.07	34.26	34.47	38.94
SVM-gfk	72.75	70.08	60.75	75.08	73.82	54.53	55.44	69.62	41.78	63.76
SVM-comgfk	74.47	70.15	59.78	75.09	73.99	54.59	55.88	**70.23**	41.85	64.00
ML-rbf	42.25	73.69	75.53	66.75	77.51	54.43	33.50	23.57	34.92	53.57
ML-comgfk	80.25	74.99	78.79	67.41	77.82	**71.68**	49.96	50.79	**53.79**	67.28
ELM-rbf	70.63	66.44	66.83	63.45	69.73	51.23	49.76	49.83	33.50	57.93
MFKS (10)	**80.79**	**80.64**	**86.75**	**79.14**	**80.69**	36.19	**68.30**	63.04	37.10	**68.07**
MFKS (20)	**85.45**	**77.96**	**88.65**	**83.61**	**89.38**	**68.80**	**84.67**	**78.66**	**42.54**	**77.75**

The boldface type denotes the best performance.

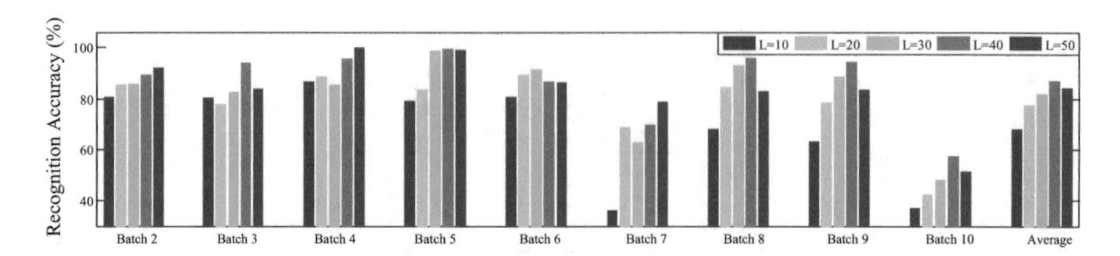

Fig. 14.3 Recognition accuracy with semi-supervised setting by using the proposed MFKS model with different number L of labeled data from batch K (K = 2,…,10)

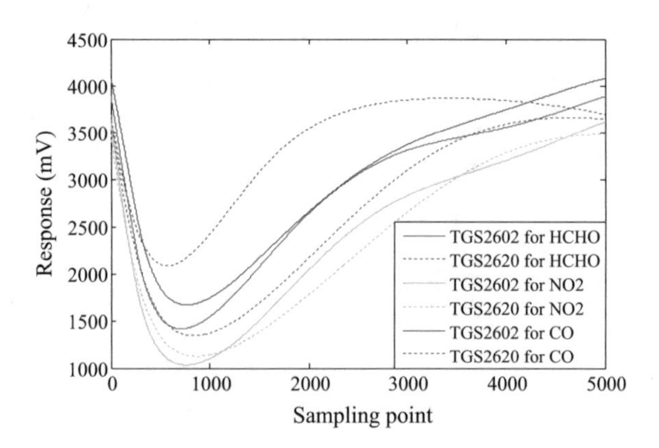

Fig. 14.4 Modulated steady state response curve for one observation under HCHO, NO_2, and CO, respectively

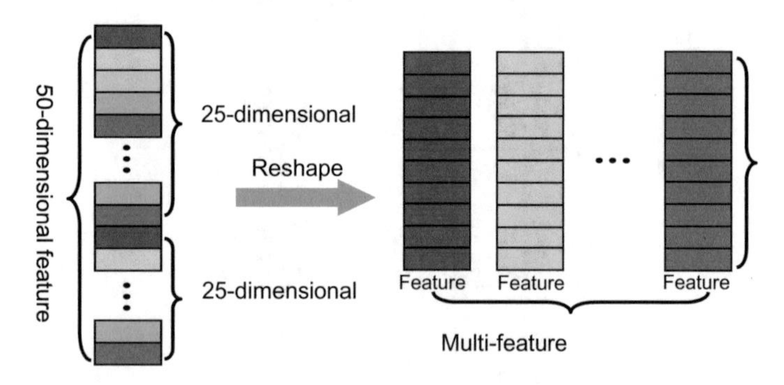

Fig. 14.5 Multi-feature representation of the extracted feature points for each observation

14.5.2 Experimental Setup

In this experiment, we adopt the following experimental setting for evaluating different methods.

Experimental setting: for each class, different number L of labeled samples is randomly selected from the dataset of each class for supervised learning. In this chapter, $L = 1, 2, 3, \ldots, 10$ is considered. The remaining samples are recognized to be unlabeled data. We have explored the performance improvement with increasing number L of labeled data.

For this modulated artificial olfactory E-Nose data, we have compared with two popular classifiers such as support vector machine (SVM) and extreme learning machine (ELM) [27]. Due to that SVM and ELM belong to single-feature learning framework, we train SVM or ELM on feature type i ($i = 1,\ldots,m$) and obtain a classifier f_i. For fair comparison, the final result of SVM or ELM is calculated by $1/m \cdot \sum_{i=1}^{m} f_i(\mathbf{x})$, which implies that each feature makes equal contribution.

14.5.3 Parameter Setting

For ELM, two parameters including penalty coefficient C and the number H of hidden neurons are involved. Therefore, for obtaining the best performance, H is tuned from the set $\{10, 10^2, 10^3, 10^4\}$ and C is tuned from the set $\{10^{-5}, 10^{-4}, \ldots, 10^{10}\}$. For SVM, two parameters including penalty coefficient C and kernel parameter σ are involved. For pursuit of the best performance, both C and σ are adjusted from the set $\{2^{-8}, 2^{-7}, \ldots, 2^{8}\}$. For MFKS, the kernel parameter σ is adjusted from the set $\{2^{-5}, 2^{-3}, \ldots, 2^{5}\}$ and the regularization coefficient γ is adjusted from the set $\{10^{-8}, 10^{-7}, \ldots, 10\}$. The coefficient λ of graph manifold regularization is set as 1. Note that the best results for each method are reported in this chapter.

14.5.4 Results and Analysis

With the temperature-modulated E-Nose data, experimental setup and the parameter setting presented above, the average recognition accuracies of HCHO, NO_2, and CO for all methods trained on different number L of labeled data for each class are reported in Table 14.2. Note that SVM (*sl*), ELM (*sl*), and MFKS (*sl*) denote purely *supervised learning*, while MFKS (*ssl*) denotes *semi-supervised learning* with unlabeled data considered. For MFKS (*sl*), the model in Eq. 14.25 is used. From the results, we can observe that the proposed MFKS model and algorithm outperform SVM and ELM to a large extent for supervised learning. The average result of

Table 14.2 Comparisons of recognition accuracy (%) with different number of labeled data in training set

Method	$L = 1$	$L = 2$	$L = 3$	$L = 4$	$L = 5$	$L = 6$	$L = 7$	$L = 8$	$L = 9$	$L = 10$	Average
SVM (*sl*)	68.69	75.91	76.73	80.4	78.64	78.42	78.61	79.16	79.15	83.94	77.97
ELM (*sl*)	72.94	77.69	78.00	83.03	82.86	82.96	82.99	84.63	85.39	88.96	81.95
MFKS (*ssl*)	66.66	78.87	**83.33**	**88.88**	85.37	85.56	86.45	85.61	86.52	87.45	83.47
MFKS (*sl*)	**75.16**	**81.51**	82.00	83.50	**88.77**	**88.65**	**91.66**	**86.66**	**89.36**	**93.54**	**86.08**

The boldface type denotes the best performance.

MFKS in semi-supervised setting is 83.47% which is higher than SVM (77.97%) and ELM (81.95%), but lower than that of MFKS in supervised learning (86.08%).

14.6 Summary

In this chapter, we aim at proposing an integrated learning framework that tends to learn multiple features and multiple sub-classifiers simultaneously, such that the "hard" learning evolves to "soft" learning and achieves a cooperative learning between features and classifiers. Specifically, a multi-feature joint semi-supervised learning with kernel mapping is proposed, that is the so-called MFKS. The proposed model, optimization algorithm, convergence, and complexity have been formulated and discussed in this chapter. For further validating the effectiveness of MFKS, two experiments based on an existing large-scale artificial olfactory data with sensor drift of 36 months and an existing small-scale temperature-modulated E-Nose data are conducted, respectively. Experimental results on two existing artificial olfactory datasets demonstrate that the proposed MFKS model outperforms other algorithms in recognition performance.

References

1. A. Vergara, S. Vembu, T. Ayhan, M.A. Ryan, M.L. Homer, Chemical gas sensor drift compensation using classifier ensembles. Sens. Actuators B: Chem. **166–167**, 320–329 (2012)
2. M. Holmberg, F.A.M. Davide, C.D. Natale, A.D. Amico, F. Winquist, I. Lundström, Drift counteraction in odour recognition applications: lifelong calibration method. Sens. Actuators B: Chem. **42**(3), 185–194 (1997)
3. S.D. Carlo and M. Falasconi, Drift correction methods for gas chemical sensors in artificial olfaction systems: techniques and challenges. Adv. Chem. Sensors, 305–326 (2012)
4. G. Fattoruso, S. De Vito, M. Pardo, F. Tortorella, G. Di Francia, A semi-supervised learning approach to artificial olfaction. Lect. Notes Electr. Eng. **109**, 157–162 (2012)

5. S. De Vito, G. Fattoruso, M. Pardo, F. Tortorella, G. Di Francia, Semi-supervised learning techniques in artificial olfaction: a novel approach to classification problems and drift counteraction. IEEE Sens. J. **12**(11), (Nov 2012)
6. O. Chapelle, A. Zien, B. Sholkopf, *Semi-Supervised Learning* (MIT Press, Boston, MA, 2006)
7. D. Zhou, O. Bousquet, T. Navin Lal, J. Weston, and B. Schölkopf, Learning with Local and Global Consistency, NIPS, 321–328 (2004)
8. Y. Luo, D. Tao, B. Geng, C. Xu, Manifold regularized multitask learning for semi-supervised multilabel image classification. IEEE Trans. Image Process. **22**(2), 523–536 (2013)
9. Y. Yang, Z. Ma, A.G. Hauptmann, N. Sebe, Feature selection for multimedia analysis by sharing information among multiple tasks. IEEE Trans. Multimedia **15**(3), 661–669 (2013)
10. S. Roweis, L. Saul, Nonlinear dimensionality reduction by locally linear embedding. Science **290**(22), 2323–2326 (2000)
11. J. Tenenbaum, V. Silva, J. Langford, A global geometric framework for nonlinear dimensionality reduction. Science **290**(22), 2319–2323 (2000)
12. M. Belkin, P. Niyogi, Laplacian eigenmaps for dimensionality reduction and data representation. Neural Comput. **15**(6), 1373–1396 (2003)
13. S. Yan, D. Xu, B. Zhang, H.J. Zhang, Q. Yang, S. Lin, Graph embedding and extensions: a general framework for dimensionality reduction. IEEE Trans. Pattern Anal. Mach. Intell. **29** (1), 40–51 (2007)
14. T. Xia, T. Mei, Y. Zhang, Multiview spectral embedding. IEEE Trans. Syst. Man Cybern. Part B **40**(6), 1438–1446 (2010)
15. A. Ziyatdinov, S. Marco, A. Chaudry, K. Persaud, P. Caminal, A. Perera, Drift compensation of gas sensor array data by common principal component analysis. Sens. Actuators B: Chem. **146**(2), 460–465 (2010)
16. S.D. Carlo, M. Falasconi, E. Sanchez, A. Scionti, G. Squillero, A. Tonda, Increasing pattern recognition accuracy for chemical sensing by evolutionary based drift compensation. Pattern Recognit. Lett. **32**(13), 1594–1603 (2011)
17. L.J. Dang, F. Tian, L. Zhang, C. Kadri, X. Yin, X. Peng, S. Liu, A novel classifier ensemble for recognition of multiple indoor air contaminants by an electronic nose. Sens. Actuators, A **207**, 67–74 (2014)
18. H. Liu, R. Chu, Z. Tang, Metal oxide gas sensor drift compensation using a two-dimensional classifier ensemble. Sensors **15**(5), 10180–10193 (2015)
19. Q. Liu, X. Li, M. Ye, S.S. Ge, X. Du, Drift compensation for electronic nose by semi-supervised domain adaption. IEEE Sens. J. **14**(3), 657–665 (2014)
20. L. Zhang, D. Zhang, Domain adaptation transfer extreme learning machine. Proc. Adapt. Learn. Optim. **3**, 103–119 (2015)
21. L. Zhang, D. Zhang, Domain adaptation extreme learning machines for drift compensation in E-Nose systems. IEEE Trans. Instrum. Meas. **64**(7), 1790–1801 (2015)
22. L. Martin, L. Amy, Unsupervised feature learning for electronic nose data applied to bacteria identification in blood. NIPS Workshop on Deep Learning and Unsupervised Feature Learning (2011)
23. Q. Liu, X. Hu, M. Ye, X. Cheng, F. Li, Gas recognition under sensor drift by using deep learning. Int. J. Intell. Syst. **30**(8), 907–922 (2015)
24. G. Liu, Z. Lin, S. Yan, J. Sun, Y. Yu, Y. Ma, Robust recovery of subspace structure by low-rank representation. IEEE Trans. Pattern Anal. Mach. Intell. **35**(1), 171–184 (2013)
25. http://archive.ics.uci.edu/ml/datasets/Gas+Sensor+Array+Drift+Dataset+at+Different+Concentrations
26. D.A.P. Daniel, K. Thangavel, R. Manavalan, R.S.C. Boss, ELM-based ensemble classifier for gas sensor array drift dataset. Adv. Intell. Syst. Comput. **246**, 89–96 (2014)
27. G.B. Huang, H. Zhou, X. Ding, R. Zhang, Extreme learning machine for regression and multiclass classification. IEEE Trans. Syst. Man, Cybern. B, Cybern **42**(2), 513–529 (2012)

Part IV
E-Nose Disturbance Elimination: Challenge III

Chapter 15
Pattern Recognition-Based Interference Reduction

Abstract Metal oxide semiconductor (MOS) sensors have some intrinsic flaw of high susceptibility to background interference which would seriously destroy the specificity and stability of electronic nose in practical application. This chapter presents an on-line counteraction of unwanted odor interference based on pattern recognition for the first time. Six kinds of target gases and four kinds of unwanted odor interferences were experimentally studied. First, two artificial intelligence learners including a multi-class least square support vector machine (learner-1) and a binary classification artificial neural network (learner-2) are developed for discrimination of unwanted odor interferences. Second, a real-time dynamically updated signal matrix is constructed for correction. Finally, an effective signal correction method was employed for E-nose data. Experimental results in the real case-studies demonstrate the effectiveness of the presented model in E-nose based on MOS gas sensors array.

Keywords Electronic nose · Sensor array · Odor interference · Counteraction Pattern recognition

15.1 Introduction

The ideal gas sensor would exhibit reliability, robustness, sensitivity, selectivity, and reversibility [1]. Gas sensors, based on the chemical sensitivity of metal oxide semiconductors (MOS), are readily available commercially [2, 3]. They have been more widely used to make arrays for odor measurement than any other class of gas sensors.

However, MOS sensors have also some intrinsic flaws which can be concluded as four kinds of background interferences in electronic nose. First, they are susceptive to environmental parameters such as temperature which has been called direct interferences [4] for the electronic nose data in practical applications. Second, drift is defined as the temporal shift of sensor response under constant physical and chemical conditions [5]. In addition, MOS sensors have instrument-related signal shift and baseline differences in measurement data. That is, the sensor of completely

L. Zhang et al., *Electronic Nose: Algorithmic Challenges*,
https://doi.org/10.1007/978-981-13-2167-2_15

identical type has different response to analytes with the same condition (concentration, temperature, and humidity) [6]. Finally, MOS gas sensors have high susceptibility to unwanted odor interferences in human surroundings (e.g., perfume, ethanol, fruit smell, and toilet water). The kind of background odor interferences will directly trouble the metal oxide semiconductor sensors in the electronic nose system with high strength in practical environmental detection. The most serious problem is the false discrimination and concentration estimation of the measured target gases when the unwanted odors exist.

For solutions of the first three kinds of interferences (environmental fluctuation, sensor drift caused by aging, and signal shift caused by sensor replacement), fortunately, many researchers have proposed different methods. Counteraction methods for background interferences caused by environmental fluctuation (the 1st interference) have been presented in [4, 7–9]. Different compensation models of long-term sensor drift (the 2nd interference) have been fully investigated and tried in [5, 10–14]. For the signal shift (the 3rd interference) between two identical sensors, calibration transfer models [15–17] have been successfully used for sensor standardization in an electronic nose. However, there is no research demonstrating the possible on-line counteraction of unwanted odor interference (the 4th interference) in an E-nose. That is, an E-nose will lose effectiveness when exposed to surroundings with unwanted background odor (e.g., perfume). Therefore, it is meaningful to present an on-line counteraction model of the 4th interference and eliminate the sensor "fault" of an E-nose caused by odor interferences in real-time ambient air monitoring. The novelty of this chapter is that we propose an on-line counteraction model of background odor interferences based on artificial intelligence learners for real-time environmental monitoring for the first time and preliminarily resolves the problem of odor interferences. In addition, the background odor interferences are various in the real world; therefore, we seemed all unwanted odor interferences as one class in artificial intelligence learners and solved the variety problem of unwanted odor. The robust on-line counteraction model can be approximately divided into two parts: "fault" recognition and "fault" counteraction.

A number of pattern recognition algorithms have been widely used in E-nose such as artificial neural network (ANN)-based methods [18, 19], heuristic and bio-inspired methods [20, 21], linear multivariate techniques and discrimination analysis [22, 23], and support vector machine (SVM)-based methods [24, 25]. Literature comparisons demonstrate that SVM and ANN are better choices in classification than any other methods due to their higher accuracy and stronger robustness [26, 27]. Besides, the previous research [25] also demonstrates the strong nonlinearity of electronic nose data. Thus, SVM and ANN were selected as the pattern recognition methods for fault recognition. Therefore, two learners (learner-1 and learner-2) were studied with subsequent counteraction, respectively. Learner-1 is based on multi-class mechanism and learner-2 is based on binary classification rule. For clarity, each gas is seemed as an independent class and multiple classes including an unwanted odor interferences class were constructed in learner-1. Consider that the key study of this chapter is counteraction of odor interferences, we induce the multiple target gases classes into one class, and the odor interferences were induced as the second class in learner-2.

15.2 Materials and Methods

15.2.1 Experimental Platform

The electronic nose system and experimental setup developed in this chapter were described previously in Chap. 4. In preparations of each gas, formaldehyde, benzene, and toluene are volatilized in a gas-collecting bag with injection of their liquor and pure nitrogen (N_2) for dilution, respectively; while CO, NH_3, and NO_2 are diffused in a gas-collecting bag with injection of their standard gas and pure nitrogen (N_2) for dilution, respectively. In experiments, the target gas is injected into the chamber with different time length through a flow-controlled pump connected with the gas-collecting bag. The experimental target temperature was set as 15, 20, 25, 30, and 35 °C; the relative humidity was set as 40, 60, and 80% RH. The total measurement cycle time for each experiment was set to 20 min, i.e., 2 min for baseline, 8 min for steady state response, and 10 min for gas bleed. Each sample contains 350 sampling points.

15.2.2 Sensor Signal Preprocessing

Signal preprocessing in this work is used for smooth filtering and normalization of the continuous sampling points. Set the length of the smooth filter as n, the length of the observation signal vector $\mathbf{S}$ is N, and the filtered and normalized signal vector $\mathbf{X}$ can be obtained by

$$X(i) = \frac{\sum_{l=i}^{i+n-1} S(l) - \max\{S(i),\ldots,S(i+n-1)\} - \min\{S(i),\ldots,S(i+n-1)\}}{n-2}, \quad i = 1,\ldots,N-n+1 \tag{15.1}$$

$$X = X/4095 \tag{15.2}$$

The normalization of formula (15.2) is that sensor responses were simply divided by 4095. It is worthy noting that the digit of 4095 (i.e., $2^{12} - 1$) is the maximum value of the 12-bit A/D output for each sensor.

15.2.3 E-Nose Data Preparation

This section will introduce five datasets, wherein dataset 1 is used for classification model learning and testing; dataset 2 and dataset 3 are used to reflect a real-world case study and validate the background elimination model in our E-nose.

- *Dataset 1 for classification model training*

This chapter aims to build a multi-classifier with seven objects: formaldehyde, benzene, toluene, carbon monoxide, ammonia, nitrogen dioxide, and odor interferences. Four kinds of familiar odor interferences (e.g., perfume, ethanol, fruit smell (orange), and toilet water) were tested. Note that the toilet water that can be used to drive mosquito is composed of fragrance and alcohol. In classification model, the whole dataset was divided into train data and test data. The train samples were selected from the whole dataset by using KS algorithm [17], and the remaining samples were used as test data which has no relation with classification model establishment and only taken as performance validation of the classifier. The number of train samples for formaldehyde, benzene, toluene, CO, NH_3, NO_2, unwanted odor interference (ethanol and orange) are 125, 48, 44, 38, 40, 25, 40 (26 and 14); and the numbers of test samples are 63, 24, 22, 20, 20, 12, 19 (13 and 6). The concentrations of target gases for each sample are different. For formaldehyde, the concentration range is 0–10 ppm; for benzene, toluene, NH_3, and NO_2, the concentration range is 0–5 ppm; for CO, the concentration range is 1–60 ppm. The specific concentrations of the six contaminants for each experimental sample in different combinations of temperature and humidity (i.e., (15, 60) represents the experimental temperature 15 °C and relative humidity 60% RH) are described in Table 5.1 of Chap. 5. It is worth noting that the reason why we only use ethanol and orange for classifier design is to verify the robustness of the proposed on-line counteraction model when exposed to perfume and toilet water in real-time application.

- *Dataset 2 of real-time interference elimination without target gas*

The dataset 2 was obtained when the electronic nose was exposed to the environment only with unwanted odor interferences, that is, no target gases were presented. An observation vector with length of 2480 points for each sensor was obtained in continuous sampling way. This dataset is developed under two odor interferences, respectively. In detail, we present the approximation positions for each object as follows. Perfume exists in two approximated regions 95–308 and 709–958; toilet water exists in two approximated regions 1429–1765 and 2056–2265; the approximated interference regions have been illustrated by four rectangular windows as shown in Fig. 15.1a.

- *Dataset 3 of real-time odor interferences under target gas*

For the general cases of counteraction, dataset 3 without target gas may be not enough for validation of the proposed interference elimination model. Therefore, the condition when electronic nose was placed in some concentration of the target gas should also be studied. Similarly to dataset 2, a dataset 3 with length of 2480 points for each sensor was also obtained in continuous sampling way. This dataset is developed under reference formaldehyde gas and four odor interferences, respectively. In detail, we present the approximation positions for each object as follows. Formaldehyde exists in three approximated regions 102–250, 719–880, and 1380–1580; ethanol exists in region 260–410; toilet water exists in region 881–1064;

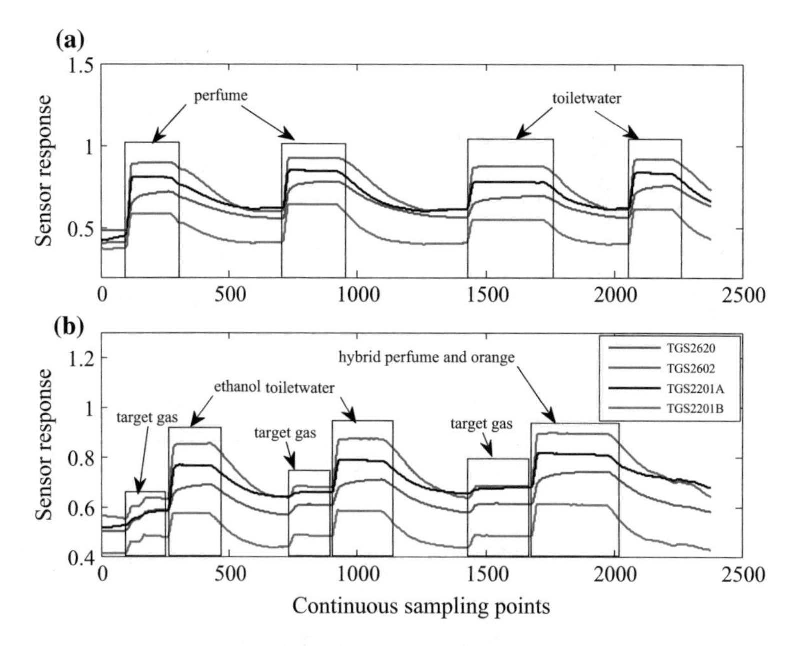

Fig. 15.1 Continuous real-time sampling curves of four sensors with odor interferences. **a** Denotes the observation signal for each sensor without target gases; **b** denotes the observation signal under some concentration of reference target gas. The rectangular windows illustrate the actual regions of target gas and odor disturbances in the two experiments

a hybrid interference of perfume and orange is located in region 1599–1899. The approximated target gas and interference regions have been illustrated by six rectangular windows in Fig. 15.1b.

15.2.4 Feature Selection of Abnormal Odor

Feature selection for representing the texture of unwanted smell being counteracted is a critical step for subsequent pattern analysis. Due to that the first six analytes are target gases in our project; thus, only the steady state responses of an array were selected as the texture of each target gas for each sample. However, for counteraction of odor interference, only steady state response is not enough because of the possible erroneous decision in adsorption and desorption process. Therefore, four features including adsorption, transient response point reflecting the sensor dynamics of the increasing/decaying transient, steady state response and desorption points were selected to express the texture of odor interference. Specifically, the adsorption and desorption positions are located at the 1/2 of the ascent stage and descent trajectory, respectively. Besides, the position of transient response is located at the maximum position of exponential moving average (*ema*) transform calculated by Muezzinoglu et al. [28]

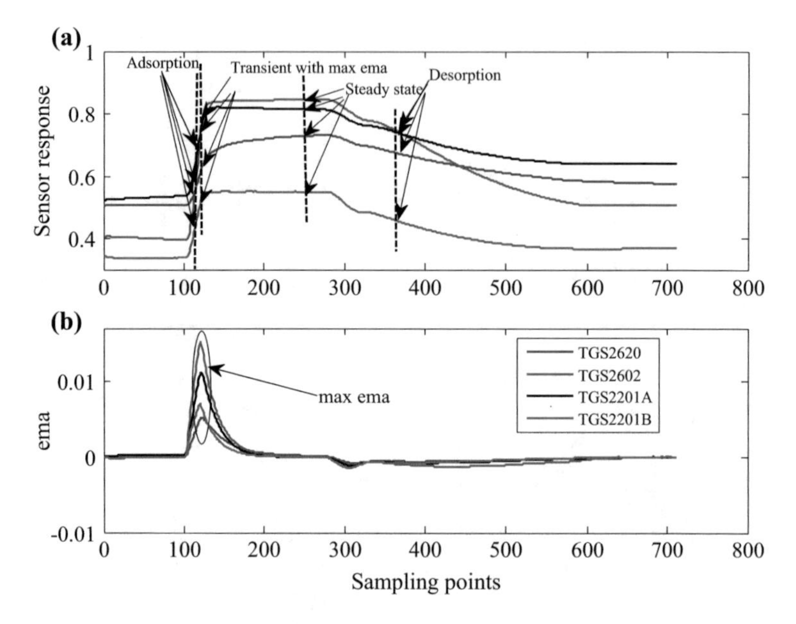

Fig. 15.2 Feature selection of odor interferences; **a** describes the specific positions (adsorption, transient response, steady state response, and desorption) of feature selection; **b** presents the *ema* curve with $\alpha = 0.05$ and determines the position of the maximum *ema* as the position of transient response

$$y[k] = (1 - \alpha) \cdot y[k-1] + \alpha \cdot (r[k] - r[k-1]), \quad k = 1, \ldots, M \tag{15.3}$$

where $\boldsymbol{r}$ denotes the sensor observation vector, $\boldsymbol{y}$ denotes the *ema* vector, α is the smooth parameter within [0, 1], and M denotes the length of the observation vector.

The reason for these selections is due to the considerations of the necessary sensitivity to unwanted odors, promptness, and accuracy of the counteraction model. However, the adsorption and desorption positions (1/2) are not arbitrary, and they can be adjusted as requirement of model sensitivity. Figure 15.2a illustrates the dynamic sensor response when exposed to perfume. The approximated positions for the selected features have been presented by vertical lines and arrows. Figure 15.2b presents the *ema* curve with $\alpha = 0.05$. The position for transient response in Fig. 15.2a is the same position of the maximum *ema* in Fig. 15.2b. The maximum *ema* of each sensor has been highlighted in an ellipse.

15.2.5 Genetic Crossover Operator for Solution of Uneven Features

In support vector multi-class problem and artificial neural network binary classification problem, uneven number of features for each class will lead to classification tendency. That is, that class with fewer samples can be easily discriminated as the class with more features, so that the accuracy and robustness of the classifier will degrade. Inspired by genetic algorithm (GA) [29], this chapter introduces a new feature generation method based on genetic crossover operator to make the number of features for each class in the train set of LSSVM equilibrium. In genetic crossover operation, parts of genes in two pairs of chromosomes exchange in a certain way and two new individuals can be produced. The arithmetic crossover operator adopted in the present study is shown by

$$\begin{cases} p_1' = p_1 \cdot \text{rand} + p_2 \cdot (1 - \text{rand}) \\ p_2' = p_1 \cdot (1 - \text{rand}) + p_2 \cdot \text{rand} \end{cases} \tag{15.4}$$

where p_1 and p_2 denote the parent features, p_1' and p_2' represent the new child features, and *rand* denotes a random value within [0, 1].

For each generation, we randomly select two parent features and get two new children features, and repeat this step until the train features for each class are balanced. To check up the characteristic of distribution between the new children features and the parent features space, the Euclidean distance d_i between the ith new child sample p_i' and the *center* of the parent features for each class was calculated by

$$d_i = \left\| p_i' - \text{center} \right\|_2 \tag{15.5}$$

The center of the parent features for each class can be calculated by

$$\text{center} = 1/m \cdot \sum_{i=1}^{m} X_i \tag{15.6}$$

Through comparison with the tolerated *threshold* which was defined as the maximum Euclidean distance between each parent sample p_i and the center of parent features.

$$\text{threshold} = \max_i \{ \|p_i - \text{center}\|_2 \}, \quad i = 1, \ldots, m \tag{15.7}$$

The trade-off of the new child sample p_i' for each class can be shown by

$$\text{if } d_i \leq \text{threshold}, \text{accept } p_i'; \text{else, refuse } p_i' \tag{15.8}$$

15.2.6 Description of Learner-1 Under Multi-class Condition

SVM, as a novel machine learning method-based statistical theory, has advantages of strong robustness, dimension insensitivities, and structural risk minimum. Consider the high complexity of the optimization solution of SVM based on inequality constraint, the least square support vector machine (LSSVM) algorithm which transforms the inequality constraint into equality constraint by introducing a slack variable is adopted for multi-class classification (learner-1) in this chapter. ANN is developed for a binary classification problem (learner-2).

- *Multi-class LSSVM*

Learner-1 is designed for counteraction under a multi-class condition [25]. That is, the unwanted odors were discriminated in a complex multi-class problem. Therefore, LSSVM was used for consideration of its strong classification ability. Multi-class problem has often been transformed into a binary classification problem. Two methods for commonly solving multi-class problems are "one-against-all (OAA)" and "one-against-one (OAO)." OAO strategy has become a better recommendation choice when the number of classes $k \leq 10$. Therefore, OAO strategy was selected for $k = 7$ classes problem in this presented study. That is, $k(k - 1)/2 = 21$ binary classifiers would be designed for final decision based on a simple voting scheme. The class with the most polls would be the final decision.

- *Nonlinear kernel function*

In most cases, two classes cannot be linearly separated. In order to make the linear learning machine work well in nonlinear cases, the original input space can be mapped through a nonlinear kernel function into some higher-dimensional feature space where the training set is linearly separable. In general, any positive semi-definite functions that satisfy the Mercer's condition can be kernel functions. Some most widely used kernels contain linear kernel, polynomial kernel, Gaussian radial basis function (RBF), and sigmoid kernel. In this study, RBF kernel shown below has been used for high dimension transformation.

$$K(x_i, x) = \exp(-\|x - x_i\|^2/\sigma^2) \tag{15.9}$$

where σ_2 is kernel parameter. For each binary classifier, the nonlinear decision function is shown as

$$f(x) = \mathrm{sign}\left[\sum_{i=1}^{l} \alpha_i \cdot K(x_i, x) + b\right] \tag{15.10}$$

where α and b are trained by LSSVM.

- *Tuning of LSSVMs' parameters*

LSSVMs have parameters to be tuned, which influence their performance. They are regularization parameters γ and the kernel parameters σ^2. As multi-class problem, each classifier has a pair of γ and σ^2; thus, 21 pairs of tuning parameters were obtained through simulated annealing (SA) optimization and a grid-search method based on a minimum leave-one-out (LOO) cross-validation error as cost function.

15.2.7 Description of Learner-2 Under Binary Classification Condition

For comparison with learner-1 under a multi-class condition, learner-2 is designed through a simple binary classification with only two classes (target gases class and odor interferences class). That is, we combine the six kinds of target gases together and recognize them as the same class. For convenient analysis, labels "000" and "100" are assigned to target gases class and odor interferences class, respectively. For such binary classification, a two-hidden-layered back-propagation neural network (BPNN) was trained and tested in dataset 1. In the present work, 10 neurons in each hidden layer were used. The training goal for BP convergence is set as 0.05. The structure of the ANN is $6 \times 10 \times 10 \times 3$. The active functions of the hidden layers ant output layer are selected as log-sigmoid and pure linear function. The log-sigmoid function can be shown by

$$f(x) = 1/(1 + \exp(-x)) \tag{15.11}$$

The outputs need a further process of 0 and 1 through a simple comparison with the threshold 0.5. It can be seen that learner-2 is easier to understand and be implemented than learner-1. Learner-2 also implies that any observation signal which is not correlated with target gases would also be recognized as odor interference. Therefore, learner-2 is still reliable to the unwanted interferences not presented in this work.

15.2.8 Adaptive Counteraction Model

- *Dynamical target signal matrix storage and updating*

Before signal correction, a dynamical target signal matrix **P** with size of $m \times n$ would be first built, where m is the dimension of sensor array and n is the length of **P**. The matrix **P** is generated in real-time E-nose application. The storage and updating process of **P** can be illustrated as follows

Step 1: Set a null matrix **P** with size of $m \times n$.
Step 2: A group of sensor signal column vector **x** in an E-nose is obtained.
Step 3: Analyze the **x** using the well-trained learner and a class label T of **x** is obtained from the learner.
Step 4: If $T = 0$ (target gas), **x** is stored into **P**, then go to step 5; else abandon **x**, return to step 2.
Step 5: If **P** is full ($\mathbf{x}_1$, $\mathbf{x}_2$, …, $\mathbf{x}_n$ have been fully stored), then the new $\mathbf{x}_{n+1}$ with $T = 0$ will hold the position of $\mathbf{x}_n$, the $\mathbf{x}_n$ will hold the position of $\mathbf{x}_{n-1}$,

the $\mathbf{x}_{n\text{-}1}$ will hold the position of $\mathbf{x}_{n-2}$, …, the $\mathbf{x}_2$ will hold the position of $\mathbf{x}_1$, and $\mathbf{x}_1$ will be erased.

Step 6: Return to step 2, a dynamical target signal matrix **P** is then produced and updated on-line

- *On-line signal correction*

On the basis of the classification learner-1 and learner-2 presented in this study, the on-line correction model of odor interferences to MOS sensors can be illustrated as

$$y_i = \begin{cases} f(x_i), & \text{if } x_i \in \text{unwanted odor disturbances} \\ x_i, & \text{if } x_i \notin \text{unwanted odor disturbances} \end{cases}, \quad i = 1, 2, 3, 4 \quad (15.12)$$

where x_i denotes the real-time observation response of the *i*th sensor (four MOS sensors presented), y_i represents the response of the *i*th sensor after counteraction of interference, and $f(\cdot)$ defines the counteraction function which can be presented as a linear model $f(\boldsymbol{x}) = h \cdot \boldsymbol{x}$ for simplification. The coefficient h ($0 < h \leq 1$) can be defined in terms of the current signal $\boldsymbol{x}_{i,curr}$ (with interference) and the previous signal vector $\mathbf{P}_{i,pre}$ of n sampling points with the nearest time interval (without interference) as follows

$$h_i = \min_n P_{i,pre} / x_{i,curr}, \quad i = 1, 2, 3, 4 \quad (15.13)$$

Note that the length of the previous signal vector $\mathbf{P}_{i,\ pre}$ is n and the sampling frequency is 1 Hz. The length of n is set to 50 in this chapter, which can be adjusted as requirement of actual application. For clear understanding of the background counteraction model, the structure of the presented method has been illustrated in Fig. 15.3.

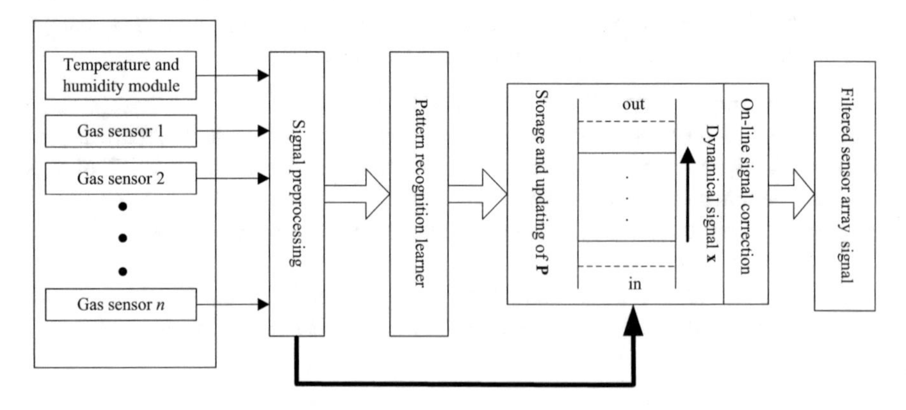

Fig. 15.3 Structure of the presented background interferences counteraction method

All algorithms for interference counteraction in this chapter were implemented in MATLAB 2009a, operating on a laboratory computer equipped with Inter Core (TM) i3 CPU 530, 2.93 GHz processors, and 2 GB of RAM. The LSSVM classification was operated using LS-SVMlab Toolbox version 1.8 [36].

15.3 Results and Discussion

15.3.1 Recognition Accuracy of Learner-1 and Learner-2

The LSSVM classifier (learner-1) was trained and tested on dataset 1 with genetic crossover operator for solution of uneven features including seven classes: formaldehyde, benzene, toluene, CO, NH_3, NO_2, and interferences. Totally, 875 (125 × 7) training samples were obtained. Thus, 21 binary classifiers have been designed based on LOO strategy. In LSSVM training process, we run the program 20 times on the train set and validate the classifier using the test set. Table 15.1 presents the average classification accuracy (ACA), standard deviation (±), and maximum classification accuracy (MCA) of the 20 runs including train accuracy and test accuracy. From the low standard deviation of the train data and test data, LSSVM classifier has a good stability. The designed classifier has a stable recognition of odor interferences with classification accuracy 97.3% for each run. For other target gases except NO_2, the classification accuracy can satisfy recognition in this work. The classification accuracy of NO_2 is lower than other gases because it is oxidizing and produces negative response to MOS sensor of this study which is different with other gases.

Consider that the learner-2 is only trained by BPNN using two classes (target gases class and odor interferences class), the number of original training samples of target gas class is 320, and thus 320 samples were also obtained for odor interferences through the new sample generator presented in this chapter. Totally, 640 training samples were obtained for BPNN. Table 15.2 presents the average classification accuracy and maximum classification accuracy of 20 runs including train

Table 15.1 Recognition accuracy of learner-1 for interferences detection in E-nose

Classes	ACA (%)		MCA (%)	
	Train data	Test data	Train data	Test data
Formaldehyde	97.16 ± 1.0575	96.35 ± 0.7274	98.40	96.83
Benzene	98.04 ± 0.8570	91.67 ± 1.3176	99.20	95.83
Toluene	99.68 ± 0.4665	100.0 ± 0.0000	100.0	100.0
Carbon monoxide	99.36 ± 0.8237	94.50 ± 2.1794	100.0	100.0
Ammonia	98.84 ± 0.9625	97.25 ± 2.4875	100.0	100.0
Nitrogen dioxide	99.24 ± 0.8570	76.92 ± 5.4393	100.0	84.62
Stimulus disturbance	97.72 ± 0.8495	97.30 ± 0.0000	98.40	97.30

accuracy and test accuracy. From the results, we find that the accuracies for target gases class and interferences class in test data are 99.38 and 97.30%. The variances are also low, and learner-2 is better than learner-1. From the results of the test samples in dataset 1, we cannot evaluate the stand or fall of learner-1 and learner-2. However, only through benchmark analysis (model learning and testing) may be not enough. Therefore, we present a real-world case study on the background elimination model based on learner-1 and learner-2, respectively.

15.3.2 Abnormal Odor Counteraction: Case Study

- *Experiment without target gas*

For validation of the effectiveness of the presented on-line counteraction model, dataset 2 with real-time collection has been analyzed. Considering the similar and repetitive counteraction work of all sensors, we just present the results of TGS2620 sensor as an example. The similar counteraction results of other three gas sensors (TGS2602, TGS2201A/B) based on learner-1 and learner-2 were given in supplementary data. The discrimination and counteraction tasks are first studied based on learner-1 (Fig. 15.4a) and learner-2 (Fig. 15.4b) on the dataset 4 without target gas. Figure 15.4a illustrates the validation results based on learner-1 and the presented counteraction model. It can be seen that the four actual regions of interferences shown in Fig. 15.1a have been recognized in Fig. 15.4 (the rectangular windows). However, the region between the third and fourth interference regions in Fig. 15.1a has also been recognized as interference region using learner-1 which is over-positive. Figure 15.4b describes the validation results based on learner-2 and the presented counteraction model. Four regions have been recognized as interference regions which are consistent with the true regions shown in Fig. 15.1a. From the interference recognitions, learner-2 is better than learner-1. However, from the interference counteraction curves (dashed lines) based on both models, the interference can be effectively eliminated.

- *Experiment with target gas*

The counteraction on the dataset 2 which has an abrupt change of sensor response may be a special case. Therefore, the counteraction should be studied in a general case that the interference appears under some concentration of target gas. The response curves of each sensor in dataset 3 have been presented in Fig. 15.1b.

Table 15.2 Recognition accuracy of learner-2 for interferences detection in E-nose

Classes	ACA (%)		MCA (%)	
	Train data	Test data	Train data	Test data
Target gases	98.84 ± 0.0028	98.64 ± 0.0066	99.38	99.38
Interferences	98.85 ± 0.0031	95.41 ± 0.0124	99.07	97.30

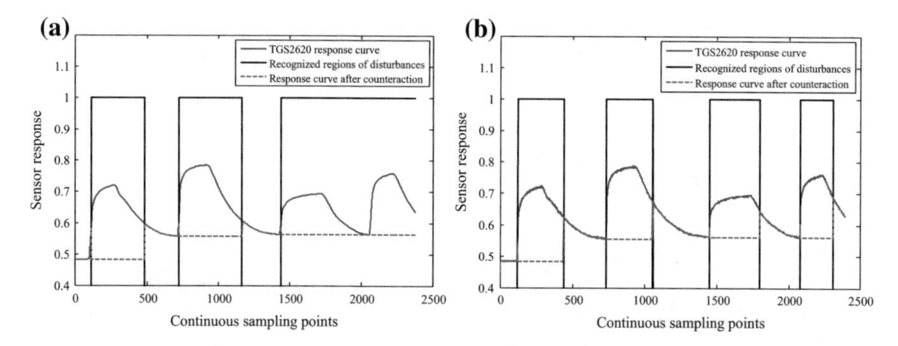

Fig. 15.4 Counteraction of stimulating interferences based on classification learner-1 (**a**) and learner-2 (**b**) without target gas. The rectangular windows illustrate the recognized regions of odor interferences; the dashed curve denotes the sensor response after counteraction

Both target gas and interference appeared in dataset 3. The actual regions of target gas and interference were presented in rectangular windows of Fig. 15.1b. Figure 15.5a illustrates the classification and counteraction of interferences based on learner-1 and the counteraction model. First, the regions in three Rectangular windows are represented as interferences discriminated by learner-1. However, the third region of target gas was incorrectly recognized as interference. Besides, the second recognized interference region has been expanded compared with the true interference regions in Fig. 15.1b. The discrimination regions in Fig. 15.5 also demonstrate that the learner-1 is over-positive. Similarly, Fig. 15.5b presents the analysis based on learner-2 and the counteraction model. Also, three regions in Rectangular windows are recognized as interference and keep consistent with the true regions shown in Fig. 15.1b. The dashed lines in Fig. 15.5 denote the TGS2620 response curve after interference counteraction. From the comparisons of both Fig. 15.5, the learner-2 is better than learner-1 from the interference recognition regions. However, the counteraction results based on both learners can be

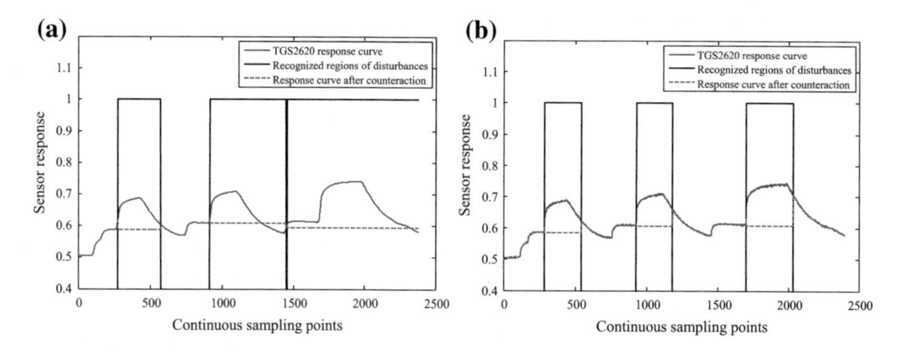

Fig. 15.5 Counteraction of odor interference based on learner-1 (**a**) and learner-2 (**b**) under some concentration of target gas

accepted. From the response curve (dashed line) after counteraction, we can see that the signal with interference has been effectively reduced and counteracted.

15.3.3 Discussion

However, there is still a problem about the location of the interference boundaries. For the correctly recognized regions of interference, the interference (desorption) in the descent process is recognized as target gas prematurely. Therefore, the response curves (dashed lines) after counteraction seem not to be very smooth. In terms of the solution of this problem, a possible way is to change the positions of adsorption and desorption in feature selection of interference in Fig. 15.2a. For example, the positions of adsorption and desorption points (1/2 is presented in this work) can be moved forward and backward, respectively, so that the recognition region for interference can be magnified. Consider the accuracy of recognition, we suggest the special positions for adsorption and desorption be determined at the middle position (1/2) in the adsorption and desorption processes so that both accuracy and robustness of the E-nose instrument can be promised in practical environmental monitoring. From the construction of learner-1 and learner-2, learner-1 based on multi-class mechanism trained by LSSVM is more complex than learner-2 trained by BPNN based on binary classification rule. However, learner-2 performs better than learner-1 from the discrimination of target gas and interference. Since both of them can help the counteraction model realize the counteraction purpose, learner-2 based on binary rule is easier to be operated on-line by the FPGA processor in an E-nose.

Unwanted odor interference in an E-nose cannot be tolerated because it can completely destroy the E-nose system for monitoring, and an E-nose will lose effect in target gases quantification monitoring when exposed to such interference. This presented research on interference counteraction of E-nose is a powerful compensation for our previous work in environmental gases quantification monitoring [21] and MOS sensors calibration transfer [17] in a large-scale manufacture of E-nose instruments. In addition, the presented study is also novel and meaningful in electronic nose development.

15.4 Summary

In the present study, we successfully demonstrated the real-time application of the on-line unwanted odor interference counteraction model based on two artificial intelligence learners in electronic nose for the first time. With respect to the advantages of this machine learning technique in combination with MOS sensors of weak selectivity to target gases, we resolved the intrinsic high susceptibility to odor interferences problem of metal oxide semiconductor sensors and strengthen the prediction accuracy and interference counteraction characteristic of E-nose. To meet the requirements of real-time environmental monitoring of pollutant gases, low complexity of model and feature selection of interference should be necessary. In this chapter, learner-2 based on binary classification may be a better choice in background elimination of E-nose.

References

1. S.M. Scott, D. James, Z. Ali, Data analysis for electronic nose systems. Microchim. Acta **156**, 183–207 (2007)
2. S.M. Kanan, O.M. El-Kadri, I.A. Abu-Yousef, M.C. Kanan, Semiconducting metal oxide based sensors for selective gas pollutant detection. Sensors **9**, 8158–8196 (2009)
3. C. Di Natale, R. Paolesse, A. Macagnano, A. Mantini, C. Goletti, A. D'Amico, Characterization and design of porphyrins-based broad selectivity chemical sensors for electronic nose applications. Sens. Actuators B **52**, 162–168 (1998)
4. C. Di Natale, E. Martinelli, A.D. Amico, Counteraction of environmental disturbances of electronic nose data by independent component analysis. Sens. Actuators B **82**, 158–165 (2002)
5. M. Holmberg, F.A.M. Davide, C. Di Natale, A. D'Amico, F. Winquist, I. Lundström, Drift counteraction in odour recognition applications: lifelong calibration method. Sens. Actuators B **42**, 185–194 (1997)
6. E.J. Wolfrum, R.M. Meglen, D. Peterson, J. Sluiter, Calibration transfer among sensor arrays designed for monitoring volatile organic compounds in indoor air quality. IEEE Sens. J. **6**, 1638–1643 (2006)
7. S.K. Jha, R.D.S. Yadava, Denoising by singular value decomposition and its application to electronic nose data processing. IEEE Sens. J. **11**, 35–44 (2011)
8. F. Hossein-Babaei, V. Ghafarinia, Compensation for the drift-like terms caused by environmental fluctuations in the responses of chemoresistive gas sensors. Sens. Actuators B **143**, 641–648 (2010)
9. J.H. Sohn, M. Atzeni, L. Zeller, G. Pioggia, Characterisation of humidity dependence of a metal oxide semiconductor sensor array using partial least squares. Sens. Actuators B **131**, 230–235 (2008)
10. M. Holmberg, F. Winquist, I. Lundström, F. Davide, C. DiNatale, A. D'Amico, Drift counteraction for an electronic nose. Sens. Actuators B **35–36**, 528–535 (1996)
11. M. Padilla, A. Perera, I. Montoliu, A. Chaudry, K. Persaud, S. Marco, Drift compensation of gas sensor array data by orthogonal signal correction. Chemometr. Intell. Lab. Syst. **100**, 28–35 (2010)
12. A. Ziyatdinov, S. Marco, A. Chaudry, K. Persaud, P. Caminal, A. Perera, Drift compensation of gas sensor array data by common principal component analysis. Sens. Actuators B **146**, 460–465 (2010)
13. H. Ding, J.H. Liu, Z.R. Shen, Drift reduction of gas sensor by wavelet and principal component analysis. Sens. Actuators B **96**, 354–363 (2003)
14. M. Paniagua, E. Llobet, J. Brezmes, X. Vilanova, X. Correig, E.L. Hines, On-line drift counteraction for metal oxide gas sensor arrays. Electron. Lett. **39**, 40–42 (2003)
15. O. Tomic, H. Ulmer, J.E. Haugen, Standardization methods for handling instrument related signal shift in gas sensor array measurement data. Anal. Chim. Acta **472**, 99–111 (2002)
16. E.J. Wolfrum, R.M. Meglen, D. Peterson, J. Sluiter, Calibration transfer among sensor arrays designed for monitoring volatile organic compounds in indoor air quality. IEEE Sens. J. **6**, 1638–1643 (2006)

17. L. Zhang, F.C. Tian, C. Kadri, B. Xiao, H. Li, L. Pan, H. Zhou, On-line sensor calibration transfer among electronic nose instruments for monitor volatile organic chemical in indoor air quality. Sens. Actuators B **160**, 899–909 (2011)
18. A.K. Pavlou, N. Magan, J.M. Jones, J. Brown, P. Klatser, A.P.F. Turner, Detection of Mycobacterium tuberculosis (TB) in vitro and situ using an electronic nose in combination with a neural network system. Biosens. Bioelectron. **20**, 538–544 (2004)
19. B. Dębska, B. Guzowska-Świder, Application of artificial neural network in food classification. Anal. Chim. Acta **705**, 283–291 (2011)
20. B. Podola, M. Melkonian, Genetic Programming as a tool for identification of analyte-specificity from complex response patterns using a non-specific whole-cell biosensor. Biosens. Bioelectron. **33**, 254–259 (2012)
21. L. Zhang, F.C. Tian, C. Kadri, G. Pei, H. Li, L. Pan, Gases concentration estimation using heuristics and bio-inspired optimization models for experimental chemical electronic nose. Sens. Actuators B **160**, 760–770 (2011)
22. C. Di Natale, A. Macagnano, E. Martinelli, R. Paolesse, G. D'Arcangelo, C. Roscioni, A.F. Agrò, A. D'Amico, Lung cancer identification by the analysis of breath by means of an array of non-selective gas sensors. Biosens. Bioelectron. **18**, 1209–1218 (2003)
23. O.F. Canhoto, N. Magan, Potential for detection of microorganisms and heavy metals in potable water using electronic nose technology. Biosens. Bioelectron. **18**, 751–754 (2003)
24. K. Brudzewski, S. Osowski, T. Markiewicz, Classification of milk by means of an electronic nose and SVM neural network. Sens. Actuators B **98**, 291–298 (2004)
25. L. Zhang, F.C. Tian, H. Nie, L. Dang, G. Li, Q. Ye, C. Kadri, Classification of multiple indoor air contaminants by an electronic nose and a hybrid support vector machine. Sens. Actuators B **174**, 114–125 (2012)
26. S.J. Dixon, R.G. Brereton, Comparison of performance of five common classifiers represented as boundary methods: Euclidean distance to centroids, linear discriminant analysis, quadratic discriminant analysis, learning vector quantization and support vector machines, as dependent on data structure. Chemometr. Intell. Lab. Syst. **95**, 1–17 (2009)
27. H.L. Chen, D.Y. Liu, B. Yang, J. Liu, G. Wang, A new hybrid method based on local fisher discriminant analysis and support vector machines for hepatitis disease diagnosis. Expert Syst. Appl. **38**, 11796–11803 (2011)
28. M.K. Muezzinoglu, A. Vergara, R. Huerta, N. Rulkov, M.I. Rabinovich, A. Selverston, H.D. Abarbanel, Acceleration of chemo-sensory information processing using transient features. Sens. Actuators B **137**(2), 507–512 (2009)
29. D.E. Goldberg, *Genetic algorithms in search, optimization, and machine learning* (Addison Wesley, Boston, 1989)

Chapter 16
Pattern Mismatch Guided Interference Elimination

Abstract Electronic nose composed of MOS sensors cannot be used when there are unwanted gases; therefore, it is urgent to solve the problem of interferences elimination. The presented method in Chap. 15 tends to discriminate the interference gases and target gases and depends on the number of types of interference gases. However, there are numerous interferences in real-world application scenario, which is impossible to be sampled in laboratory experiments. Considering that the target gases rather than interferences can be fixed as invariant information, a novel and effective *pattern mismatch-based interference elimination* (PMIE) method is proposed in this chapter. It contains two parts: discrimination (i.e., pattern mismatch) and correction (i.e., interference elimination). Specifically, the principle behind is that the interference discrimination is achieved by deciding whether a new pattern violates the rules established on the invariant target gases information (i.e., interference gas) or not (i.e., target gas). If the current pattern of the sensor array is interference, orthogonal signal correction (OSC) algorithm is used for interference correction. Experimental results prove that the proposed PMIE method is very effective for interferences elimination in electronic nose.

Keywords Electronic nose · Pattern mismatch · Interference elimination
Interference discrimination

16.1 Introduction

The performance of an E-nose depends largely on the selected sensors, which should have good cross-sensitivity, selectivity, reliability, and robustness [1]. Objectively, the cross-sensitivity of sensor array is either an advantage or a disadvantage. Specifically, the cross-sensitivity is beneficial for detecting many kinds of gases with a limited number of gas sensors. However, this also results in that the sensor array in an E-nose would produce significant responses to some undesired interferences. The disadvantage is that the interferences will have a serious impact

L. Zhang et al., *Electronic Nose: Algorithmic Challenges*,
https://doi.org/10.1007/978-981-13-2167-2_16

on the detection of the target gases. Therefore, in this chapter, we aim at solving the problem of interference elimination by leveraging machine learning idea.

Generally, there are two kinds of interferences in E-nose. One is the unwanted (non-target) odors that do not belong to the target gases being detected. The other one is the environmental factors, such as temperature and humidity [2]. In conventional electronic noses, the latter, i.e., temperature and humidity, have been compensated by integrating the temperature and humidity sensor into the gas sensor array. Currently, the methods for eliminating the impacts from environmental temperature and humidity are based on temperature and humidity compensation model [3–8]. A physical way, where SiO_2 was used as sensing material with a titanium thermistor being used to maintain a constant temperature, was used to compensate temperature and humidity influences [9]. Independent component analysis was also used for reduction of environmental impacts [10], in which the component with the maximum correlation coefficient w.r.t. the reference vector is removed as interference. However, the former (i.e., non-target odor interference) cannot be handled effectively so far, which seriously limits the real-time usage of E-nose in real-world scenario, such as the mall with perfumes and the kitchen with smoke flavor, because the metal oxide semiconductor gas sensors are very sensitive to the pungent odors like perfumes and smoke flavor.

The problem of non-target odors is urgent to solve in E-nose, where the conventional method that tends to solve this problem can be divided into two steps: interferences discrimination and interferences correction [11]. Briefly, the first step is to determine that whether the current E-nose signal is from some interference. If yes, the second step is to correct the sensor responses with inferences eliminated. Therefore, a classifier that is used to discriminate the interference should be first setup by leveraging limited number of types of interference samples, such that the learned classifier largely depends on the limited number of interferences and it may only be effective to the limited types of interference. However, there are many possible kinds of interferences that cannot be collected in laboratory experiments. Therefore, the conventional method cannot discriminate all the unwanted odor interferences in real-world application scenarios. For the correction step, the interference signal depends on the previous target signal, such that the E-nose cannot work in real time when interferences exist.

As shown in [12] for interferences elimination in wound detection, it tends to remove the background interference according to the spatial correlation coefficients by performing wavelet transform on the collected odor samples between the infected and healthy mice. In [13], the sensor response is analyzed by independent component analysis (ICA), the sensor response of healthy mice body odor measured independently was regard as the reference vector of background interference, and then the correlation coefficients between the components of ICA and the reference vector were calculated in which the components with the highest coefficient are recognized to be background. It is known that, in this method, the reference vector is very important for determining the interference component. However, in this study the interferences are various in complex real-world scenario, such that it is difficult to obtain the so-called reference vector. Additionally, the ICA algorithm shows a good effect in eliminating the environment interference, but it does not fit the non-target gas interferences. One reason may be that the target gas and non-target gas have similar nature and effect to the sensor array, and cannot be effectively separated from the signal of mixtures. The orthogonal signal correction (OSC) algorithm can remove the information that is orthogonal to target signal, where the orthogonal information is recognized to be the irrelevant information; therefore, it has been used to correct signals and remove the interference information in [14, 15]. However, the problem of determining the reference vector still exists as well as ICA. Moreover, the interference in [15] is fixed in the whole detection. In this chapter, various interferences in real-world environment that are independent of the target gases are considered in our proposed method.

From the analysis above, many appropriate methods were found to compensate and eliminate the interference from environmental factors. But the interferences of non-target gases are still troubling. In order to solve this problem, a novel pattern mismatch-based Interference Elimination (PMIE) method consisting of discrimination and correction is proposed in this work.

16.2 Data Acquisition

The E-nose systems and the data collection are following the same settings as Chap. 15. Therefore, the experimental setup and datasets are not presented repeatedly. The dataset is shown in Table 16.1.

16.3 Proposed PMIE Method

16.3.1 Main Idea

Suppose the dataset $\mathbf{S_1}$ to be $(\mathbf{X}, \mathbf{Y})$, where $\mathbf{X} = (\mathbf{x}_1, \mathbf{x}_2, \ldots, \mathbf{x}_N)^{\mathrm{T}} \in \Re^{N \times d}, \mathbf{x}_i = (x_{i1}, \ldots, x_{\mathrm{id}})$ and $\mathbf{Y} \in \Re^N$. N denotes the number of training samples, and d is the feature dimension. The prediction function $f(\bullet)$ can be defined as follows.

$$\min_f \sum_{i=1}^{N} \|f(\mathbf{x}_i) - Y_i\|_2^2 \tag{16.1}$$

Table 16.1 Target gas samples for models setting up

Target gas	Formaldehyde	Toluene	Benzene	NO_2	NH_3	CO
Number of samples	504	264	288	152	116	232
Number of train/test samples	336/168	176/88	192/96	101/51	77/39	155/77

The absolute prediction error of another dataset $\mathbf{S_2}$ can be calculated by,

$$\text{error}_i = |f(\mathbf{x}_i) - Y_i|, i = 1, \ldots, M \tag{16.2}$$

where M denotes the number of dataset $\mathbf{S_2}$. We can imagine that a larger prediction error will be obtained by using the learned sensor prediction function $f(\bullet)$ on some sample with different pattern from the patterns $\mathbf{S_1}$. Therefore, an unknown sample can be discriminated whether it belongs to the patterns from $\mathbf{S_1}$ or not by finding an appropriate threshold T. For example, if some sample $(\mathbf{U}, V)$ with different pattern from $\mathbf{S_1}$ is used as the input of $f(\bullet)$, we can obtain that $|f(\mathbf{U}) - V| > T$. This idea implies that the invariant information of the fixed dataset $\mathbf{S_1}$ is memorized in the learned prediction f. The new patterns that have different space distribution from $\mathbf{S_1}$ will have a conflict, such that it is easy to recognize all types of different patterns that are uncorrelated with the invariant information in $\mathbf{S_1}$.

In this work, this idea is motivated by the fact that many kinds of interferences in real-world scenarios may happen in E-nose applications. From the viewpoint of experiments, it is impossible to collect all the interference samples. However, the target gases, as fixed patterns, can be recognized as invariant information, such that the proposed idea is reasonable for interference elimination. Motivated by the information invariance and the proposed idea, a pattern mismatch-based interference elimination (PMIE) method is proposed. Note that we nominate "pattern mismatch" instead of "pattern match" because we aim at recognizing the interference by the mismatch degree (i.e., prediction error), instead of recognizing the target gases.

16.3.2 Pattern Mismatch-Based Interference Elimination (PMIE)

The proposed PMIE method includes two parts: interference discrimination and interference elimination. The details for each part have been presented as follows.

- PMIE-based Interference Discrimination

In E-nose, the responses of temperature (x_{tem}), humidity (x_{hum}), and three MOS sensors (x_{s1}, x_{s2}, x_{s3}) are used to predict the response of the fourth sensor (x_{s4}). In the proposed PMIE method, the pattern vector $\mathbf{x}$ is described as (x_{tem}, x_{hum}, x_{s1}, x_{s2}, x_{s3}) and the predicted variable Y is described as x_{s4}. In this chapter, the existing regression techniques such as regularized least square (RLS), back-propagation neural network (BPNN), and support vector regression (SVR) have been used to learn the prediction function f, by assuming that $x_{s4} = f(x_{\text{tem}}, x_{\text{hum}}, x_{s1}, x_{s2}, x_{s3})$. We call them PMIE-RLS, PMIE-BP, and PMIE-SVR, respectively. Note that in this chapter, four gas sensors have been used as predicted variables, respectively.

For discrimination of an unknown sample **x**, the *label* of **x** can be obtained by calculating the *error* between its predicted output $y = f(x_{\text{tem}}, x_{\text{hum}}, x_{s1}, x_{s2}, x_{s3})$ and the true value x_{s4}, as follows.

$$\text{label} = \begin{cases} 1, & \text{if} \quad \text{error} > T \\ 0, & \text{if} \quad \text{error} \leq T \end{cases} \tag{16.3}$$

where the *threshold* can be determined by ROC curve and misjudgment rate (*mR*). Note that the *label* 1 denotes the interference pattern (i.e., non-target gas) and 0 denotes target gas pattern.

Specifically, the interference discrimination process of the PMIE method is shown in Fig. 16.1. Note that the training process of the sensor prediction function *f* in Fig. 16.1 is conducted by using dataset 1.

Threshold determination: The dataset 2 is used to determine the threshold *T* for pattern mismatch. Specifically, the receiver operating characteristic (ROC) curve is used to determine the threshold, which is a popular index in reflecting the sensitivity and specificity of recognition. The area under the curve can reflect the overall recognition performance [16–18]. For a binary classification problem, four items such as true positive (TP), true negative (TN), false positive (FP), and false negative (FN) are included. In ROC, the true positive rate is calculated by,

$$\text{TPR} = \frac{\text{TP}}{\text{TP} + \text{FN}} \tag{16.4}$$

The false positive rate is calculated by,

$$\text{FPR} = \frac{\text{FP}}{\text{FP} + \text{TN}} \tag{16.5}$$

In Eqs. 16.4 and 16.5, TP is the number of true positive samples, FN is the number of false negative samples, FP is the number of false positive samples, and TN is the number of true negative samples.

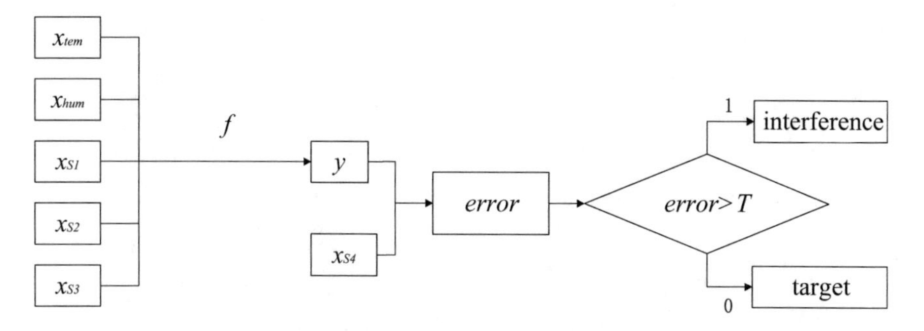

Fig. 16.1 Structure of PMIE for interference discrimination

First, the prediction error vector of dataset 2 can be obtained by using the learned function *f*, then the maximum and minimum errors denoted as $E_{\max}$ and $E_{\min}$ are obtained. The interval $[E_{\min}, E_{\max}]$ is divided into 100, such that 100 potential thresholds can be defined. For each threshold, the sensitivity (TPR, true positive rate) and specificity (FPR, false positive rate) can be calculated for ROC curve. Then, the misjudgment rate is calculated for each threshold, in which the threshold with the smallest misjudgment rate is chosen as the final threshold *T* in Eq. 16.3.

- *PMIE-based Interference Elimination*

After interference discrimination, the proposed interference elimination is formulated as follows.

$$\mathbf{z}_{\mathrm{cor}} = \begin{cases} \mathbf{z} - \mathbf{z} \cdot \mathbf{h}, & \text{label} = 1 \\ \mathbf{z}, & \text{label} = 0 \end{cases} \tag{16.6}$$

where **z** is a sensor array response and **h** is the correction matrix being determined. In this chapter, orthogonal signal correction (OSC) algorithm is used to establish the signal correction matrix **h**. The OSC has been used to preprocess the near-infrared spectrum data [19–21], which tends to remove the irrelevant information (orthogonal part) in the projection matrix **R** from the spectral matrix **S**. Therefore, in this chapter, the OSC is used to remove the interference information that is irrelevant to target signal. The projection matrix **R** is the mean vector of the 20 target samples.

- *PMIE Algorithm*

The training and testing stages of PMIE algorithm have been illustrated in Algorithms 16.1 and 16.2, respectively.

Algorithm 16.1 PMIE training algorithm

Input:

Dataset 1 ($\mathbf{X}_1$);
Dataset 2 ($\mathbf{X}_2$);

Output:

Prediction function *f*;
Threshold *T*.

Procedure:

1. Data preprocessing;
2. Obtain the prediction function *f* by training on $\mathbf{X}_1$;
3. For $i = 1$ to N

Predict the fourth sensor response: $y = f(x_{\mathrm{tem}}, x_{\mathrm{hum}}, x_{s1}, x_{s2}, x_{s3})$ by using $\mathbf{X}_2$;

Calculate the prediction error: $\text{error}_i = |y - x_{s4}|$

end

4. Calculate the misjudgment rates of the 100 thresholds;
5. Get the threshold T with the smallest misjudgment rate.

Algorithm 16.2 PMIE testing algorithm

Input:

The prediction function f;
The threshold T;
Dataset 3 ($\mathbf{X}_3$).

Output:

The interference *label*;
The interference eliminated sensor response $\mathbf{Z}_{\text{cor}}$.

Procedure:

1. Data preprocessing;
2. Predict the fourth sensor response: $y = f(x_{\text{tem}}, x_{\text{hum}}, x_{s1}, x_{s2}, x_{s3})$ by using $\mathbf{X}_3$;
3. Calculate the prediction error: $\text{error} = |y - x_{s4}|$;
4. Predict the interference *label* based on pattern mismatch:

if *error* > T

label = 1;

else *label* = 0;

5. Correct the original signal $\mathbf{X}_3$ using the OSC:

if label = 1

$\mathbf{Z}_{\text{cor}} = \mathbf{X}_3 - \mathbf{X}_3 \cdot \mathbf{h}$

else $\mathbf{Z}_{\text{cor}} = \mathbf{X}_3$;

16.4 Results and Discussion

16.4.1 Data Preprocessing

Smooth filtering and the vector standardization are used for preprocessing and normalization, respectively, in this chapter. Set the length of the smooth filter as 20, and the length of the signal vector **S** is n. The filtered signal vector **S** can be calculated by,

$$\mathbf{X}(i) = \frac{\sum_{l=i}^{i+20-1} \mathbf{S}(l) - \max(\mathbf{S}(i), \ldots, \mathbf{S}(i+20-1)) - \min(\mathbf{S}(i), \ldots, \mathbf{S}(i+20-1))}{18}$$
$$i = 1, \ldots, n-20+1 \tag{16.7}$$

Considering that the information of concentration is not important in discrimination, the vector standardization can be used to compensate the concentration difference.

The relationship between gas sensors and gas concentration can be described by,

$$x_{ij} = \alpha_i \left[C_j\right]^{\beta} \tag{16.8}$$

where x_{ij} is the j-th response of i-th MOS sensor for the gas; α_i is the sensitivity of i-th sensor; C_j is the gas concentration; β is the characteristic constant related to gas concentration. The signal normalization can be calculated by,

$$y_{ij} = \frac{x_{ij}}{\sqrt{\sum_i (x_{ij})^2}} = \frac{\alpha_i}{\sqrt{\sum_i \alpha_i^2}}, \quad i = 1, \ldots, 4; j = 1, \ldots, n \tag{16.9}$$

Equation 16.9 indicates that vector standardization can be used to compensate the concentration difference, such that each gas sample is uncorrelated with concentration. Therefore, this method can effectively avoid the influence of concentration.

16.4.2 PMIE Training on Dataset 1

Three models such as PMIE-RLS, PMIE-BP and PMIE-SVR were used in this chapter. The average prediction errors of each sensor for three models were described in Table 16.2, from which we can see that the prediction accuracy of the PMIE-SVR model shows the best performance. However, the proposed PMIE method in this chapter is not strict to the prediction accuracy, as long as the boundary points which discriminate between target gas signal and interference gas signal can be found.

Table 16.2 Average training and testing prediction errors for three models

Predicted sensor	TGS2620	TGS2602	TGS2201A	TGS2201B
	train/test error	train/test error	train/test error	train/test error
PMIE-RLS	0.0269/0.0242	0.0545/0.055	0.0475/0.0499	0.0213/0.0182
PMIE-BP	0.0106/0.0112	0.015/0.016	0.019/0.021	0.0171/0.02
PMIE-SVR	2.2e−5/2.0e−5	1.6e−5/1.4e−5	2.3e−5/1.7e−5	1.0e−5/9.4e−6

16.4.3 Threshold Analysis Based on Dataset 2

The Dataset 2 is used to determine the threshold. First, the prediction errors can be observed from dataset 2 for different prediction models, and then the maximum error E_{max} and minimum error E_{min} can be found. In this chapter, the interval [E_{min}, E_{max}] is divided into 100 thresholds uniformly. For each threshold, the sensitivity (TPR) and specificity (FPR) can be calculated. The ROC curves are shown in Fig. 16.2, which shows that the areas under curve (AUC) of the ROC for three models are more than 0.5. The PMIE-BP model is better than PMIE-SVR and PMIE-RLS models for pattern prediction of TGS2620 and TGS2201B. The PMIE-SVR model is the best model for pattern prediction of TGS2602 and TGS2201A. In general, the PMIE-BP model and PMIE-SVR model have their own advantages in discrimination, and the PMIE-RLS is the worst due to the linear nature.

Second, the misjudgment rate can be calculated for each threshold. The threshold with the smallest misjudgment rate will be used as the threshold for interference discrimination. Relationships between threshold and misjudgment rate were shown in Fig. 16.3, from which the threshold and misjudgment rate of each model for each sensor can be observed. And the red point is the threshold point. The details for each sensor are described in Table 16.3, respectively.

In terms of the misjudgment rate, the PMIE-BP model with respect to the TGS2201B sensor as the dependent variable in PMIE has the smallest misjudgment rate for interference discrimination.

In summary, this section denotes the training process of the discrimination model f based on Dataset 1 and the search for the best threshold T based on Dataset 2. The obtained f and T will be tested in the following section by using Dataset 3.

16.4.4 Interference Discrimination on Dataset 3

- Discrimination based on Single Model

In this section, the prediction results of three models for each sensor were described in Table 16.4. It can be seen that for each sensor, the misjudgment rates of PMIE-BP model and PMIE-SVR model are smaller. Therefore, the PMIE-BP model and PMIE-SVR model are more suitable for interference discrimination.

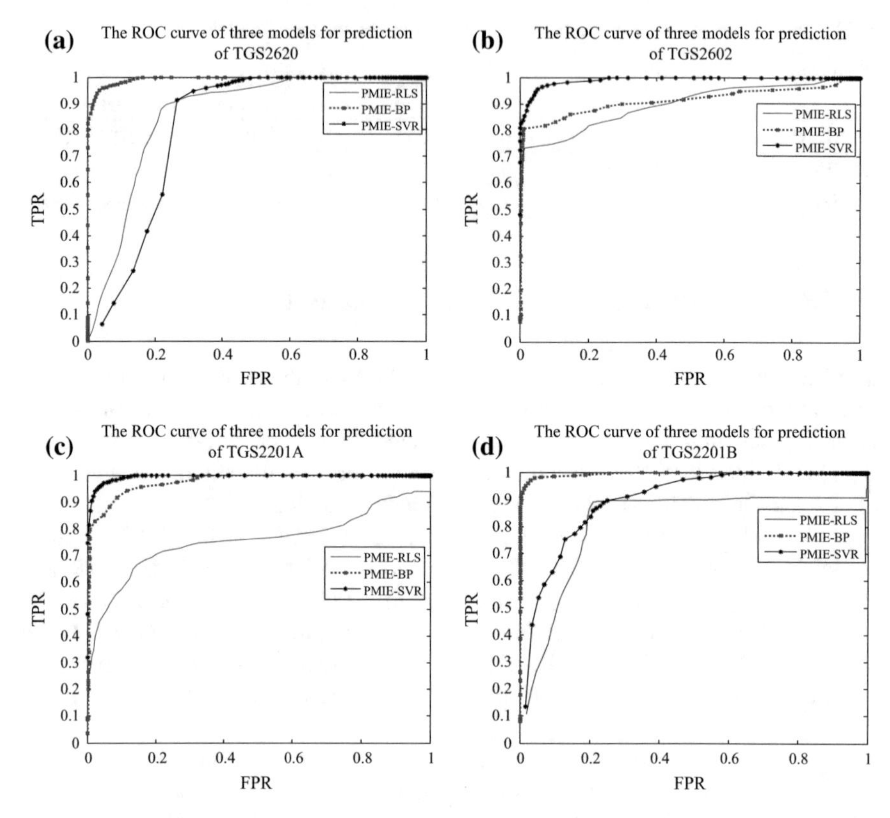

Fig. 16.2 ROC curves of the sensor TGS2620 (**a**), TGS2602 (**b**), TGS2201A (**c**), and TGS2201B (**d**) based on the three prediction models, respectively

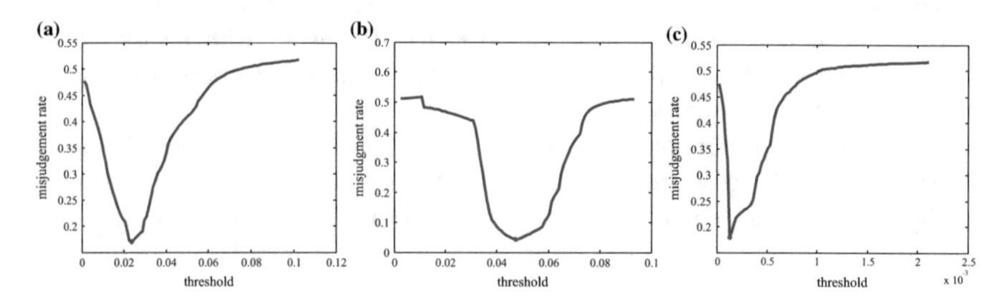

Fig. 16.3 Misjudgment rate (*mR*) with threshold (*T*) for TGS2620 sensor by using PMIE-RLS (**a**), PMIE-BP (**b**), and PMIE-SVR (**c**), respectively

- Discrimination based on Multiple Models

For each model, the responses of four sensors are predicted, respectively, i.e., $S_1 = f_1(S_2, S_3, S_4)$, $S_2 = f_2(S_1, S_3, S_4)$, $S_3 = f_3(S_1, S_2, S_4)$, and $S_4 = f_4(S_1, S_2, S_3)$. So there are four prediction functions for each model (e.g., SVR). In order to prevent

Table 16.3 Threshold (T) and misjudgment rate (mR) of four sensors for three models

Method	PMIE-RLS		PMIE-BP		PMIE-SVR	
	T	mR (%)	T	mR (%)	T	mR (%)
TGS2620	0.024	17.38	0.0464	7.46	0.00015	18.75
TGS2602	0.188	13.58	0.0905	10.54	0.0012	5.08
TGS2201A	0.0183	24.8	0.1342	8.58	0.0004	9.67
TGS2201B	0.0189	16.9	0.1802	4	0.00018	17.83

Table 16.4 Misjudgment rate (mR) of four sensors for three models

Method	PMIE-RLS	PMIE-BP	PMIE-SVR
	mR (%)	mR (%)	mR (%)
TGS2620	17.75	3.63	12.04
TGS2602	6	3.91	2.33
TGS2201A	28.40	3.83	6.70
TGS2201B	34.70	8.38	8.50

Table 16.5 Comparison of the misjudgment rate (mR) between the proposed PMIE based methods and the traditional method in [11]

	The proposed method									Method [11]		
Method k	PMIE-RLS			PMIE-BP			PMIE-SVR			RLS	BP	SVR
	1	2	3	1	2	3	1	2	3			
mR	42.10	15.25	10.45	5.67	3.41	3.25	5.41	5.62	5.29	6.55	5.75	5.08

the recognition error caused by a single sensor prediction, it uses multiple sensor predictions to discriminate between target signals and interference signals in the same model. As shown in Table 16.5, k ($k = 1, 2, 3$) denotes that a signal is discriminated as an interference response when the signal is predicted as interference by k prediction function. Table 16.5 illustrates that the misjudgment rate is smallest with $k = 3$ under the same model.

The PMIE method proposed in this chapter is different with the pure classification-based method [11]. The PMIE method proposed in this chapter only relies on the target samples when it sets up, and it is irrelevant to interference gases. So the method is effective to all interference gases. However, the method in [11] needs some interference samples to train classifier, so it depends on the type of interference gas to some extent. Thus, it is effective to limited types of interference. Table 16.5 shows that the PMIE-BP method is superior to the method in [11] in discrimination performance.

16.4.5 PMIE-Based Interference Elimination Result

In this chapter, the OSC algorithm was used to correct the interference signals. The curves of TGS2260 responses which were discriminated by PMIE-BP prediction model and corrected by OSC were described in Fig. 16.4b, from which we can find that the misjudgment positions are the boundaries between the target gases and interferences. We have to admit that the boundary problem is still an unsolved issue, which is a cost-sensitive problem.

The independent component analysis (ICA) algorithm and orthogonal signal correction (OSC) algorithm can be used in components correction [22–25]. In the E-nose based on MOS sensor array, they are used to solve the interference problem and drift problem [26–29] caused by sensor aging. For comparison, we have implemented ICA for interference elimination. The TGS2620 response curve with interference elimination is showed in Fig. 16.4a. Additionally, we also apply single OSC method, which can remove the information that is irrelevant to target signal. The corrected curve of TGS2620 was shown in Fig. 16.4c. From the results, we can find that the proposed PMIE method can successfully discriminate and eliminate the

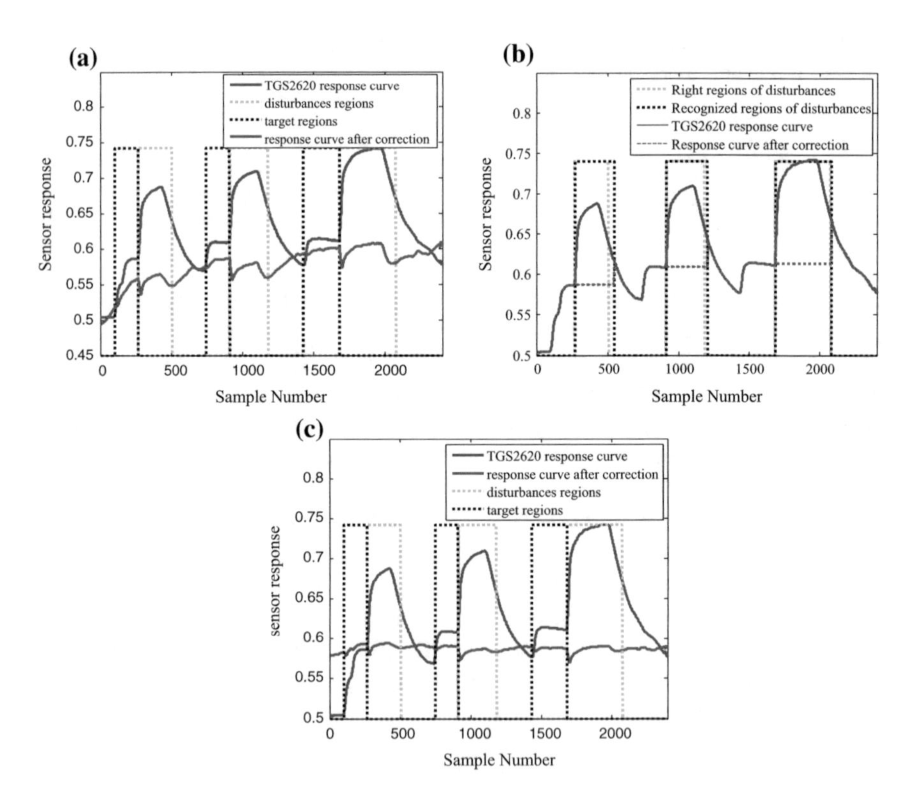

Fig. 16.4 Curve of sensor 2620 after correction by ICA algorithm (**a**), OSC with PMIE-BP model discrimination (**b**), and OSC algorithm (**c**), respectively

interferences in real-time application. This also implies that ICA or OSC may not effectively eliminate the interferences that are independent of target gases in real-time application from their principle. Therefore, for real-time use of an E-nose in real-world scenarios, our proposed method can well address the issue in E-nose community.

16.5 Summary

In this chapter, a pattern mismatch-based interference elimination (PMIE) method is proposed for addressing the key and unsolved issue of interferences in E-nose. This method consists of two parts: interferences discrimination and interferences elimination. The discrimination model of PMIE in this chapter only relies on the target gas samples, and it is irrelevant to interference gases, such that the PMIE method is effective to all interference gases. In experiments, the proposed PMIE method in this chapter shows the best results and improves the interferences problem of E-nose caused by unwanted gases.

References

1. S.M. Scott, D. James, Z. Ali, Data analysis for electronic nose systems. Microchim. Acta **156**, 183–207 (2007)
2. L. Zhang, F. Tian, Performance study of multilayer perceptrons in a low-cost electronic nose. IEEE Trans. Instrum. Meas. **63**, 1670–1679 (2014)
3. X. Tian, Y. Yin, H. Liu, Research on artificial olfactory sensor technology for liquor identification. Food Sci. **2**, 29–32 (2004)
4. B. Mumyakmaz, A. Özmen, M.A. Ebeoğlu, C. Taşaltın, İ. Gürol, A study on the development of a compensation method for humidity effect in QCM sensor responses. Sens. Actuators, B **1**, 277–282 (2010)
5. K.R. Kashwan, M. Bhuyan, in *Robust electronic-nose system with temperature and humidity drift compensation for tea and spice flavour discrimination, Sensors and the International Conference on new Techniques in Pharmaceutical and Biomedical Research*, Asian Conference on, 2005-07-20 (2005)
6. J.W. Gardner, E.L. Hines, F. Molinier, P.N. Bartlett, T.T. Mottram, Prediction of health of dairy cattle from breath samples using neural network with parametric model of dynamic response of array of semiconducting gas sensors, Science, Measurement and Technology, IEEE Proceedings—2, 102–106 (1999)
7. X. Xiao-Liang, Q. Jun-Na and C. Chun, A study on local sensor fusion of wireless sensor networks based on the neural network, Machine Learning and Cybernetics, International Conference on, Kunming, 2008-01-01 (2008)
8. S. Jianfang, T. Hongbiao, G. Haiyan, Application of wavelet neural network and multi-sensor data fusion technique in intelligent sensor, Intelligent Control and Automation. WCICA 2008. 7th World Congress on, Chongqing, 2008-01-01 (2008)
9. T.A. Emadi, C. Shafai, M.S. Freund, D.J. Thomson, D.S. Jayas, N.D.G. White, Development of a polymer-based gas sensor—humidity and CO2 sensitivity, Microsystems and Nanoelectronics Research Conference. MNRC 2009. 2nd, Ottawa, ON, Canada, 2009-01-01 (2009)

10. C. Di Natale, E. Martinelli, A. D'Amico, Counteraction of environmental disturbances of electronic nose data by independent component analysis. Sens. Actuators, B. **82**(2–3), 158–165 (2002)
11. L. Zhang, F. Tian, L. Dang, G. Li, X. Peng, X. Yin, S. Liu, A novel background interferences elimination method in electronic nose using pattern recognition. Sens. Actuators, A **201**, 254–263 (2013)
12. J. Feng, F. Tian, J. Yan, Q. He, Y. Shen, L. Pan, A background elimination method based on wavelet transform in wound infection detection by electronic nose. Sens. Actuators, B **2**, 395–400 (2011)
13. F. Tian, J. Yan, S. Xu, J. Feng, Q. He, Y. Shen, P. Jia, Background interference elimination in wound infection detection by electronic nose based on reference vector-based independent component analysis. Inf. Technol. J. 7 (2012)
14. N.G. Yee, G.G. Coghill, Factor selection strategies for orthogonal signal correction applied to calibration of near-infrared spectra. Chemometr. Intell. Lab. Syst. **67**, 145–156 (2003)
15. J. Feng, F. Tian, P. Jia, Q. He, Y. Shen, S. Fan, Improving the performance of electronic nose for wound infection detection using orthogonal signal correction and particle swarm optimization. Sens. Rev. **34**, 389–395 (2014)
16. X. Zhang, X. Li, Y. Feng, Z. Liu, The use of ROC and AUC in the validation of objective image fusion evaluation metrics. Sig. Process. **115**, 38–48 (2015)
17. V. Nykänen, I. Lahti, T. Niiranen, K. Korhonen, Receiver operating characteristics (ROC) as validation tool for prospectivity models—A magmatic Ni–Cu case study from the Central Lapland Greenstone Belt. Northern Finl. Ore Geol. Rev. **71**, 853–860 (2015)
18. M. Thomas, K. De Brabanter, J.A.K. Suykens, B. De Moor, Predicting breast cancer using an expression values weighted clinical classifier. BMC Bioinform. 15 (2014)
19. S. Wold, H. Antti, F. Lindgren, J. Öhman, Orthogonal signal correction of near-infrared spectra. Chemometr. Intell. Lab. Syst. **44**, 175–185 (1998)
20. Z. Talebpour, R. Tavallaie, S.H. Ahmadi, A. Abdollahpour, Simultaneous determination of penicillin G salts by infrared spectroscopy: evaluation of combining orthogonal signal correction with radial basis function-partial least squares regression. Spectrochim. Acta Part A Mol. Biomol. Spectrosc. **76**, 452–457 (2010)
21. L. Laghi, A. Versari, G.P. Parpinello, D.Y. Nakaji, R.B. Boulton, FTIR spectroscopy and direct orthogonal signal correction preprocessing applied to selected phenolic compounds in red wines. Food Anal. Methods **4**, 619–625 (2011)
22. D.J. Bouveresse, A. Moya-González, F. Ammari, D.N. Rutledge, Two novel methods for the determination of the number of components in independent components analysis models. Chemometr. Intell. Lab. Syst. **112**, 24–32 (2012)
23. S. Balasubramanian, S. Panigrahi, C.M. Logue, C. Doetkott, M. Marchello, J.S. Sherwood, Independent component analysis-processed electronic nose data for predicting Salmonella typhimurium populations in contaminated beef. Food Control **19**, 236–246 (2008)
24. T. Aguilera, J. Lozano, J.A. Paredes, F.J. Alvarez, J.I. Suarez, Electronic nose based on independent component analysis combined with partial least squares and artificial neural networks for wine prediction. Sens. Basel **6**, 8055–8072 (2012)
25. M. Padilla, A. Perera, I. Montoliu, A. Chaudry, K. Persaud, S. Marco, Drift compensation of gas sensor array data by orthogonal signal correction. Chemometr. Intell. Lab. Syst. **100**, 28–35 (2010)
26. L. Zhang, F. Tian, S. Liu, L. Dang, X. Peng, X. Yin, Chaotic time series prediction of E-nose sensor drift in embedded phase space. Sens. Actuators, B **182**, 71–79 (2013)
27. M. Holmberg, F.A.M. Davide, C. Di Natale, A. D'Amico, F. Winquist, I. Lundström, Drift counteraction in odour recognition applications: lifelong calibration method. Sens. Actuators, B **42**, 185–194 (1997)
28. L. Zhang, D. Zhang, Domain adaptation extreme learning machines for drift compensation in E-nose systems. IEEE Trans. Instrum. Meas. **64**, 1790–1801 (2015)
29. L. Zhang, F.C. Tian, C. Kadri, B. Xiao, H. Li, L. Pan, H. Zhou, On-line sensor calibration transfer among electronic nose instruments for monitor volatile organic chemical in indoor air quality. Sens. Actuators, B **160**, 899–909 (2011)

Chapter 17
Self-expression-Based Abnormal Odor Detection

Abstract Abnormal odors (e.g., perfume, alcohol) show strong sensor response, such that they deteriorate the usual usage of E-nose for target odor analysis. An intuitive idea is to recognize abnormal odors and remove them online. A known truth is that the kinds of abnormal odors are countless in real-world scenarios. Therefore, general pattern classification algorithms lose effect because it is expensive and unrealistic to obtain all kinds of abnormal odors data. In this chapter, we propose two simple yet effective methods for abnormal odor (outlier) detection. (1) A self-expression model (SEM) with l_1/l_2-norm regularizer is proposed, which is trained on target odor data for coding and then a very few abnormal odor data is used as prior knowledge for threshold learning. (2) Inspired by self-expression mechanism, an extreme learning machine (ELM)-based self-expression (SE^{2}LM) is presented. Experiments on several datasets by an E-nose system fabricated in our laboratory prove that the proposed SEM and SE^{2}LM methods are significantly effective for real-time abnormal odor detection.

Keywords Extreme learning machine · Self-expression · Odor detection
Electronic nose

17.1 Introduction

17.1.1 Background

Currently, there are commonly three challenging problems in E-nose community, which are summarized as 3D (i.e., discreteness, drift, and disturbance) issue in [1]. Specifically, the discreteness issue has been well handled in recent years by using calibration transfer methods [2–5]. The drift issue is currently a hot problem in E-nose, which is recognized to be time-varying noise and difficult to be described by some deterministic models. A number of different methods have been proposed by researchers to compensate and process the drift [6–10], and a big progress has been achieved by using transfer learning techniques. However, for the disturbance

L. Zhang et al., *Electronic Nose: Algorithmic Challenges*,
https://doi.org/10.1007/978-981-13-2167-2_17

issue (i.e., abnormal odors), there is little work in E-nose [11–13]. Specifically, in [11, 13], the authors follow a general classification route and simply take abnormal odors as one class, but neglect that there are thousands of abnormal odors which are impossible to collect. In [12], the authors attempt to establish a self-correspondence by using regression idea, i.e., predicting one sensor by using other sensors based on the target samples, but neglect the intrinsic independence between sensors. This disturbance issue is closely related to the cross-sensitivity characteristics of gas sensors. Generally speaking, during the target gases sensing by an E-nose system, the gas sensors show strong response when exposed to the disturbances (abnormal odors, e.g., perfume, alcohol). Consequently, the sensors are seriously deteriorated and the target odor detection by an E-nose comes to a failure in such application scenarios. In this chapter, we would focus on the abnormal odor detection and improve the E-nose performance in complex application scenarios (i.e., with abnormal odors).

17.1.2 Problem Statement

As claimed above, we target at solving the disturbance (i.e., abnormal odors) problem in E-nose. An intuitive idea is to recognize the abnormal odors, because the abnormal odors should be with large inter-variance by comparing to target odors (i.e., normal odors) that are detected by an E-nose. With this idea, it may not be difficult to have a rational strategy that appropriate pattern recognition algorithms can be used to train a model for classification, by treating abnormal odors as one class and target odors as another class. However, we have to face the fact that there are so many kinds of disturbances (countless) appeared in real-world air scenarios, such that the discrimination between target odors and abnormal odors cannot be simply recognized as a general pattern recognition problem, because it is expensive and unrealistic to acquire all kinds of abnormal odor data. Therefore, abnormal odor detection without prior knowledge of abnormal odor patterns is currently an urgent problem being solved.

17.1.3 Motivation

By thinking about the above problem from scratch, in our E-nose system, although the prior knowledge of abnormal odor detection is deficient, the prior knowledge of target odor data (six kinds of contaminants) can be easily obtained. Therefore, the problem becomes how to accurately detect abnormal odors by using data of target odors. Our motivations are as follows.

- For abnormal odor detection, the prior knowledge of target odor can be recognized as some invariant information, which is used for modeling some self-correspondence. Once the established self-correspondence when fed into some input is violated, it will be categorized as abnormal odors.

- To establish a self-correspondence, the prior knowledge of target odors can be modeled by using self-expression in representation-based machine learning theory.
- A fast learning algorithm for solving a single hidden layer feed-forward neural network (SLFN), known as extreme learning machine (ELM) proposed by Huang et al. [14, 15], has turned out to be the remedy for biological learning. ELM is with rather simple structure, and its speed can be thousands of times faster than the traditional network learning algorithms. Recently, ELM has been explored efficiently in hierarchal learning [16], transfer learning [17], and deep learning [18]. A deep insight into ELM theory about its learning mechanism and biological learning idea can be found in [19, 20]. Inspired by self-expression and ELM, we would like to model the self-correspondence of target odor data as a SLFN network with nonlinear activation.

With the above motivations, the research on abnormal odor detection in an E-nose system by using self-expression learning and extreme learning machine is expanded. The idea and motivation can be briefly described in Fig. 17.1, which clearly show the abnormality detection process by an E-nose system. Some other interesting applications in vision and tactile perception can be referred to as [21–25].

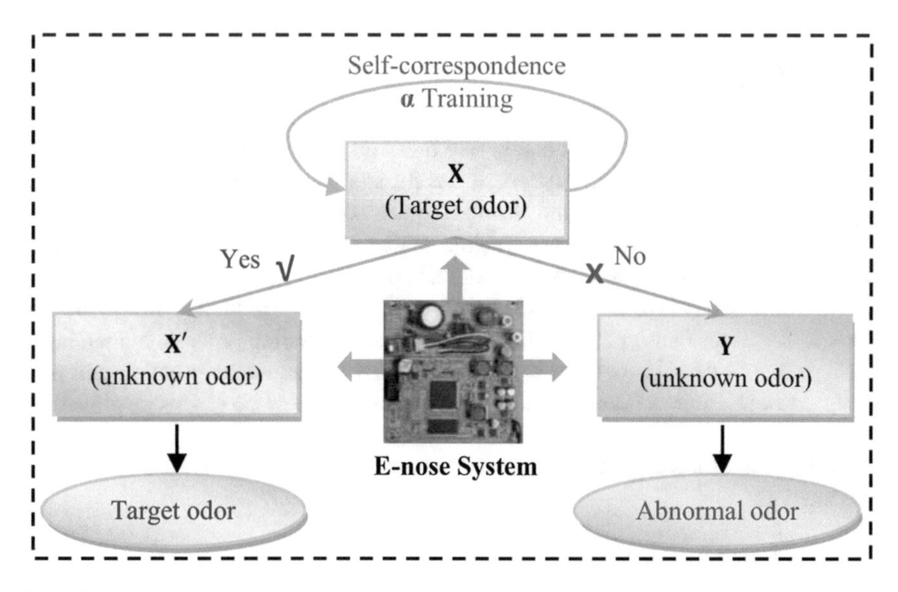

Fig. 17.1 Schematic diagram of abnormal odor detection in our E-nose system. "Yes" indicates that the unknown odor $\mathbf{X}'$ complies with the self-correspondence $\boldsymbol{\alpha}$. "No" indicates that the unknown odor $\mathbf{Y}$ violates the self-correspondence. The odor data acquisition and detection are implemented in our E-nose system

In this chapter, we propose two methods including self-expression model (SEM) and extreme learning machine-based self-expression (SE^{2}LM), for abnormal odor detection in E-nose. The contribution of this chapter can be summarized as threefold, as follows.

(1) We propose a self-correspondence concept based on the prior knowledge of target odors for abnormal odor detection, without using the prior information of abnormal odors in model training.
(2) With the representation-based learning mechanism, a self-expression model (SEM) with l_1/l_2-norm regularization is proposed in our E-nose system.
(3) Inspired by biological learning concept of extreme learning machine, a heuristic self-expression method (SE^{2}LM) is proposed in our biological olfaction (E-nose) system for abnormal odor detection.

The basic idea of self-correspondence is illustrated in Fig. 17.1, which is simply divided into two steps. First, the training is conducted for self-correspondence establishment, and the coefficient $\boldsymbol{\alpha}$ describes the self-correspondence. Second, for abnormality detection, each new pattern is represented by using the self-correspondence coefficient matrix, and representation error is computed for abnormality detection. As shown in Fig. 17.2, in testing phase, the instances $\mathbf{y}_1$, $\mathbf{y}_2$, and $\mathbf{y}_3$ indicate target odor, and therefore small errors are observed. However, the instance $\mathbf{y}_4$ indicates abnormal odor, and a big error is observed, which can then be correctly recognized as abnormal odor.

17.2 Related Work

ELM is closely related to this chapter and therefore resented as related works. The magic of ELM is that the parameters of weight and bias can be assigned randomly independent of training data and do not require computationally intensive tuning upon the data. Besides, the output weights can be solved with different constraints. ELM has been for regression and classification problems, and the activation function can be any type of piecewise continuous nonlinear hidden neurons, such as sigmoid function, Fourier function, RBF function. In learning process, hidden layer nodes (number of neurons) can be tuned in terms of the actual situation and naturally do not require an iterative adjustment. Briefly, the ELM [14] is described as follows.

In the case of clean data, the output function of ELM for generalized SLFNs is presented as

$$f(\mathbf{x}) = \sum_{i=1}^{L} \boldsymbol{\beta}_i G(\mathbf{a}_i, b_i, \mathbf{x}) \tag{17.1}$$

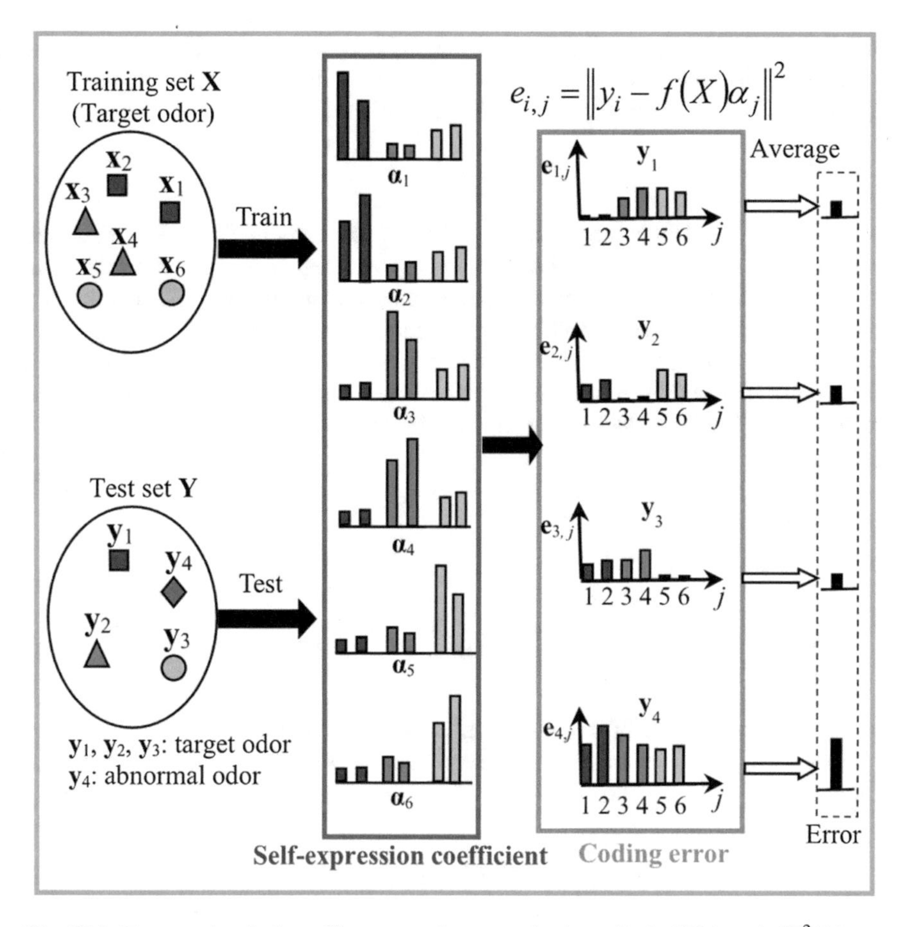

Fig. 17.2 Framework of the self-correspondence mechanism. Both SEM and SE^{2}LM are proposed based on this framework. In the training data X (target odor data), six samples of three target classes are shown for obtaining the self-correspondence coefficients ($\boldsymbol{\alpha}_1,\ldots,\ \boldsymbol{\alpha}_6$). In the testing data Y, four samples of three target classes and one abnormal class are used to calculate the coding error ($\mathbf{e}_1,\ldots,\ \mathbf{e}_4$)

where $\mathbf{x}$ is the input vector, L is the number of hidden nodes, $\mathbf{a}_i$ is the input weights, b_i is the bias of the hidden nodes, $\boldsymbol{\beta}_i$ is the output weights between the hidden layer with L nodes and the output nodes, $f(\mathbf{x})$ is the respective target output vectors, and $G(\mathbf{a}_i,\ b_i,\ \mathbf{x})$ is the output vector of the ith hidden neuron, respectively. Equation 17.1 can also be compactly written as

$$f(\mathbf{x}) = \mathbf{h}(\mathbf{x})\boldsymbol{\beta} \tag{17.2}$$

where $h_i(\mathbf{x}) = G(\mathbf{a}_i,\ b_i,\ \mathbf{x})$ is the output vector of the ith hidden neuron, thus $\mathbf{h}(\mathbf{x}) = [h_1(\mathbf{x}),\ h_2(\mathbf{x}),\ldots,\ h_L(\mathbf{x})]$ is the output vector of the hidden layer, and $\boldsymbol{\beta} = [\boldsymbol{\beta}_1,\ \boldsymbol{\beta}_2,\ \ldots,\ \boldsymbol{\beta}_L]$ is the output weights. In order to minimize the norm of the output weights, the minimal norm least square method is employed in ELM instead of the standard optimization methods. Thus, the output weights vector $\boldsymbol{\beta}$ is determined analytically using Moore–Penrose (MP) generalized inverse as

$$\boldsymbol{\beta} = \mathbf{h}(\mathbf{x})^{+}\mathbf{T} \tag{17.3}$$

where $\mathbf{T}$ is the labeled hypothesis, $\mathbf{h}(\mathbf{x})^{+}$ is the Moore–Penrose generalized pseudo-inverse of the hidden layer output matrix, and $\boldsymbol{\beta}$ has the smallest norm among all the optimization solutions, and this is the reason why ELM has better generalization performance and higher learning accuracy. According to Bartlett's neural network generalization theory, in addition to achieving smaller training error, the smaller the norms of weights are, the better generalization performance the networks tend to have, and the regularized ELM is expressed as

$$\min_{\boldsymbol{\beta}} \|\boldsymbol{\beta}\|_F^2 + C\|\mathbf{H}\boldsymbol{\beta} - \mathbf{T}\|_F^2 \tag{17.4}$$

Then, the solution can be written as

$$\boldsymbol{\beta} = \mathbf{H}^{\mathrm{T}}\left(\frac{\mathbf{I}}{C} + \mathbf{H}\mathbf{H}^{\mathrm{T}}\right)^{-1}\mathbf{T}, \quad \text{if } N \leq L \tag{17.5}$$

where N is the number of training samples and L is the number of hidden nodes.

When the number of training samples N is larger than that of nodes L, then one can have

$$\boldsymbol{\beta} = \left(\frac{\mathbf{I}}{C} + \mathbf{H}^{\mathrm{T}}\mathbf{H}\right)^{-1}\mathbf{H}^{\mathrm{T}}\mathbf{T}, \quad \text{if } N > L \tag{17.6}$$

where $\mathbf{I}$ is an identity matrix.

17.3 Self-expression Model (SEM)

17.3.1 Model Formulation

There are numerous types of abnormal odors in real-world application scenarios, which can seriously deteriorate the performance of E-nose systems. Obviously, it is expensive and unrealistic for researchers to obtain all of them in experiments as training samples. Therefore, we attempt to use the prior information of the target odors in modeling the self-correspondence. Specifically, the prior knowledge of target odors is invariant information, and thus for constructing a self-correspondence model, it is rational to imagine that a self-expression model can be designed for capturing the internal relationship (i.e., self-correspondence) among target odors. The relationship within target odors can be used to detect the abnormality if "violation" of this relationship is caused. The proposed SEM framework includes two phases: *self-correspondence learning* and *violation threshold learning*.

- *Self-correspondence* $\boldsymbol{\alpha}$ *learning*

Instinctively, the relationship can be modeled by satisfying

$$\mathbf{X} = \mathbf{X}\boldsymbol{\alpha} \tag{17.7}$$

where $\boldsymbol{\alpha} \in \Re^{N \times N}$ describes the self-correspondence and $\mathbf{X} \in \Re^{D \times N}$ denotes the training set of target odors. It is important to find a robust $\boldsymbol{\alpha}$ based on Eq. 17.7. Generally, we propose to solve $\boldsymbol{\alpha}$ by minimizing the following objective

$$\min_{\boldsymbol{\alpha}, \boldsymbol{\alpha}_{ii=0}, \forall i} \|\mathbf{X} - \mathbf{X}\boldsymbol{\alpha}\|_F^2 + \lambda \cdot R(\boldsymbol{\alpha}) \tag{17.8}$$

where $0 < \lambda \leq 1$ denotes the regularization coefficient and $R(\boldsymbol{\alpha})$ represents an appropriate regularizer formulated as

$$R(\boldsymbol{\alpha}) = \|\boldsymbol{\alpha}\|_p \tag{17.9}$$

where $\|\cdot\|_p$ indicates l_p-norm. Specifically, $p = 1$ denotes sparse constraint imposed on $\boldsymbol{\alpha}$, and $p = 2$ shows smoothness of the self-correspondence. Therefore, the SEM-sparse model is formulated as follows

$$\min_{\boldsymbol{\alpha}, \boldsymbol{\alpha}_{ii=0}, \forall i} \|\mathbf{X} - \mathbf{X}\boldsymbol{\alpha}\|_F^2 + \lambda \cdot \|\boldsymbol{\alpha}\|_1 \tag{17.10}$$

The SEM-smooth model is formulated as follows

$$\min_{\boldsymbol{\alpha}, \boldsymbol{\alpha}_{ii=0}, \forall i} \|\mathbf{X} - \mathbf{X}\boldsymbol{\alpha}\|_F^2 + \lambda \cdot \|\boldsymbol{\alpha}\|_F^2 \tag{17.11}$$

- *Violation threshold T learning*

After obtaining the self-correspondence $\boldsymbol{\alpha}$, the coding error E_X of target odor pattern $\mathbf{x}$ is calculated as

$$E_X(\mathbf{x}_j) = \frac{1}{N} \sum_{i=1}^{N} \|\mathbf{x}_j - \mathbf{X}\alpha_i\|^2, \quad j = 1, \ldots, N \tag{17.12}$$

Similarly, the coding error E_Y of abnormal odor pattern $\mathbf{y}$ is

$$E_Y(\mathbf{y}_j) = \frac{1}{N}\sum_{i=1}^{N} \|\mathbf{y}_j - \mathbf{X}\alpha_i\|^2, \quad j = 1, \ldots, n \tag{17.13}$$

The violation threshold T can be determined by uniform search between the minimum E_X (i.e., $E_{X,\min}$) and the maximum E_X (i.e., $E_{X,\max}$), until the average classification accuracy of $\mathbf{X}$ and $\mathbf{Y}$ is maximized.

$$T^* = \underset{E_{X,\min} \leq T \leq E_{X,\max}}{\arg\max} \frac{1}{2}(\text{Accuracy}(X) + \text{Accuracy}(Y)) \tag{17.14}$$

Note that for simplification, the target odors are categorized as one class (i.e., normal class). The classification accuracy is easy to be computed by using the popular coding error. Additionally, other strategies other than the average accuracy can also be used in Eq. 17.14 for determining the threshold. Once the optimal T is determined, the abnormal odor detection can be made by comparing T^* with the coding error E_z computed in Eq. 17.12 or Eq. 17.13 when given a new instance z. Without loss of generality, if $E_z \geq T^*$, then z is discriminated as some kind of abnormal odor. Otherwise, z is recognized to be one kind of target odors.

17.3.2 Algorithm

According to the SEM framework, two steps in training phase are included as follows.

For the first step, two models in Eqs. 17.10 and 17.11 are presented based on l_1/l_2-norm regularizer.

When l_1-norm constraint on $\boldsymbol{\alpha}$ is considered, Eq. 17.10 is a sparse optimization problem and can be easily solved by a standard *Lasso* solver [26]. Generally, the update strategy of $\alpha_{i,j}$ is shown as

$$\alpha_{i,j} = \text{sign}(\alpha_{i,j})\left(|\alpha_{i,j}| - \frac{\lambda}{2}\right)_+ \tag{17.15}$$

where $\left(|\alpha_{i,j}| - \frac{\lambda}{2}\right)_+ = \max\left(|\alpha_{i,j}| - \frac{\lambda}{2}, 0\right)$.

When l_1-norm constraint on $\boldsymbol{\alpha}$ is considered, Eq. 17.11 is a least square optimization problem, and a closed-form solution can be induced as follows.

$$\boldsymbol{\alpha} = \left(\mathbf{X}^{\mathrm{T}}\mathbf{X} + \lambda \cdot \mathbf{I}\right)^{-1}\mathbf{X}^{\mathrm{T}}\mathbf{X} \tag{17.16}$$

Specifically, the detailed implementation of the whole SEM framework for abnormality detection (abnormal odor) is summarized as Algorithm 17.1.

Algorithm 17.1. SEM (SEM-sparse vs. SEM-smooth)

Algorithm 1. SEM (SEM-sparse vs. SEM-smooth)

Input:

The training data $\mathbf{X} \in \mathbb{R}^{D\times N}$ and $\mathbf{Y} \in \mathbb{R}^{D\times n}$;

Parameter λ;

Procedure:

- *Phase 1: self-correspondence* $\boldsymbol{\alpha}$ *learning*

  if l_1-norm constraint is used (p=1), solve Eq. 17.10 by using *Lasso* operator (SEM-sparse)

  for i, j=1 to N

  Initialize $\alpha_{i,j} = \mathbf{x}_j^{\mathrm{T}}\mathbf{x}_i$;

  Update $\alpha_{i,j}$ by using Eq. 17.15;

  end

  else if l_2-norm constraint is used (p=2), solve Eq. 17.11 by using *least-square* operator (SEM-smooth)

  Compute the close-form solution $\boldsymbol{\alpha}$ by using Eq. 17.16;

- *Phase 2: violation threshold T learning*

  Compute E_{X} and E_{Y} using Eq. 17.12 and 17.13;

  Compute the optimal T^* by solving Eq. 17.14.

Output: $\boldsymbol{\alpha}$ and T^*.

17.4 Extreme Learning Machine-Based Self-expression Model (SE^{2}LM)

17.4.1 Model Formulation

In Eq. 17.7, the self-expression is purely linear. Inspired by ELM theory, we propose to establish the self-correspondence by using the nonlinearity activated data in SE^{2}LM space (i.e., hidden layer output). Suppose the random hidden layer output of SE^{2}LM to be $\mathbf{H} \in \Re^{D\times L}$, then Eq. 17.7 can be further written as

$$\mathbf{X} = \mathbf{H}\boldsymbol{\alpha} \tag{17.17}$$

where $\boldsymbol{\alpha} \in \Re^{L\times N}$ denotes the output weights between hidden layer and output layer, and $\mathbf{H}$ is represented as follows

$$\mathbf{H}=\begin{bmatrix} h(\mathbf{w}_1\mathbf{x}_1+b_1) & h(\mathbf{w}_2\mathbf{x}_1+b_2) & \cdots & h(\mathbf{w}_L\mathbf{x}_1+b_L) \\ h(\mathbf{w}_1\mathbf{x}_2+b_1) & h(\mathbf{w}_2\mathbf{x}_2+b_2) & \cdots & h(\mathbf{w}_L\mathbf{x}_2+b_L) \\ \vdots & \vdots & \vdots & \vdots \\ h(\mathbf{w}_1\mathbf{x}_D+b_1) & h(\mathbf{w}_2\mathbf{x}_D+b_2) & \cdots & h(\mathbf{w}_L\mathbf{x}_D+b_L) \end{bmatrix} \tag{17.18}$$

where $h(\cdot)$ indicates the activation function, such as sigmoid, Gaussian function, L denotes the number of hidden nodes, $\mathbf{W}=[\mathbf{w}_1,\ldots,\mathbf{w}_L]\in\Re^{N\times L}$ is the randomly generated weights between input layer and hidden layer, and $\mathbf{B}=[b_1,\ldots,b_L]^{\mathrm{T}}\in\Re^L$ is the randomly generated bias for hidden nodes.

The structure of SE^{2}LM is shown in Fig. 17.3, which is similar to ELM, yet there are some differences in nodes design.

Specifically, the proposed SE^{2}LM model is formulated as

$$\begin{aligned} &\min_{\boldsymbol{\alpha},\xi_i,\forall i}\frac{1}{2}\|\boldsymbol{\alpha}\|_F^2+\frac{1}{2}\mu\cdot\sum_{i=1}^{N}\|\xi_i\|^2 \\ &\text{s.t. } \xi_i=\mathbf{x}_i-\mathbf{H}\boldsymbol{\alpha}_i,\quad i=1,\ldots,N \end{aligned} \tag{17.19}$$

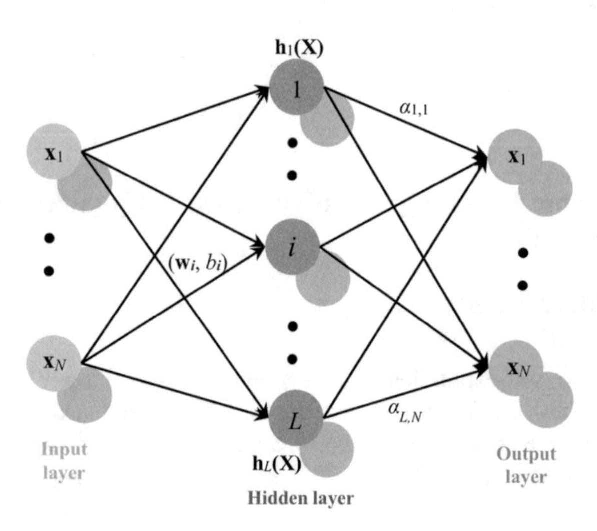

Fig. 17.3 Network structure of SE^{2}LM. The difference between this structure and ELM lies in that the number of input nodes is associated with the dimensionality. In SE^{2}LM, it serves for representing each sample by using hidden layer output and the analytically determined output weights $\boldsymbol{\alpha}$. It implies that the ELM space represents the dictionary space. Note that an interesting aspect is that in ELM, the number of input nodes and output nodes can be the dimension D, which would become a transformation problem. If N is set, it is expression problem focused on this chapter. Also, in the proposed structure, each node is composed of D sub-nodes (i.e., shadow nodes)

The model in Eq. 17.19 can be compactly written as

$$\begin{aligned} &\min_{\boldsymbol{\alpha},\boldsymbol{\xi}} \frac{1}{2}\|\boldsymbol{\alpha}\|_F^2 + \frac{1}{2}\mu \cdot \|\boldsymbol{\xi}\|_F^2 \\ &\text{s.t. } \boldsymbol{\xi} = \mathbf{X} - \mathbf{H}\boldsymbol{\alpha} \end{aligned} \tag{17.20}$$

The model can also be explained as that each sample can be represented by the dictionary $\mathbf{H}$ through the coefficient $\boldsymbol{\alpha}$. Additionally, the proposed SE^{2}LM inherits the advantages of ELMs. The objective is to learn the self-correspondence coefficients $\boldsymbol{\alpha}$, based on the fixed dictionary $\mathbf{H}$. That is, the SE^{2}LM is only proposed in training process.

17.4.2 Algorithm

Similar to SEM framework, in SE^{2}LM framework, the same two phases are included.

- *Self-correspondence $\boldsymbol{\alpha}$ learning*

The optimization of SE^{2}LM model Eq. 17.20 can be easily conducted, by following similar induction with ELM. Specifically, the closed-form solution of $\boldsymbol{\alpha}$ can be described as follows.

$$\boldsymbol{\alpha} = \begin{cases} \mathbf{H}^{\mathrm{T}}\left(\frac{\mathbf{I}}{\mu} + \mathbf{H}\mathbf{H}^{\mathrm{T}}\right)^{-1}\mathbf{X}, & \text{if } D \le L \\ \left(\frac{\mathbf{I}}{\mu} + \mathbf{H}^{\mathrm{T}}\mathbf{H}\right)^{-1}\mathbf{H}^{\mathrm{T}}\mathbf{X}, & \text{if } D > L \end{cases} \tag{17.21}$$

The deduction of Eq. 17.21 is similar to the standard ELM framework, by considering the property of hidden matrix $\mathbf{H}$.

- *Violation threshold T learning*

After obtaining the self-correspondence $\boldsymbol{\alpha}$, similar to Eqs. 17.12 and 17.13, the coding error E_X of target odor pattern $\mathbf{x}$ is calculated as Eq. 17.22

$$E_X(\mathbf{x}_j) = \frac{1}{N}\sum_{i=1}^{N} \left\|\mathbf{x}_j - \mathbf{H}\alpha_i\right\|^2, \quad j = 1, \ldots, N \tag{17.22}$$

Similarly, the coding error E_Y of abnormal odor pattern $\mathbf{y}$ is

$$E_Y(\mathbf{y}_j) = \frac{1}{N}\sum_{i=1}^{N} \left\|\mathbf{y}_j - \mathbf{H}\alpha_i\right\|^2, \quad j = 1, \ldots, n \tag{17.23}$$

The search process of the optimal T is similar to Eq. 17.14.

Specifically, the whole process for abnormality detection of SE^{2}LM framework is summarized as Algorithm 17.2.

Algorithm 17.2. SE^{2}LM
Input:
The training data $\mathbf{X} \in \mathbb{R}^{D \times N}$ and $\mathbf{Y} \in \mathbb{R}^{D \times n}$;
Parameter μ;
Procedure:

- *Phase 1: self-correspondence* $\boldsymbol{\alpha}$ *learning*

  Generate the input weights $\mathbf{W}$ and hidden bias $\mathbf{B}$ randomly.
  Compute the hidden layer matrix $\mathbf{H}$ by using Eq. 17.18.
  Compute the output weights $\boldsymbol{\alpha}$ by using Eq. 17.21.

- *Phase 2: violation threshold T learning*

  Compute E_X and E_Y using Eqs. 17.22 and 17.23.
  Compute the optimal T^* by solving Eq. 17.14.

Output: $\boldsymbol{\alpha}$ and T^*.

17.5 Experiments

17.5.1 Electronic Nose and Data Acquisition

Our electronic nose system and experimental setup developed in this chapter have been described previously in [9]. The E-nose system is composed of an array of metal oxide semiconductor sensors, which includes TGS2602, TGS2620, TGS2201A, and TGS2201B as shown in Fig. 11.4. Additionally, the gas sensors are also sensitive to the environmental variables, such as temperature and humidity, and result in an impact on concentration measure and discrimination of gases. Therefore, a module of temperature and humidity (i.e., STD2230-I2C), which is used to measure the ambient temperature and humidity, is also integrated into our E-nose system. The real-time response of this module has been used as feature variables in our algorithms for environmental compensation. In this chapter, six kinds of target odors/gases including formaldehyde (HCHO), benzene (C_6H_6), toluene (C_7H_8), carbon monoxide (CO), ammonia (NH_3), and nitrogen dioxide (NO_2) are being detected by our E-nose.

That is, other odors except the six target odors will be uniformly recognized to be abnormal odors. In addition to computing the recognition accuracy, we also collect two extra real-time sequences for validating the effectiveness of the proposed frameworks in real-time application scenarios. The data acquisition

Table 17.1 Target odor dataset for training and testing

Target gases	Formaldehyde	Benzene	Toluene	CO	NO_2	NH_3	Total
Number of total samples	188	72	66	58	38	60	482
Number of samples for training $\boldsymbol{\alpha}$	75	29	27	23	15	24	193
Number of samples for training T	75	29	27	23	15	24	193
Number of test samples	38	14	12	12	8	12	96

experiments were measured in a gas chamber, where the E-nose system was fixed. The odor sample (target odor and abnormal odor) mixed with pure nitrogen (i.e., N_2) is collected in a gas bag, and an air pump is used to transfer the odor from the bag to the chamber, controlled by a flowmeter for different concentrations. During the measurements, the temperature and relative humidity of the gas chamber are within 10–40 °C and 40–80% RH.

The E-nose systems and the data collection are following the same settings as Chap. 15. Therefore, the experimental setup and the three datasets are not presented repeatedly. The target odor dataset for training and testing is shown in Table 17.1.

17.5.2 Abnormal Odor Detection Based on Dataset 1

The training performance of the proposed SEM and SE^{2}LM frameworks is relevant to the data amount during training $\boldsymbol{\alpha}$ and T. In experiments, to observe the performance impact with respect to the number of training samples in training $\boldsymbol{\alpha}$, 10, 15, 20, 25, 30, 35, and 40 samples per class of the training set $\mathbf{X}$ are explored for sample balance, respectively. Because the number of samples for some target odor shown in Table 17.2 is less than the maximum value (i.e., 40), we repeat the sample selection randomly for sample balance. The recognition accuracy is shown in Table 17.2. For SE^{2}LM method, sigmoid function and Gaussian (RBF) function are used as activation function separately. From the results, we can see the best average performance when 30 samples per class are used in training set. The recognition accuracy of target odors is 90.91%, and the accuracy of abnormal odors is 91.67%. Note that, we show the average performance, because in Eq. 17.14 the average accuracy is used as criteria in searching the optimal violation error threshold T. Additionally, we could observe that SE^{2}LM method outperforms SEM method for different settings. The SEM with sparse l_1-norm constraint achieves 89.78%, which is much better than SEM with smooth l_2-norm constraint (84.73%). This demonstrates that the self-correspondence coefficients $\boldsymbol{\alpha}$ should be sparse for robust self-expression.

Similarly, for observing the impact with respect to the number of training samples in searching T, 10, 15, 20, 25, 30, 35, and 40 training samples per class in training set $\mathbf{X}$ are explored, respectively. The recognition accuracies are shown in

Table 17.2 Recognition accuracy (%) of abnormal odor detection under a different number of training samples per class for training **α**

Number of samples per class		40	35	30	25	20	15	10
SEM-sparse (l_1-norm)	Target odor	98.99	98.99	87.88	59.6	52.53	17.17	0
	Abnormal odor	75	75	91.67	100	100	100	100
	Average	87	87	**89.78**	79.8	76.27	58.59	50
SEM-smooth (l_2-norm)	Target odor	98.99	96.97	77.78	48.48	3.03	2.51	0
	Abnormal odor	50	58.33	91.67	100	100	100	100
	Average	74.5	77.65	**84.73**	74.24	51.52	51.75	50
SE^{2}LM (sigmoid)	Target odor	100	92.93	90.91	80.81	72.73	34.34	15.15
	Abnormal odor	66.67	83.33	91.67	100	100	100	100
	Average	83.34	88.13	**91.29**	90.4	86.37	67.17	57.57
SE^{2}LM (Gaussian)	Target odor	100	91.92	90.91	76.77	57.58	52.53	5.05
	Abnormal odor	66.67	83.33	91.67	100	10	100	100
	Average	83.34	87.63	**91.29**	88.39	78.77	76.27	52.53

The boldface type denotes the best performance.

Table 17.3. We can observe that the best average accuracy is 91.29% when 25 samples per class are used. Also, it turns out to be that SE^{2}LM not only outperforms SEM but also shows better stability when fewer training samples are used. Additionally, SEM-based methods show imbalanced recognition between target odor and abnormal odor. Specifically, we have shown the performance variation curves w.r.t. the threshold T in searching process in Fig. 17.4. We can observe that with increasing of T, the recognition rate of target odor is decreasing because the rejection rate of target odor is increasing. In contrast, the recognition rate of abnormal odor is increasing. Clearly, the near-optimal T appears in their cross-point region. From Fig. 17.4, SE^{2}LM-based method shows better detection performance and lower bias for both target and abnormal odors. Note that the scale of T may be different which depends on the method. Also, the model parameters λ and μ are tuned in the range of 10^{-4} and 10^4. For different tasks, the optimal model parameters may be different during the learning process.

Through the comparisons shown in Tables 17.2 and 17.3, we can observe that the results based on SE^{2}LM are better than that of SEM-based methods. Generally, if we simply treat the target/abnormal odor recognition as a binary classification problem, the recognition accuracy by using conventional ELM classifier is shown in Table 17.4. Note that ELM (sigmoid) denotes the ELM classifier based on sigmoid kernel function. The principle of pattern mismatch-based interference elimination (PMIE) [12] is that a similar self-correspondence is established by regression between sensors based on target odor data, which is a regression idea. Specifically, PMIE uses five sensors to predict the remaining sensor for target odors and search an optimal prediction error threshold. As shown in Table 17.4, we can observe that the results with general binary classification method between target

Table 17.3 Recognition accuracy (%) of abnormal odor detection under a different number of training samples per class for training **T**

Number of samples per class		40	35	30	25	20	15	10
SEM-sparse (l_1-norm)	Target odor	98.99	98.99	78.79	75.76	74.75	94.95	98.99
	Abnormal odor	41.67	41.67	100	100	100	83.33	50
	Average	70.33	70.33	**89.4**	87.88	87.38	89.14	74.5
SEM-smooth (l_2-norm)	Target odor	85.86	89.9	51.52	79.8	94.95	86.42	74.75
	Abnormal odor	91.67	83.33	100	100	58.33	56.83	50
	Average	88.77	86.62	75.76	**89.9**	76.64	71.63	62.38
SE^{2}LM (sigmoid)	Target odor	81.82	81.82	89.9	90.91	81.82	81.82	80.81
	Abnormal odor	100	100	91.67	91.67	100	100	100
	Average	90.91	90.91	90.79	**91.29**	90.91	90.91	90.41
SE^{2}LM (Gaussian)	Target odor	81.82	81.82	85.86	80.81	81.82	80.81	80.81
	Abnormal odor	100	100	100	100	100	100	100
	Average	90.91	90.91	**92.93**	90.41	90.91	90.41	90.41

The boldface type denotes the best performance.

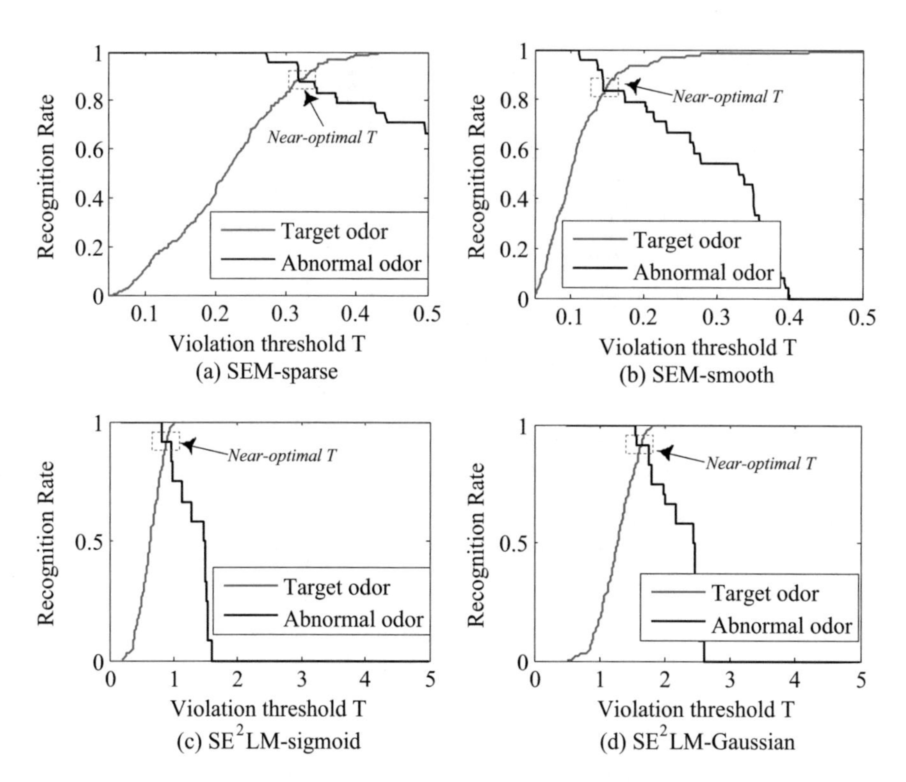

Fig. 17.4 Performance variation with respect to the violation threshold T for target and abnormal odor. The near-optimal T is labeled in rectangle region (the cross-point)

Table 17.4 Detection accuracy (%) of different methods

Number of samples per class		ELM (Gaussian)	PMIE (TGS2602)	PMIE (TGS2620)	PMIE (TGS2201A)	PMIE (TGS2201B)	SEM-sparse (l_1-norm)	SEM-smooth (l_2-norm)	SE^{2}LM (Gaussian)
10	Target	84.67	82.42	85.93	68.88	74.72	98.99	74.75	80.81
	Abnormal	82.86	86.71	88	58.33	81.19	50	50	100
	Average	83.77	84.57	86.97	63.61	77.96	74.5	62.38	**90.41**
20	Target	77.45	85.93	86.34	92.12	78.34	74.75	94.95	81.82
	Abnormal	62.47	88.26	92.88	66.67	75	100	58.33	100
	Average	69.96	87.10	89.61	79.39	76.67	87.38	76.64	**90.91**
30	Target	52.84	87.14	87.37	75.55	84.27	78.79	51.52	85.86
	Abnormal	37.6	89.0	94.12	81.67	66.67	100	100	100
	Average	45.22	88.07	90.75	78.61	75.47	89.4	75.76	**92.93**

The boldface type denotes the best performance.

odor (positive class) and abnormal odor (negative class) are much worse than the proposed SEM and SE^{2}LM methods. Besides, binary classification-based method should rely on all kinds of abnormal odors in real-world scenarios, which is expensive and unrealistic in electronic nose. Therefore, both the results and reality demonstrate that abnormal odor detection cannot be simply recognized as a binary classification problem. The truth also confirms the difficulty of problem, significance of our motivations, and the proposed methods. As shown in Table 17.4, the proposed SE^{2}LM method still outperforms other binary classification-based abnormal odors detection methods.

17.5.3 Validation of Real-Time Sequence on Dataset 2

As expressed in experimental data (i.e., dataset 2), this dataset was collected in real time and used for validating the proposed SEM and SE^{2}LM methods. The abnormal odor region recognition results for different methods are shown in Fig. 17.5, in which the rectangular windows are represented as detected abnormal odor regions (i.e., disturbance). In addition to the qualitative recognition of regions, we have described the receiver operating characteristic curve (ROC) on this validation dataset 2 in Fig. 17.6a, by computing true positive rate and false positive rate by adjusting the threshold *T*.

17.5.4 Validation of Real-Time Sequence on Dataset 3

The abnormal odor region recognition results for different methods are shown in Fig. 17.7, where the regions labeled by rectangular windows are indicated as abnormal odor regions. The effectiveness of the proposed methods is clearly demonstrated. The ROC curves are shown in Fig. 17.6b, and it shows that SE^{2}LM is better.

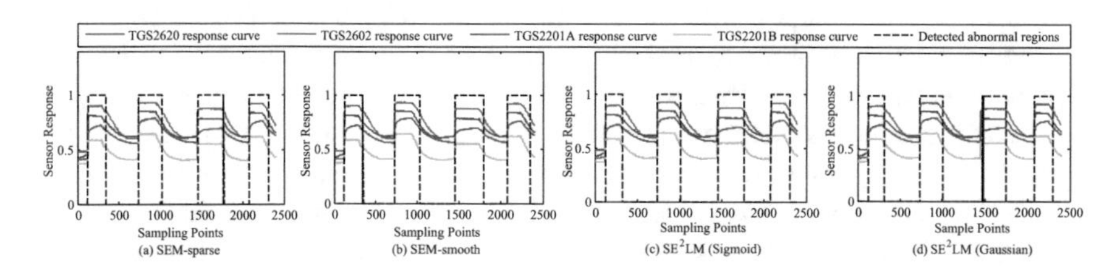

Fig. 17.5 Detected abnormal regions based on dataset 2

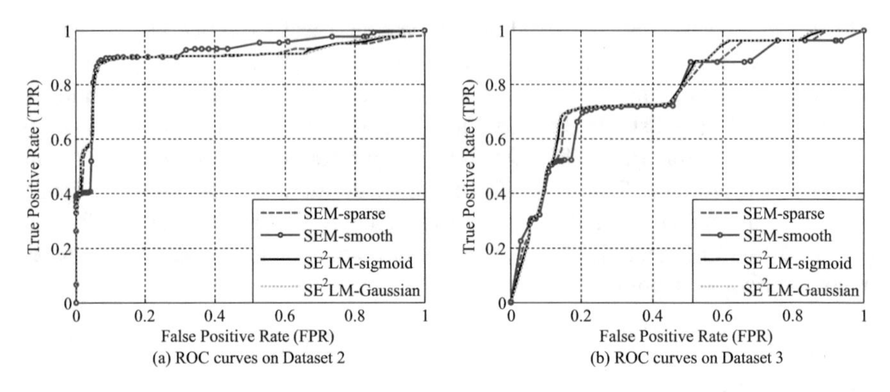

Fig. 17.6 ROC curves on real-time validation odor sequences

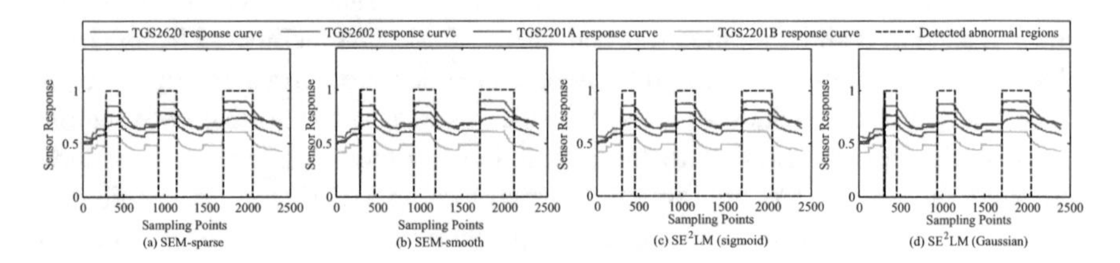

Fig. 17.7 Detected abnormal regions based on dataset 3

17.6 Discussion

The key idea behind the proposed methods is to construct an internal relationship (i.e., self-correspondence) based on target odor data, such that the abnormal odor (i.e., disturbance) can be detected once the established relationship is violated. The rationality and motivation behind are that it is hard and even impossible to collect all kinds of abnormal odors (countless) in real-world application scenarios by using an E-nose system. That is, the detection of abnormal odors cannot be simply recognized to be a binary classification problem. Therefore, we have to rely on the known prior knowledge of the target odors and establish a self-correspondence. During the compared method, the PMIE method [12] actually relies on a regression idea, which attempt to construct a similar self-correspondence between sensors (i.e., five sensors are used to approximate the remaining one sensor for target odors), but neglect the assumption of internal independence between sensors. In our methods, a very necessary step is also to search an optimal violation threshold T (i.e., an appropriate decision bound). Actually, the value of bound T is important to decide rejection rate or acceptance rate of an unknown odor, which is also an important and difficult problem in machine learning.

17.7 Summary

In this chapter, we focus on the challenge of abnormal odor detection (i.e., disturbance) in E-nose. We propose two frameworks, i.e., SEM and SE^2LM for abnormal odor detection, which consist of two general phases: self-correspondence establishment (i.e., self-expression $\boldsymbol{\alpha}$) and violation threshold T search. The strength of the proposed methods is twofold: (1) The self-correspondence α is easily implemented by using target odor data as invariant information; (2) the search of violation threshold T is conducted by using a very few abnormal odor data as prior knowledge, without considering countless kinds of abnormal odors in surroundings. Numerous experiments on our E-nose system were conducted, and the results demonstrate the effectiveness of the proposed methods. Particularly, in comparisons, the SE^2LM method shows a superior performance in real-world application scenarios.

References

1. L. Zhang, D. Zhang, Efficient solutions for discreteness, drift and disturbance (3D) in electronic olfaction, in *IEEE Transactions on Systems, Man, and Cybernetics: Systems* (2016). https://doi.org/10.1109/tsmc.2016.2597800
2. K. Yan, D. Zhang, Calibration transfer and drift compensation of e-noses via coupled task learning. Sens. Actuators B: Chem. **225**, 288–297 (2016)
3. L. Zhang, F.C. Tian, X.W. Peng, X. Yin, A rapid discreteness correction scheme for reproducibility enhancement among a batch of MOS gas sensors. Sens. Actuators A: Phy. **205**, 170–176 (2014)
4. J. Fonollosa, L. Fernandez, A. Gutierrez-Galvez, R. Huerta, S. Marco, Calibration transfer and drift counteraction in chemical sensor arrays using direct standardization. Sens. Actuators B: Chem. **236**, 1044–1053 (2016)
5. L. Zhang, F. Tian, C. Kadri, B. Xiao, H. Li, L. Pan, H. Zhou, On-line sensor calibration transfer among electronic nose instruments for monitoring volatile organic chemicals in indoor air quality. Sens. Actuators B: Chem. **160**, 899–909 (2011)
6. M. Padilla, A. Perera, I. Montoliu, A. Chaudry, K. Persaud, S. Marco, Drift compensation of gas sensor array data by orthogonal signal correction. Chemo. Intelli. Lab. Syst. **100**(1), 28–35 (2010)
7. A. Ziyatdinov, S. Marco, A. Chaudry, K. Persaud, P. Caminal, A. Perera, Drift compensation of gas sensor array data by common principal component analysis. Sens. Actuators B: Chem. **146**(2), 460–465 (2010)
8. H. Ding, J.-H. Liu, Z.-R. Shen, Drift reduction of gas sensor by wavelet and principal component analysis. Sens. Actuators B: Chem. **96**(1–2), 354–363 (2003)
9. L. Zhang, D. Zhang, Domain adaptation extreme learning machines for drift compensation in E-nose systems. IEEE Trans. Instrum. Meas. **64**(7), 1790–1801 (2015)
10. K. Yan, D. Zhang, Correcting instrumental variation and time-varying drift: a transfer learning approach with autoencoders. IEEE Trans. Instrum. Meas. **65**(9), 2012–2022 (2016)
11. L. Zhang, F. Tian, L. Dang, G. Li, X. Peng, X. Yin, S. Liu, A novel background interferences elimination method in electronic nose using pattern recognition. Sens. Actuators, A **201**, 254–263 (2013)

12. F. Tian, Z. Liang, L. Zhang, Y. Liu, Z. Zhao, A novel pattern mismatch based interference elimination technique in E-nose. Sens. Actuators B: Chem. **234**, 703–712 (2016)
13. E. Phaisangittisagul, H.T. Nagle, Enhancing multiple classifier system performance for machine olfaction using odor-type signatures. Sens. Actuators B Chem. **125**(1), 246–253 (2007)
14. G.B. Huang, Q.Y. Zhu, C.K. Siew, Extreme learning machine: theory and applications. Neurocomputing **70**(1–3), 489–501 (2006)
15. G.B. Huang, H. Zhou, X. Ding, R. Zhang, Extreme learning machine for regression and multiclass classification. IEEE Trans. Syst. Man Cybern. B Cybern. **42**(2), 513–529 (2012)
16. J. Tang, C. Deng, G.B. Huang, Extreme learning machine for multilayer perceptron. IEEE Trans. Neural Netw. Learn. Syst. **27**(4), 809–821 (2015)
17. L. Zhang, D. Zhang, Robust visual knowledge transfer via extreme learning machine based domain adaptation, in *IEEE Transactions Image Processing* (2016). https://doi.org/10.1109/tip.2016.2598679
18. G.B. Huang, Z. Bai, L.L.C. Kasun, C.M. Vong, Local receptive fields based extreme learning machine. IEEE Comput. Intell. Mag. **10**(2), 18–29 (2015)
19. G.B. Huang, What are extreme learning machines? Filling the gap between Frank Rosenblatt's dream and John von Neumann's puzzle. Cogn. Comput. **7**, 263–278 (2015)
20. G.B. Huang, An insight into extreme learning machines: random neurons, random features and kernels. Cogn. Comput. **6**, 376–390 (2014)
21. F. Sun, C. Liu, W. Huang, J. Zhang, Object classification and grasp planning using visual and tactile sensing. IEEE Trans. Syst., Man, Cybern: Syst. **46**(7), 969–979 (2016)
22. H. Kaindl, M. Vallee, E. Arnautovic, Self-representation for self-configuration and monitoring in agent-based flexible automation systems. IEEE Trans. Syst., Man, Cybern.: Syst. **43**(1), 164–175 (2013)
23. V. Bruni, D. Vitulano, An improvement of kernel-based object tracking based on human perception. IEEE Trans. Syst. Man Cybern.: Syst. **44**(11), 1474–1485
24. C. Aytekin, Y. Rezaeitabar, S. Dogru, I. Ulusoy, Railway fastener inspection by real-time machine vision. IEEE Trans. Syst. Man Cybern.: Syst. **45**(7), 1101–1107
25. L. Zhang, W.M. Zuo, D. Zhang, LSDT: latent sparse domain transfer learning for visual adaptation. IEEE Trans. Image Process. **25**(3), 1177–1191 (2016)
26. R. Tibshirani, Regression shrinkage and selection via the Lasso. J. Roy. Stat. Soc. B **58**, 267–288 (1996)

Part V
E-Nose Discreteness Correction: Challenge IV

Chapter 18
Affine Calibration Transfer Model

Abstract Reproducibility of E-nose is an important issue. An inherent problem of MOS sensors is that the signal is different when exposed to the same surroundings. This is caused by the discreteness of MOS sensors. Therefore, this chapter shows how to solve this challenge of discreteness issue in E-nose. In terms of the homogeneous linearity between multi-sensors systems, an on-line calibration transfer model based on global affine transformation (GAT) and Kennard–Stone sequential (KSS) algorithm is presented and evaluated in this chapter. GAT is achieved in terms of one single sensor by a robust weighted least square (RWLS) algorithm, and KSS is studied for representative transfer sample subset selection from a large sample space. This chapter consists of two aspects: calibration step (for responses of sensors) and prediction step (for gas concentration). Prediction is developed to evaluate the performance of calibration transfer. In prediction step, three artificial neural networks for concentration prediction of three analytes were trained based on back-propagation algorithm. Experimental results show that the reproducibility can be significantly improved by using GAT method, and the discreteness problem is solved.

Keywords Electronic nose · Calibration transfer · Affine transformation
Kennard–Stone sequential algorithm · Robust weighted least square

18.1 Introduction

The inherent variability during the sensor manufacturing process leads to slight differences in the reactivity of the tin oxide substrate of individual sensor [1]. For instance, when two identical sensors are exposed to the same environment, slightly different responses would be produced. One reason is that state-of-the-art gas sensors do not maintain their sensitivity profile over time. Even brand new gas sensors, coming from the same production batch, may not generate identical response values when measuring identical samples. Other possible reasons are

L. Zhang et al., *Electronic Nose: Algorithmic Challenges*,
https://doi.org/10.1007/978-981-13-2167-2_18

related to the physical environment (e.g., temperature, humidity, and pressure). The whole purpose and why we do calibration in this chapter would be fully explained as follows.

As we know, traditional E-nose instruments which have a big volume are inconvenient and expensive to the users. Our project is devoted to design of portable E-nose used indoor so that many instruments should be produced for extensive users. However, in mass production of E-nose products such as air quality monitors, it is impossible to train an individual prediction model (e.g., ANN) on each E-nose product (instrument). However, it is convenient for us to construct an effective model on one standard instrument only (e.g., master) through a large amount of analytical chemical experiments. Unfortunately, due to the baseline differences of identical sensors, the developed prediction model using the chemical datasets on the master E-nose instrument may not be fitted with the datasets on other slave instruments even with completely the same sensor array and electrical components as the master instrument. Thereby, in real-time monitoring of environmental chemicals indoor, the displayed concentrations would not be identical and even false discriminations of gas concentrations for multiple instruments become possible. So, a well-developed ANN with high prediction accuracy would lose adaptability with new instruments. Considering the intensive complexity of repeated ANN training process for each instrument, it is impractical to construct one individual ANN for each instrument, respectively. Sensor calibration is therefore the first step during the mass developments of E-nose instruments. A good promotion of E-nose instruments for air quality monitoring should be based on a high performance of calibration transfer. Thus, on-line calibration transfer among E-nose instruments in mass production, with subsequent prediction of concentrations in unknown samples using electronic nose instruments, has become increasingly important for manufacturers and researchers in artificial olfactory field. Manufacturers aim to construct instruments that generate exactly the same gas concentrations when exposed to identical environments.

Until now, so many literature on spectrophotometers instrument standardization based on transferring near-infrared (NIR) has been published, such as classical direct standardization (DS) and piece direct standardization (PDS) [2, 3], orthogonal signal correction (OSC) [4], wavelet [5], and principle component regression (PCR) [6]. But these methods can only be used in the multi-dimensional E-nose system directly for off-line instrument calibration, because they should be implemented based on a dataset, and also the robustness can also be promised when the responses of the sensors measured on a single instrument change over a period of time because of temperature and humidity fluctuation. That is to say, when the model studied based on a known dataset is used to another new dataset, the model would lose its effect. So, these methods cannot be used for on-line calibration. However, so a few literature [7–9] on calibration transfer between electronic nose instruments with multi-dimensional sensor array in artificial olfactory fields inspire us to have a deep research on this issue toward attempting to solve the E-nose instrument-related signal shift for on-line calibration. In the existent literature, neural methods were used for sensor array calibration. Because a generalized neural network model should be based on a large number of transfer samples or datasets, it would certainly increase the experimental calibration complexity, especially in mass production. In this case, the neural methods would also become worthless. The on-line mass standardization is now still a bottleneck problem because of the multi-dimensional nonlinear characteristic of E-nose system.

The goal of this chapter is to develop a tin oxide sensor device (E-nose) capable of detecting and quantifying volatile organic chemicals present at typical indoor environments. In the previous publications [10–12], the E-nose sensor array and circuit device, feature selection, and probabilistic neural network for wound classification have been studied through theoretical and experimental analysis. With knowledge of the approximated relation of linearity (homogeneous linearity) between two E-nose instruments with the same types of sensors, and the inherent flaws of the previous methods discussed above, a brand new method for on-line mass E-nose instruments calibration is proposed to realize the high-accuracy standardization between instruments in this chapter. The applied methodology for calibration is global affine transformation (GAT) and Kennard–Stone sequential (KSS) algorithm. Therein, the solutions of affine transformation coefficients are studied using a robust weighted least square (RWLS) algorithm and the sample selection is on the basis of Euclidean distance. Affine transformation was widely used for pattern matching [13, 14]. The concentration prediction step is also developed for validation of the calibration transfer. Considering the strong non-linearity of multi-sensors system internal (e.g., response and concentration), a feed-forward multilayer perceptron neural network based on a back-propagation algorithm [15] is used for organic gas concentration prediction. As we know, artificial neural networks have been widely used for concentration prediction with electronic noses [16–20].

18.2 Method

18.2.1 Calibration Step

- Affine Transform based on RWLS

Affine transformation is a map F: $\mathfrak{R}^n \rightarrow \mathfrak{R}^n$ of the form $F(x) = \boldsymbol{A}^{\mathrm{T}}\boldsymbol{x} + \boldsymbol{t}$, for all $x \in \mathfrak{R}^n$, where $\boldsymbol{A}$ is a linear transformation of $\mathfrak{R}^n$ [21] and "T" denotes transpose of matrix. Scaling, translation, and rotation are included in affine transformation, but the first two (scaling and translation) are used here. Let $\boldsymbol{x}$ denote the dataset measured on the slaved E-nose instrument and $\boldsymbol{y}$ the calibrated dataset from $\boldsymbol{x}$ to the master E-nose instrument. The calibration transfer model is shown as follows

$$\boldsymbol{y}_{i,n} = \boldsymbol{a}_i \times \boldsymbol{x}_{i,n} + \boldsymbol{b}_i, \quad i = 1,\ldots,k, \quad n = 1,\ldots,N \tag{18.1}$$

in which index i indicates the ith sensor, k denotes the number of sensors in the sensor array, index n represents the nth sample, N refers to the number of being calibrated samples, and parameters $\boldsymbol{a}_i$ and $\boldsymbol{b}_i$ represent the on-line calibration transfer coefficients slope and intercept of the ith sensor, respectively. The coefficients are studied using hybrid robust weighted least square algorithm and KSS algorithm.

Robust weighted least square (RWLS) in this work is first applied to electronic nose datasets for accurately mapping one instrument (called "slave") to another (called "master") to realize instruments standardization. Suppose the number of transfer samples to be V, with knowledge of that ordinary least square (OLS) which is based on the minimum of square sum of error (SSE) is sensitive to outliers and therefore results in a failure in fuzzy chemical dataset measured using the nonlinear multivariate system. Interestingly, RWLS, which aims to reduce the sensitivity of SSE, can avoid the disadvantages of OLS through minimizing a weighted square sum of error (WSSE) [22] shown by

$$\min \sum_{v=1}^{V} w_v (y_v - \hat{y}_v)^2 \tag{18.2}$$

The reweighted function "bi-square" is used in this work shown as follows

$$w_v = \begin{cases} (1 - u_v^2)^2, & |u_v| < 1 \\ 0, & |u_v| \geq 1 \end{cases} \quad v = 1,\ldots,V \tag{18.3}$$

where u_v is the adjusted residuals with standardization. This method minimizes a weighted sum of squares, in which the weight assigned to each sample point depends on how far the point is from the fitted line. The detailed procedure of *iterative RWLS algorithm* is presented as follows

Step 1: Fit the model using ordinary least square (OLS), and compute the initial error residual $\boldsymbol{r}_{1\times V}$.

Step 2: Compute the adjusted residuals using $r_{\text{adj}} = r_v / \sqrt{1 - h_v}, \quad v = 1,\ldots,V$, where $\boldsymbol{r}$ is from *step* 1 and $\boldsymbol{h}$ are leverages that adjust the residuals by down-weighting high leverage sample points that have a large effect on the least square fit. The leverages are the elements situated on the diagonal of the prediction matrix (hat matrix), defined as $\mathbf{H} = \boldsymbol{x}(\boldsymbol{x}^{\text{T}}\boldsymbol{x})^{-1}\boldsymbol{x}^{\text{T}}$; thus, the leverage $h_v = x_v(\boldsymbol{x}^{\text{T}}\boldsymbol{x})^{-1}x_v^{\text{T}}$, where x_v denotes the vth transfer sample point.

Step 3: Standardize the adjusted residuals from *step* 2 using $\boldsymbol{u} = \boldsymbol{r}_{\text{adj}}/K \cdot s$, where K is a tuning constant commonly set to 4.685, s is the robust variance given by MAD/0.6745, and MAD is the median absolute deviation of the residuals.

Step 4: Compute the updated robust weights in terms of the bi-square function of $\boldsymbol{u}$ described in Eq. 18.3.

Step 5: If the fit converges, algorithm terminated; otherwise, return to the *step* 1 for next iteration.

- KSS Algorithm for Sample Subset Selection

In calibration, the selection of most representative samples that can reflect the whole sample space becomes necessary for building the model of Eq. 18.1. Therefore, the Kennard–Stone sequential algorithm [23] for sample selection is used in this work. Let z denote the dataset measured on the master instrument. In order to assure a uniform distribution of such a sample subset along the sensor response dataset space, KSS follows a stepwise procedure that new selections are taken in regions of the space far from the samples already selected in terms of the multivariate Euclidean distances $d_z(p, q)$ between the response z-vectors of each pair (p, q) of samples calculated as

$$d_z(p,q) = \sqrt{\sum\nolimits_{i=1}^{k} [x_p(i) - x_q(i)]^2} \quad p,q \in [1,N] \tag{18.4}$$

The selection starts by taking the pair (p_1, p_2) of samples for which the distance $d_z(p_1, p_2)$ is the largest. For clear understanding of this algorithm, the flow of *KSS algorithm* is described as follows

Step 1: Set the desired number of transfer samples.

Step 2: Select the two furthest samples from each other in the whole sample space.

Step 3: Calculate the distances between other remaining samples and the selected ones; the nearest one for each pair is retained from all the pairs of distances.

Step 4: The sample with the furthest distance from these nearest distances retained in *step* 2 is selected.

Step 5: Repeat *steps* 3 and 4 until the required number of transfer samples is achieved.

- Application of the Calibration Model

The measurement datasets were divided into transfer set ($\mathbf{X_{tran}}$ for one slave instrument and $\mathbf{Z_{tran}}$ for the master instrument) and validation set ($\mathbf{X_{val}}$ for slave instrument and $\mathbf{Z_{val}}$ for master instrument). In calibration, $\mathbf{X_{tran}}$ and $\mathbf{Z_{tran}}$ with m transfer samples are used to design the standardization models (m = 5). The remaining p samples excluding that m transfer samples are used to validate the models (p = 120, in this chapter). The transfer sets that contain m samples with k sensors (k = 6) are shown as follows

$$\mathbf{X_{tran}} = \begin{pmatrix} x_{11} & \cdots & x_{1k} \\ \vdots & \ddots & \vdots \\ x_{m1} & \cdots & x_{mk} \end{pmatrix}_{m \times k} \tag{18.5}$$

And

$$\mathbf{Z_{tran}} = \begin{pmatrix} z_{11} & \cdots & z_{1k} \\ \vdots & \ddots & \vdots \\ z_{m1} & \cdots & z_{mk} \end{pmatrix}_{m \times k} \tag{18.6}$$

The validation sets that contain p samples with k sensors are shown as follows

$$\mathbf{X_{val}} = \begin{pmatrix} x_{11} & \cdots & x_{1k} \\ \vdots & \ddots & \vdots \\ x_{p1} & \cdots & x_{pk} \end{pmatrix}_{p \times k} \tag{18.7}$$

And

$$\mathbf{Z_{val}} = \begin{pmatrix} z_{11} & \cdots & z_{1k} \\ \vdots & \ddots & \vdots \\ z_{p1} & \cdots & z_{pk} \end{pmatrix}_{p \times k} \tag{18.8}$$

After calibration on $\mathbf{X_{val}}$ of the slave instrument, the new corrected response matrix is indicated as

$$\mathbf{X_{val}} \xrightarrow{\text{After corrected}} \mathbf{Y_{val}} = \begin{pmatrix} y_{11} & \cdots & y_{1k} \\ \vdots & \ddots & \vdots \\ y_{p1} & \cdots & y_{pk} \end{pmatrix}_{p \times k} \tag{18.9}$$

The mean relative difference (MRD) of the ith sensor shown below between the sensor response of the slaves ($\boldsymbol{x}$, before calibration; $\boldsymbol{y}$, after calibration) and the sensor response z of the master instrument is performed as the evaluation measure of the proposed calibration model.

$$\mathrm{MRD}_i = \frac{1}{N}\sum_{n=1}^{N}\left|\frac{x_{i,n} - z_{i,n}}{z_{i,n}}\right| \infty \frac{1}{N}\sum_{n=1}^{N}\left|\frac{y_{i,n} - z_{i,n}}{z_{i,n}}\right| \tag{18.10}$$

For the purpose of clarity, as we know, when the calibration model is used in other new systems, some prior information of each new system should be known. So, five transfer samples should be obtained through experiments using that new instrument being calibrated. Then, the proposed calibration model can be used to determine the calibrated coefficients of the new instrument.

18.2.2 Prediction Step

Multilayer perceptron feed-forward neural network based on error back-propagation algorithm is applied for organic chemicals concentration prediction in our project. The prediction benefits from the strong generalization ability for approximation of artificial neural network and its low computation complexity for a large number of multi-dimensional experimental datasets. For predicting concentration, we adopt the multi-input and single-output neural network to solve the approximation between responses of the sensors and concentrations of the odor. Three-layered neural network *m-h-o* (one input layer with *m* neurons, one hidden layer with *h* neurons, and one output layer with *o* neurons) is used in experience. The neural network can be illustrated through weight matrix, $\mathbf{W_1}$ $(m \times h)$, $\mathbf{W_2}$ $(h \times o)$, and bias vector, $\mathbf{B_1}$ $(h \times 1)$, $\mathbf{B_2}$ $(o \times 1)$ $(o = 1$, in this chapter). The architecture of the neural network is illustrated in Fig. 18.1.

Given the responses z (training samples measured on the master instrument) and the training targets $\mathbf{T}$ (actual concentrations of analytes), the training process of ANN is employed for the weight matrix and bias vectors learning. The training weights $\mathbf{W}_1$, $\mathbf{B}_1$, $\mathbf{W}_2$, and $\mathbf{B}_2$ are first achieved through the back-propagation algorithm. Detailed description of neural networks is out of the scope of this present study; for that, we refer the reader to [15, 19] to clearly understand the training procedure or learning process through back-propagation algorithm. The active functions of the hidden layer and output layer for that approximation are selected as log-sigmoid and pure linear function. After training, the forward computation process for predicted concentration is shown as follows.

The output of the *j*th hidden node is calculated by the log-sigmoid transfer function

$$f(\text{node}_j) = 1 \Big/ \left[1 + \mathrm{e}^{-\left(\sum_{i=1}^{m} w_{ij} \cdot z_i - b_j\right)}\right], \quad j = 1, \ldots, h \tag{18.11}$$

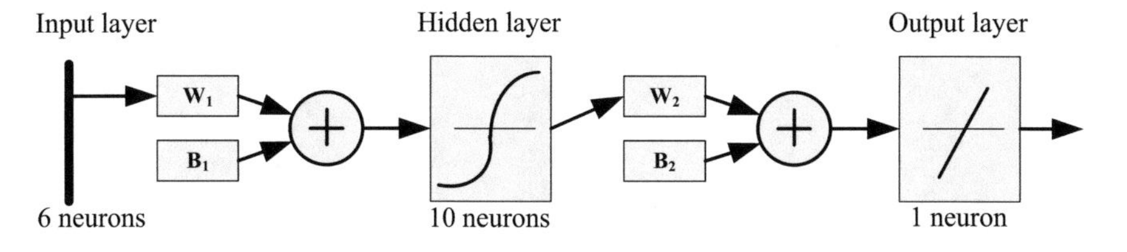

Fig. 18.1 Type of architecture of the neural network

where w_{ij} (one element of $\mathbf{W_1}$) is the connection weight from the ith node of input layer to the jth node of hidden layer and b_j (one element of $\mathbf{B_1}$) is the bias of the jth node of hidden layer.

The output concentration of the k th output node is calculated by the pure linear function

$$y_k = \sum_{j=1}^{h} w_{kj} \cdot f(\text{node}_j) - b_k, \quad k = 1, \ldots, o \tag{18.12}$$

where w_{kj} (one element of $\mathbf{W_2}$) is the connection weight from the jth hidden node to the kth output node and b_k (one element of $\mathbf{B_2}$) is the bias of the kth output layer.

The learning error E can be calculated by the following formulation

$$E = \sum_{c=1}^{q} E_c/(q \cdot o) \tag{18.13}$$

where $E_c = \sum_{k=1}^{o} \left(y_k^c - T_k^c\right)^2$, where q is the number of total training samples and E_c is the error of the actual output and desired output of the kth output unit when the cth training sample is used for training.

In performance evaluation, root mean square error of prediction (RMSEP) and mean absolute relative error of prediction (MAREP) (similar to reference [23, 24]) are used to verify the calibration effect through concentration prediction. The RMSEP is calculated as

$$\text{RMSEP} = \sqrt{\frac{1}{N} \sum_{n=1}^{N} (\varphi_n - T_n)^2} \tag{18.14}$$

The MAREP is calculated as

$$\text{MAREP} = \frac{1}{N} \sum_{n=1}^{N} \left| \frac{\varphi_n - T_n}{T_n} \right| \tag{18.15}$$

where φ_n and T_n denote the predicted and actual concentration for the nth sample, respectively.

18.3 Experiments

18.3.1 Electronic Nose Module

The metal oxide semiconductor gas sensors used in our E-nose system consist of TGS series from FIGARO, USA. They are TGS2602, TGS2620, and TGS2201 with two outputs *A* and *B* (TGS2201A/B). In addition, a module (SHT2230 of Sensirion in Switzerland) with two auxiliary sensors for temperature and humidity compensations is also used. The sensors were mounted on a custom-designed printed circuit board (PCB), along with associated electrical components. An analog–digital converter (AD) is used as interface between the FPGA processor and the sensors. The system can be connected to a PC via a JTAG port. An additional flash memory is used to save the weights and biases of the neural network trained on the PC and the calibration transfer coefficients of each sensor. The datasets for these gases are made up of samples in $\Re^6$ space, and it just means that an input vector with six independent variables was obtained in each observation. The gases measurements are implemented in the constant temperature and humidity chamber (LRH-150S) in which the temperature and humidity can be effectively controlled in terms of the desired temperatures and humidity. In this chapter, six E-nose instruments with completely identical types of sensors and electrical components are used for verifying the proposed calibration model. For convenience of subsequent analysis, one instrument is selected as the “master” which is recognized as the standard instrument and the left five are named as “slave_1, slave_2, slave_3, slave_4, and slave_5” which will be calibrated according to [2]. It is worth noting that all the six instruments would employ the experiments together at the same time in the chamber to ensure the consistency of the experimental samples on each instrument.

18.3.2 Gas Datasets

Three organic chemicals have been analyzed in calibration step and prediction step: formaldehyde, benzene, and toluene. The experiments of sample collections are developed in the chamber. The E-nose instrument should be exposed to each of the three candidates in many different concentrations separately, and the responses of the sensors are saved on PC in each measurement. Totally, 243 measurements (dataset) including 125 formaldehyde samples, 52 benzene samples, and 66 toluene samples for each instrument are measured in the same way. These samples are measured through different combinations of the target temperatures of 15, 25, 30, and 35 °C and target humidity of 40, 60, 80 RH. The total measurement cycle time for a single measurement was set to 10 min, i.e., 2 min reference air (baseline) and

8 min sampling. Between two single measurements, 10 min for cleaning the chamber through injecting clean air is also consumed. In one single measurement, the temperature and humidity have little change except the slight fluctuation. Note that the calibration and prediction variables (features) used in our work are the steady state responses. In the studies of calibration models, formaldehyde was chosen as referenced gas.

18.4 Results and Discussion

18.4.1 Sensor Response Calibration

Sensor response calibration indicates the projections from the response of slave instruments to the response of the master instrument. For on-line calibration among E-nose instruments, we select formaldehyde as reference gas and a certain number of transfer samples are also needed for developing the calibration coefficients using the hybrid RWLS and KSS algorithm. Figure 18.2 studies the calibration error (MRD in Eq. 18.10 of all sensors between slave instruments and the master after calibration) when using a different number of transfer samples. We can see that five transfer samples are enough to perform a good calibration. More transfer samples would also increase the experimental and calibration complexity in mass production. Figure 18.3 illustrates the linear regression curves of each sensor between the slave_5 and the master using hybrid RWLS and KSS algorithm on the selected five transfer samples. Each subfigure in Fig. 18.3 represents one type of sensor.

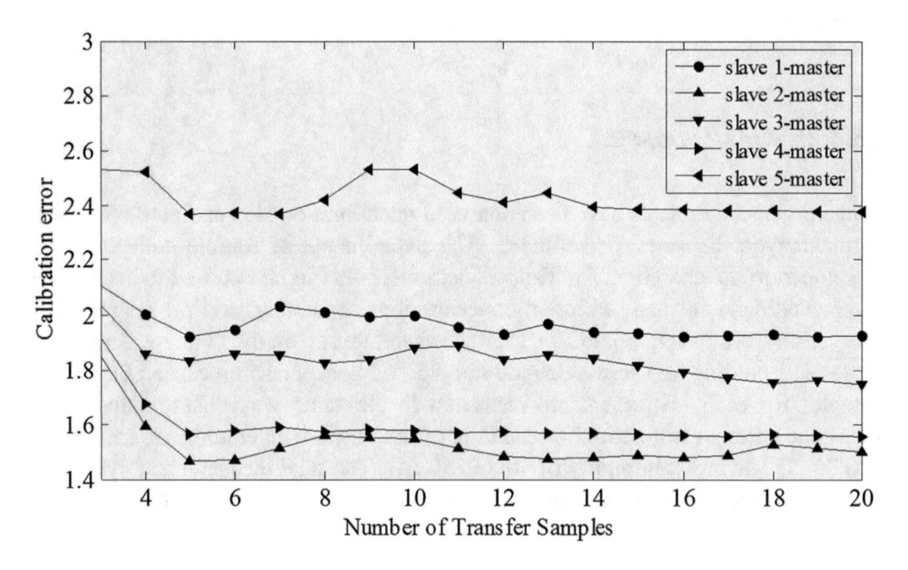

Fig. 18.2 Performance of calibration with increasing number of transfer samples

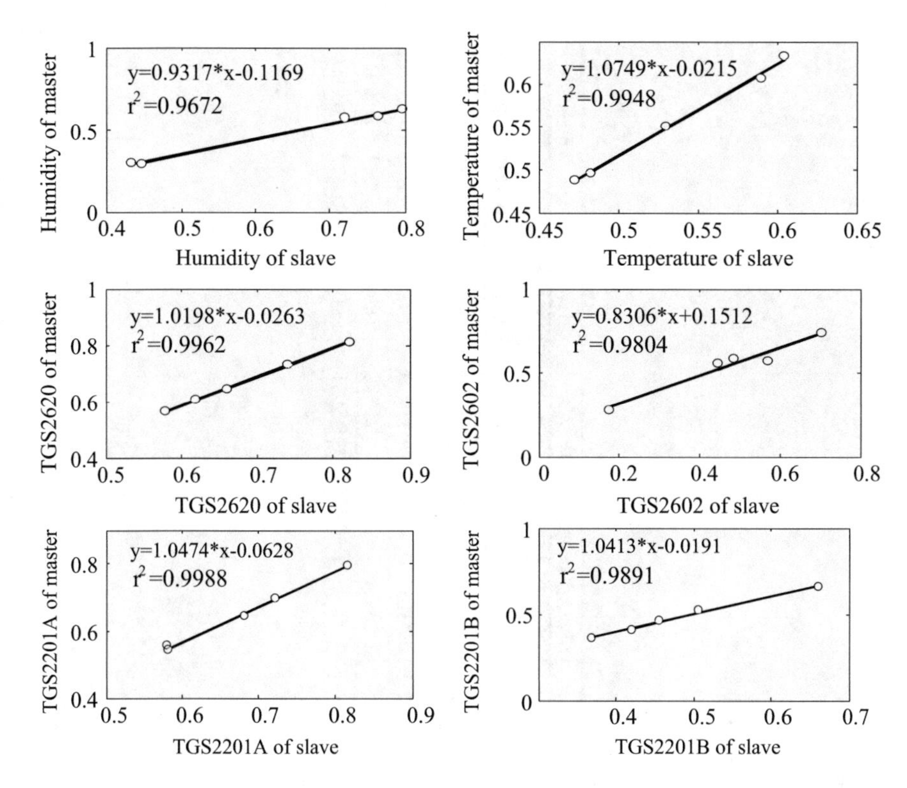

Fig. 18.3 Calibration regression equations between slave_5 and the master using RWLS estimator and five representative transfer samples selected by KSS algorithm. In each subfigure, the open symbols are the selected five transfer sample points; the line is the regression of the five points

Figure 18.3 also demonstrates the homogeneous linearity between multi-sensors systems and feasibility of the proposed calibration transfer model.

Table 18.1 also presents the regression coefficients of another four slaves (from slave_1 to slave_4). All squared correlation coefficient r^2 values are between 0.967 and 0.998, and there was no statistically significant difference ($\alpha = 0.05$) between the master instrument responses and calibrated responses from the slave instruments. It is worth noting that all the instruments should use the same transfer sample number (*index* 1,…, 5) obtained from the master dataset by calling the KSS algorithm, namely in subsequent calibration of other slaves (from slave_1 to slave_5), and the same samples number (*index* 1,…, 5) should be used.

Table 18.2 presents the performance validation error of the remaining 120 formaldehyde samples using the calibration coefficients described in Fig. 18.3 and Table 18.1. Figure 18.4 illustrates the steady state response curves (single sensor) of the 120 formaldehyde samples measured on the 6 instruments together,

Table 18.1 Calibration transfer regression coefficients (calibration equation, $y = a * x + \mathrm{b}$)

Sensor	Slave_1-master			Slave_2-master			Slave_3-master			Slave_4-master		
	a	b	r^2	a	b	r^2	a	b	r^2	a	b	r^2
Humidity	0.9280	0.0136	0.9736	1.0073	−0.1056	0.9600	0.9903	−0.0091	0.9798	0.8125	0.0472	0.9714
Temperature	0.9929	0.0032	0.9952	1.0510	−0.0213	0.9974	1.0714	−0.0399	0.9944	0.9799	0.0220	0.9963
TGS2620	1.0126	−0.0156	0.9975	0.9848	0.0157	0.9939	1.0069	−0.0071	0.9988	1.0009	−0.0097	0.9983
TGS2602	0.9291	0.0540	0.9753	0.9482	0.0409	0.9922	0.8326	0.1178	0.9725	0.9161	0.0924	0.9778
TGS2201A	1.0687	−0.0666	0.9885	1.0516	−0.0581	0.9835	1.1283	−0.1323	0.9812	0.9719	0.0032	0.9913
TGS2201B	1.0381	−0.0671	0.9876	1.0356	−0.0317	0.9958	1.0030	−0.0254	0.9931	1.0096	−0.0272	0.9880

Table 18.2 Mean relative difference (MRD) of each sensor between the master instrument and the slaves

Sensor	Master–slave_1		Master–slave_2		Master–slave_3		Master–slave_4		Master–slave_5	
	N	*Y*	*N*	*Y*	*N*	*Y*	*N*	*Y*	*N*	*Y*
Humidity	0.0550	0.0354	0.0968	0.0301	0.3539	0.0333	0.0439	0.0266	0.3038	0.0866
Temperature	0.0054	0.0047	0.0171	0.0052	0.0325	0.0055	0.0088	0.0041	0.0628	0.0162
TGS2620	0.0113	0.0032	0.0149	0.0057	0.0183	0.0047	0.0042	0.0026	0.0063	0.0038
TGS2602	0.0468	0.0242	0.1003	0.0226	0.1202	0.0324	0.0826	0.0286	0.0248	0.0132
TGS2201A	0.0347	0.0224	0.0309	0.0111	0.0535	0.0117	0.0594	0.0155	0.0245	0.0113
TGS2201B	0.0981	0.0298	0.0384	0.0110	0.0221	0.0207	0.0368	0.0171	0.0284	0.0128

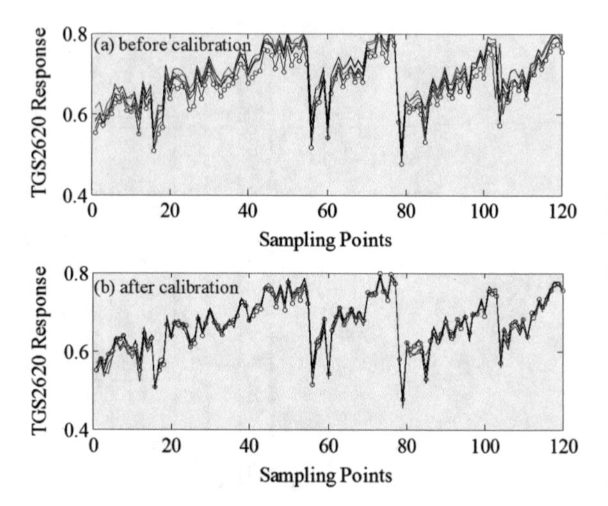

Fig. 18.4 TGS2620 responses with 120 samples measured before and after calibration; six curves including five slaves and the master

respectively; (a) denotes the 6 curves before calibration; (b) denotes the 6 curves after calibration. That line with circles is the steady state response curve of the master instrument which is recognized as the calibration targets. In Table 18.2, the columns labeled "*N*" denote the mean relative difference (MRD) of all sensors and samples between the slave instruments and the master without calibration, while the columns labeled "*Y*" are with calibration. Through Table 18.2 and Fig. 18.4, we can find that the calibration coefficients developed using the selected five transfer samples successfully realized the on-line response projections.

The responses of the master instrument are denoted by the line with circles, and other five slaves aim to approximate the master instrument by the calibration transfer coefficients. In actual applications of these five slave instruments, the calibration coefficients $\boldsymbol{a}$, $\boldsymbol{b}$ would be saved on each slave instrument for on-line automated calibration. In mass production, all the slaves should be first located in specific experimental environment in accordance with the selected five transfer samples (*index* 1,…,5) measured on the master instrument for determining the projection parameters of each slave instrument.

18.4.2 Concentration Prediction

In this section, the concentration prediction model is constructed using the dataset measured on the master E-nose instrument using multilayer feed-forward artificial neural network trained by error back-propagation algorithm. The training process of ANN is automatically operated using MATLAB toolbox of neural network. The concentration prediction is also studied mainly to find out the improved fitting ability of the five slaves when still using the same prediction model as the master after using the presented on-line calibration transfer; in this case, the importance and necessity of calibration transfer have also been demonstrated. In this section, three organic chemicals datasets were used for concentration measure: formaldehyde, benzene, and toluene. Thus, three single ANNs (ANN_1, ANN_2, and ANN_3) were built for these three analytes, and the three ANNs should be trained on their gas dataset individually. Considering the generalization of ANNs, three sets of samples for each analyte have been classified: training set, monitoring set, and test set [20]. The monitoring set is used to control the training process and avoid overfitting. In this chapter, the predicted and actual concentrations of 20 test samples for each analyte are presented to test our calibration model. The validation must be performed with the test set, composed of samples not used in the ANN training and monitor. Once trained, the obtained weights $\mathbf{W_1}$, $\mathbf{W_2}$ and bias vectors $\mathbf{B_1}$, $\mathbf{B_2}$ should be saved on each E-nose instrument for calculating the concentrations of analytes in real time using Eqs. 18.11 and 18.12. The parameters of each ANN consist of six input neurons, ten hidden neurons, and one output neuron. The maximum number of epochs is set as 2000, and the convergence goal of each epoch in the training process is set randomly between 0.0005 and 0.05.

Table 18.3 presents the formaldehyde concentrations of 20 test samples including the five slave instruments, the master instrument, and the actual (reference). To each table, the columns labeled "*N*" denote the predicted concentration without calibration process, while the columns labeled "*Y*" denote the predicted concentration with calibration. It is worth noting that the predicted concentration values on each slave instrument are calculated using the same ANN_1, ANN_2, and ANN_3 as the master instrument. In visual, Fig. 18.5 illustrates the predicted concentration curves of 20 test samples (slave_1, 2, 3, 4, and 5, master, and actual) of three analytes, respectively; (a) denotes the predicted and actual concentrations before calibration, and (b) denotes the same results as (a) after calibration.

The line with circles denotes the actual concentrations (reference); the line with triangles indicates the predicted concentrations of the master. Through Table 18.3 and Fig. 18.5, we can find that the predicted concentrations of the five slaves after calibration become more close to the master and actual concentrations than before. Also, we can believe that the calibration transfer coefficients employed from slaves to the master can be fitted with formaldehyde, benzene, and toluene. For error quantifications, Tables 18.4 and 18.5 present the RMSEP and MAREP values of the three measured analytes. Therein, the columns labeled "*N*" and "*Y*" denote the results without and with calibration, respectively.

From Tables 18.4 and 18.5, the RMSEP and MAREP of predicted concentrations on the three analytes are reduced significantly after calibration. The fitting ability of the slaves to the prediction model trained on the master instrument has been strengthened obviously. Thus, all the E-nose instruments for air quality monitoring can share the same prediction model (weights and biases of ANN) as the

Table 18.3 Test of formaldehyde concentration estimation (concentration in ppm)

Sampling number	Slave_1		Slave_2		Slave_3		Slave_4		Slave_5		Master	Actual
	N	*Y*	*N*	*Y*	*N*	*Y*	*N*	*Y*	*N*	*Y*	–	–
1	0.084	0.066	0.086	0.067	0.558	0.064	0.123	0.067	0.026	0.064	0.067	0.064
2	0.068	0.076	0.070	0.076	0.109	0.080	0.111	0.081	0.116	0.084	0.074	0.074
3	0.666	0.090	0.364	0.100	0.959	0.098	0.114	0.085	0.734	0.120	0.079	0.081
4	0.114	0.082	0.126	0.083	0.800	0.086	0.118	0.091	0.477	0.095	0.080	0.090
5	0.061	0.094	0.075	0.087	0.115	0.105	0.139	0.094	0.113	0.090	0.096	0.103
6	0.061	0.163	0.084	0.159	0.129	0.140	0.204	0.130	0.158	0.138	0.140	0.140
7	1.030	0.300	0.163	0.115	0.439	0.126	0.813	0.190	0.813	0.195	0.188	0.172
8	0.221	0.164	0.096	0.200	0.236	0.198	0.348	0.182	0.583	0.166	0.181	0.174
9	0.012	0.123	0.088	0.172	0.148	0.162	0.293	0.162	0.137	0.180	0.180	0.196
10	0.033	0.117	0.084	0.241	0.109	0.118	0.179	0.201	0.239	0.201	0.185	0.206
11	0.418	0.236	0.447	0.422	1.040	0.230	0.233	0.244	1.130	0.266	0.256	0.234
12	0.079	0.207	0.004	0.234	0.577	0.206	1.236	0.312	1.687	0.200	0.216	0.250
13	1.268	1.497	0.598	1.361	2.274	1.193	1.684	1.529	1.594	1.300	1.387	1.419
14	1.243	1.892	0.172	1.581	0.959	1.512	2.000	1.484	2.618	1.523	1.816	1.753
15	1.574	2.125	2.322	2.132	2.380	1.950	1.984	2.100	0.153	2.237	2.127	2.191
16	2.012	2.113	2.267	2.559	3.251	1.964	2.031	2.160	1.668	1.938	2.481	2.456
17	2.464	2.765	0.176	2.503	1.442	2.382	2.019	2.133	3.152	2.582	2.461	2.615
18	2.763	3.340	2.547	3.219	2.467	2.557	2.752	2.833	3.323	3.146	3.216	3.169
19	3.035	3.894	1.311	4.211	3.151	4.094	4.746	4.612	4.057	4.837	4.539	4.529
20	5.283	5.307	5.099	5.320	4.425	5.314	5.309	5.314	5.302	5.315	5.318	5.322

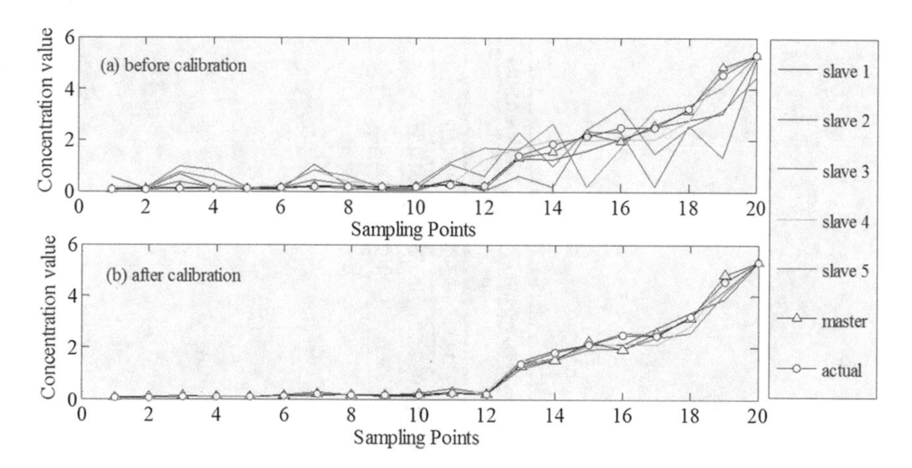

Fig. 18.5 Performance of predicted concentrations on test samples (before and after calibration) of formaldehyde

master instrument and avoid the repeated and complex ANN training process on each instrument. More results can be referred to as [25].

In mass production, the on-line sensor calibration method proposed here is based on a global affine transformation. The difference between global and local affine transformation is that the slave instruments are calibrated in the whole sample space without considering the piecewise linearity. Local transformation (e.g., piecewise) can also be used in terms of different conditions (e.g., types of hazardous gases and concentrations of gases). Through experimental analysis, we find that the calibration model constructed by formaldehyde can also be directly applied to benzene and toluene. For these three analytes, low, middle, and high concentration predictions have been tested for calibration based on global affine transformation. Note that acquirement of representative transfer samples using KSS algorithm should be built in a big enough sample space so that the selected transfer samples can represent the whole range of concentrations being monitored. However, calibration accuracy may be reduced slightly. The experimental results demonstrate the good performance of global calibration transfer. Considering the unknown concentration of hazardous gas in real time, it is difficult for local affine transformation to determine different calibration models using piecewise linearity in terms of different stages of concentrations, and it needs a further research in the future work.

Table 18.4 RMSEP of the three measured analytes

RMSEP	Slave_1–actual		Slave_2–actual		Slave_3–actual		Slave_4–actual		Slave_5–actual		Master–actual
	N	*Y*	*N*	*Y*	*N*	*Y*	*N*	*Y*	*N*	*Y*	–
Formaldehyde	0.4743	0.1782	1.0048	0.1014	0.6695	0.2282	0.3440	0.1641	0.7113	0.1484	0.0439
Benzene	0.1580	0.0299	0.0880	0.0140	0.1782	0.0363	0.1386	0.0358	0.0828	0.0535	0.0269
Toluene	0.0655	0.0101	0.0136	0.0041	0.0294	0.0093	0.0214	0.0163	0.0866	0.0078	0.0059

Table 18.5 MAREP of the three measured analytes

MAREP	Slave_1–actual		Slave_2–actual		Slave_3–actual		Slave_4–actual		Slave_5–actual		Master–actual
	N	*Y*	*N*	*Y*	*N*	*Y*	*N*	*Y*	*N*	*Y*	–
Formaldehyde	18.744	2.8550	12.000	2.6549	37.100	2.5408	13.253	1.7031	32.094	1.9712	0.9720
Benzene	7.0101	1.0626	4.0642	0.6454	8.5268	1.5997	4.3465	1.4660	4.1445	1.7059	1.2842
Toluene	11.211	1.7730	2.4016	0.8488	7.0995	1.8126	5.1516	2.6932	14.517	1.4781	1.1507

18.5 Summary

This chapter addresses the critical issue of calibration transfer among electronic nose instruments based on our finding of the homogeneous linearity among E-nose multi-sensors system. An on-line sensor calibration transfer model used for indoor air quality monitoring through global affine transformation based on robust weighted least square algorithm and Kennard–Stone sequential algorithm is proposed. Six E-nose instruments, including one master instrument and five slave instruments equiped with the same types of sensors and other electrical components, were used to evaluate the performance of the proposed model. Experimental results for concentration estimation in real time confirmed the efficiency of the proposed calibration transfer models.

References

1. E.J. Wolfrum, R.M. Meglen, D. Peterson, J. Sluiter, Calibration transfer among sensor arrays designed for monitoring volatile organic compounds in indoor air quality. IEEE Sens. J. **6**, 1638–1643 (2006)
2. E. Bouveresse, D.L. Massart, Standardisation of near-infrared spectrometric instruments: a review. Vib. Spectrosc. **11**, 3–15 (1996)
3. E. Bouveresse, C. Hartmann, D.L. Massart, Standardization of near-infrared spectrometric instruments. Anal. Chem. **68**(6), 982–990 (1996)
4. J. Sjoblom, O. Svensson, M. Josefson, H. Kullberg, S. Wold, An evaluation of orthogonal signal correction applied to calibration transfer of near infrared spectra. Chemometr. Intell. Lab. Syst. **44**, 229–244 (1998)
5. B. Walczak, E. Bouveresse, D.L. Massart, Standardization of near-infrared spectra in the wavelet domain. Chemometr. Intell. Lab. Syst. **36**, 41–51 (1997)
6. K.S. Park, Y.H. Ko, H. Lee, C.H. Jun, H. Chung, M.S. Ku, Near-infrared spectral data transfer using independent standardization samples: a case study on the trans-alkylation process. Chem. Intel. Lab. Syst. **55**, 53–65 (2001)
7. C. Dinatale, F.A. Davide, A. Damico, W. Gopel, U. Weimar, Sensor array calibration with enhanced neural networks. Sens. Actuators B: Chem. **18**, 654–657 (1994)
8. S. Osowski et al., Neural methods of calibration of sensors for gas measurements and aroma identification system. J. Sens. Stud. **23**(4), 533–557 (2008)
9. P. Laurent, O.B. Jacques, T. Raphael, Data transferability between two MS-based electronic noses using processed cheeses and evaporated milk as reference materials. Eur. Food Res. Technol. **214**, 160–162 (2002)
10. F.C. Tian, S.X. Yang, K. Dong, Circuit and noise analysis of odorant gas sensors in an E-nose. Sensors **5**, 85–96 (2005)
11. X.T. Xu, F.C. Tian, S.X. Yang, Q. Li, J. Yan, J.W. Ma, A solid trap and thermal desorption system with application to a medical electronic nose. Sensors **8**, 6885–6898 (2008)
12. F.C. Tian, X.T. Xu, Y. Shen, J. Yan, Q.H. He, J.W. Ma, T. Liu, Detection of wound pathogen by an intelligent electronic nose. Sens. Mater. **21**, 155–166 (2009)
13. L.H. Zhang, W.L. Xu, C. Chang, Genetic algorithm for affine point pattern matching. Pattern Recogn. Lett. **24**, 9–19 (2003)
14. J. Heikkilä, Pattern matching with affine moment descriptors. Pattern Recogn. **37**, 1825–1834 (2004)
15. S. Haykin, *Neual Networks, a Comprehensive Foundation* (Macmillan, New York, NY, 2002)

16. D. Gao, M. Chen, J. Yan, Simultaneous estimation of classes and concentrations of odors by an electronic nose using combinative and modular multilayer perceptrons. Sens. Actuators B **107**, 773–781 (2005)
17. B. Yea, T. Osaki, K. Sugahara, R. Konishi, The concentration estimation of inflammable gases with a semiconductor gas sensor utilizing neural networks and fuzzy inference. Sens. Actuators B **41**, 121–129 (1997)
18. S. De Vito, A. Castaldo, F. Loffredo, E. Massera, T. Polichetti, I. Nasti, P. Vacca, L. Quercia, G. Di Francia, Gas concentration estimation in ternary mixtures with room temperature operating sensor array using tapped delay architectures. Sens. Actuators B **124**, 309–316 (2007)
19. M. Pardo, G. Sberveglieri, Remarks on the use of multilayer perceptrons for the analysis of chemical sensor array data. Sens. J., IEEE. **4**(3), 355–363 (2004)
20. D.L.A. Fernandes, M. Teresa, S.R. Gomes, Development of an electronic nose to identify and quantify volatile hazardous compounds. Talanta **77**, 77–83 (2008)
21. Š. Obdržálek, J. Matas, Object recognition using local affine frames on distinguished regions, in BMVC (2002), pp. 113–122
22. R.M. Heiberger, R.A. Becker, Design of an S function for robust regression using iteratively reweighted least squares. J. Comput. Graphical Stat. **1**, 181–196 (1992)
23. F. Sales, M.P. Callao, F.X. Rius, Multivariate standardization for correcting the ionic strength variation on potentiometric sensor arrays. Analyst **125**, 883–888 (2000)
24. Y.H. Huang, D. Jiang, D.F. Zhuang, J.Y. Fu, Evaluation of hyperspectral indices for chlorophyll-a concentration estimation in Tangxun Lake (Wuhan, China). Int. J. Environ. Res. Public Health. **7**, 2437–2451 (2010)
25. L. Zhang, F. Tian, C. Kadri, B. Xiao, H. Li, L. Pan, H. Zhou, On-line sensor calibration transfer among electronic nose instruments for monitoring volatile organic chemicals in indoor air quality. Sens. Actuators B: Chem. **160**, 899–909 (2011)

Chapter 19
Instrumental Batch Correction

Abstract Metal oxide semiconductor (MOS) gas sensors have been widely reported in machine olfaction system (i.e., electronic nose/tongue) for rapid detection of gas mixture components due to their positive characteristics of cross-sensitivity, broad-spectrum response, and low cost. However, as described in Chap. 18, the discreteness of MOS gas sensors caused by inherent sensor variability during the manufacturing process results in the failure of the batch-oriented applications of MOS gas sensors due to their weak reproducibility. Therefore, the contribution of this chapter is to solve the discreteness and improve the reproducibility of sensors by designing an effective and easily realized scheme for large-scale calibration, which is an extension of Chap. 18. Experimental results demonstrate that the proposed scheme can effectively and rapidly realize the calibration of the sensors' discreteness in batch of electronic noses production and the proposed scheme can also be used in industry. Besides, this chapter also proves that one sensor's discreteness is constant and keeps unchanged when the sensor is exposed to different kinds of gas components.

Keywords Metal oxide semiconductor gas sensor · Electronic nose
Reproducibility · Discreteness correction · Large-scale application

19.1 Introduction

Recently, a variety of algorithms have been proposed for dealing with the sensor drift problem [1–5]. Fonollosa et al. also studied the sensor failures in discrimination of chemical substances [6]. However, most of the research in E-nose based on MOS gas sensor array focus on the pattern recognition analysis using one fixed sensor array, and without considering the problem of sensor's reproducibility that will result in the difference of electrical signal between two sensor arrays of the same type. In other words, other sensor arrays with completely the same type may not be appropriate with the learned pattern recognition model (i.e., artificial neural network) due to the weak reproducibility [7]. MOS gas sensors are operated with

L. Zhang et al., *Electronic Nose: Algorithmic Challenges*,
https://doi.org/10.1007/978-981-13-2167-2_19

the principle that volatile odor components can produce a reaction inside the sensor in contact with a catalytic metal, changing the electrical resistance of the sensor device and producing some voltage signal [8]. Generally, the sensing material is metal oxide, most typically S_nO_2. The principles can be described as follows.

When the metal oxide crystal is heated at a certain high temperature in air, oxygen is adsorbed on the crystal surface with a negative charge. In the presence of a deoxidizing gas, the surface density of the negatively charged oxygen will decrease so that the barrier height is reduced which will decrease the sensor resistance. The detection principle of MOS gas sensors is based on the chemical adsorption and desorption of chemicals on the sensor's surface. Besides, the ambient temperature and humidity will also affect the sensitivity characteristics of sensor by changing the rate of chemical reaction [9]. Therefore, from the complex sensing principle and the various factors related in sensing, the reproducibility of MOS gas sensors should be taken into consideration in the industrial production of MOS gas sensors based instruments.

In batch of E-nose production, the homogeneity of multiple electronic noses' predictions when exposed to the same gas component is very important and the week reproducibility would seriously degrade the homogeneity [10]. The homogeneity completely depends on the MOS gas sensor array embedded in E-nose, because the pattern recognition module used internally is identical among E-noses and different inputs from sensors would lead to different predictions. Unfortunately, the MOS gas sensor array of identical type reflects diverse responses to the same chemical in the same experimental condition due to the inherent sensor variability and discreteness during the manufacturing process. The kind of sensor discreteness must cause the reduction of E-nose prediction accuracy and reproducibility. To solve the problem of sensor discreteness, specific correction methods have been studied in previous publications [10, 11]. Comparatively, the GAT-RWLS method proposed in [10], for its simplification of algorithm, is easier to implement for calibration in real-time E-nose detection. However, in large-scale sensors application (i.e., production of E-nose instruments), a rapid discreteness correction scheme is very necessary to reduce the complexity and also promise the accuracy, especially for regular calibration. The problems of sensor's discreteness and reproducibility in large-scale application of MOS gas sensors have been mentioned and fully solved in this work.

Therefore, the contribution of this chapter is to present an effective implementation scheme of sensors' discreteness correction coupled with GAT-RWLS method for batch-oriented instruments production. In addition, this chapter also reveals that the sensors' discreteness has little relation with the type of measured gas.

19.2 Materials and Method

19.2.1 Sensors' Discreteness

MOS gas sensors' discreteness will cause large difficulties in batch-oriented instruments development, especially that the discreteness extremely lowers the accuracy of electronic nose instruments. The discreteness can be illustrated in two facets:

(1) Baseline difference: the sensitive resistance R_o of identical sensors in the standard environment (clean air) with temperature 20 °C and relative humidity (RH) 60% is variable which results in that the baseline of sensor with identical type is different in the same environment.
(2) Sensitivity difference: when exposed to some kind of pollutant gas, the MOS sensors with identical type also have different sensitivity which can be denoted as R_s/R_o (R_s is the sensitive resistance in the pollutant gas and R_o is the sensitive resistance in clean air). This will result in that the sensor responses with identical type are also different when exposed to the same type of gas with the same concentration in the same environment. That is, the same two sensors in the same environment have different outputs.

Therefore, it is not difficult to infer that the discreteness of MOS sensors can largely influence the accuracy of detective instruments, and rapid correction of the discreteness without changing the sensor circuits is very significant in improving the sensors' reproducibility and the performance of instruments, especially in batch-oriented application.

For studying of the sensor discreteness and its rapid correction in batch-oriented application, we have employed multiple kinds of gases experiments using six electronic nose systems embedded with identical sensor array. Our E-nose has been illustrated in Fig. 11.4 (Chap. 11). Considering the characteristics of broad spectrum and low-cost of metal oxide semiconductor gas sensors, four metal oxide semiconductor gas sensors from Figaro Inc. including TGS2602, TGS2620, TGS2201A, and TGS2201B are used in the sensor array. The heating voltage of TGS2620 and TGS2602 is 4 V (Volt), and the heating voltage of TGS2201A/B is 7 V. The supplied power voltage of system is DC12 V. The experiments of electronic noses were employed in the climate chamber (LRH-150S). Totally, 126 formaldehyde samples, 72 benzene samples, 66 toluene samples, 58 carbon monoxide samples, 27 ammonia samples, and 30 nitrogen dioxide samples were obtained. The experimental conditions and concentrations are different from each other. The discreteness of TGS2620, TGS2602, TGS2201A, and TGS2201B sensors when exposed to the same concentration of formaldehyde gas has been illustrated in Fig. 19.1. Note that TGS2620 (1–6), TGS2602 (1–6), TGS2201A

(1–6), and TGS2201B (1–6) represent six sensors with completely the same type, respectively. We can see from Fig. 19.1 that the discreteness of TGS2620 is weaker than other three MOS gas sensors. In other words, the reproducibility of TGS2620 is comparatively better. This phenomenon results from several facets. Though their sensing principles in detection are similar, their manufacturing process, electrical characteristics, and sensitivity will also influence the reproducibility. For instance, the sensor resistance (R_s) of TGS2620 is 1–5 kΩ, while 10 k–100 kΩ is for TGS2602, 250 and 25 kΩ are for TGS2201A and TGS2201B, respectively. In our experiments, TGS2620 shows the best reproducibility and stability, while TGS2602 performs the worst.

19.2.2 Review of the GAT-RWLS Method

GAT-RWLS method using reference formaldehyde gas for discreteness correction was proposed in Chap. 18. Through a large number of electronic nose experiments, we found that there exists a good linear relation between two sensors with the same type when exposed to the same environment and conditions. That is, the discreteness can be easily corrected in a linear way. For simplification, the calibration transfer model is shown by

$$y_{i,n} = a_i \cdot x_{i,n} + b_i, \quad i = 1, \ldots, k; \quad n = 1, \ldots, N \tag{19.1}$$

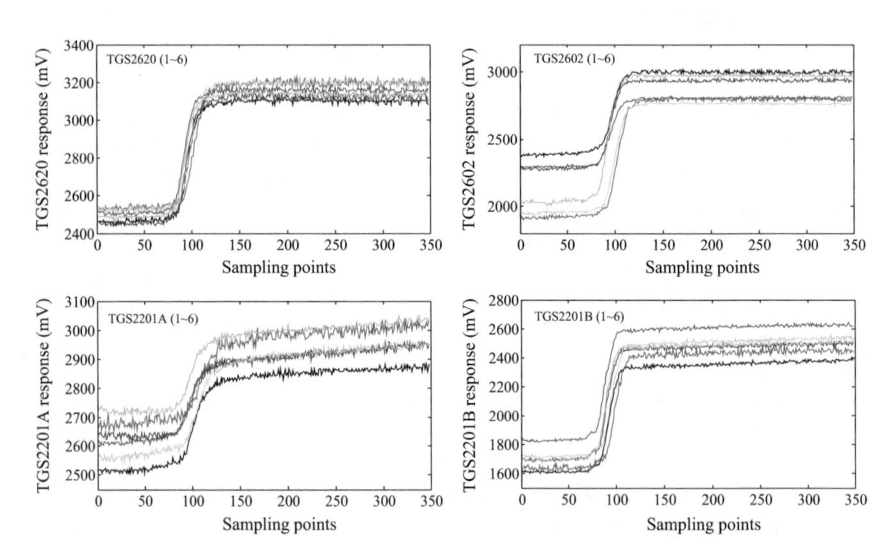

Fig. 19.1 Responses of six sensor arrays of the same type with TGS2620, TGS2602, TGS2201A, and TGS2201B when exposed to the same concentration of formaldehyde

where k denotes the number of sensors being calibrated, N denotes the number of calibration samples, x denotes the response of slave, y denotes the estimation of the master, a_i and b_i represent the calibration coefficients of the ith sensor obtained using reference gas (formaldehyde). In this model, a master should be determined as the standard electronic nose in advance, take other electronic nose as slaves and calibrate the slaves to the master.

The proposal of model Eq. 19.1 is based on a global affine transform (GAT) with scaling and translation in a special way that the sensors in a sensor array influence each other in calibration. That is, the calibration is independent for each sensor. Also, considering the experimental error of a number of samples which will result in the inaccuracy of the calibration model, a robust weighted least square (RWLS) method was used for regression Eq. 19.1 and obtaining the parameters a and b. Experimental results of the formaldehyde samples correction demonstrate that the proposed method was very effective and easy to realize. We refer readers [10] for the details of the GAT-RWLS method.

In this chapter, the calibration parameters obtained using formaldehyde as reference gas would also be validated for correction of the sensors' discreteness when exposed to other five kinds of gases in different conditions and concentrations.

19.2.3 Instrumental Batch Correction Scheme

An easily realized and fast correction scheme is very necessary in batch of more than 50 electronic noses production. Therefore, this chapter aims to propose an effective scheme for sensors' discreteness based on the GAT-RWLS calibration method.

The sensors' discreteness correction should be completed through experiments in this chapter. Considering that the uniformity of reference gas in the chamber in experiment is very necessary, the volume of the chamber should not be too large. Note that the uniformity denotes that all electronic noses placed into the climate chamber can be exposed to the same concentration of reference gas. Then, 11 electronic noses including the master and 10 slaves were employed in each batch of experiments, and the time consumption is approximately 1 h. Therefore, totally 5 batches of calibration experiments would be employed to complete all calibrations of 50 slaves to the master, and only about 5 h are needed. Considering the environmental robustness of calibration, three experimental conditions 15 °C/60% RH, 25 °C/60% RH, and 35 °C/60% RH are employed in each batch of experiments. In each experimental condition, two different concentrations of reference gas are produced for sensitivity correction.

The proposed experimental scheme for discreteness correction of 50 slaves to the master can be shown as follows.

Step 1: For differentiation, we define the standard electronic nose and 50 being calibrated electronic noses as master, slave 1, slave 2, slave 3, slave 4,…, and slave 50, respectively; then, divide the 50 slaves into 5 batches (each batch contains 10 slaves).

Step 2: Put the master and the first batch of 10 slaves into the chamber; set the temperature and relative humidity of the chamber as 15 °C and 60% RH, and turn on the humidifier;

Step 3: Wait until the setting temperature and humidity are achieved and turn off the humidifier. First, the baseline collection of 5 min is sustained. Second, inject reference gas (formaldehyde) into the chamber using a pump for 10 s, and continue the data collection for 5 min. Then, inject reference gas (formaldehyde) into the chamber using a pump for 10 s again and continue the data collection for 5 min;

Step 4: Set the temperature and relative humidity of the chamber as 25 °C and 60% RH, and turn on the humidifier; then, repeat step 3;

Step 5: Set the temperature and relative humidity of the chamber as 35 °C and 60% RH and turn on the humidifier; then, repeat step 3;

Step 6: Air exhaust and chamber cleaning. After finish the batch of experiments, air exhaust by a pump is necessary for chamber cleaning to recover the sensor response quickly and take out the electronic noses in the chamber;

Step 7: Put the master and the other batches of slaves into the chamber, respectively, and repeat step 3–step 6;

Note that step 2–step 6 denote the whole process of calibration experiments in one batch of 11 electronic noses. There are approximately 3000 sampling points collected in the whole process. In correction, we should first determine the positions of three features which correspond to the three experimental conditions 15 °C/60% RH (step 2), 25 °C/60% RH (step 4), and 35 °C/60% RH (step 5) in the steady state response in the master and obtain the calibration coefficients between each slave and the master using the model denoted by Eq. 19.1. Then, the obtained calibration coefficients would be used to correct the whole curves and verify the performance of the proposed scheme.

19.3 Results and Discussion

For validation of the calibration coefficient obtained from the reference gas (formaldehyde) on other five gases samples, we select one of sensor arrays in electronic noses as slave and calibrated it to the master. For calibration, we select one point in the steady state response as the feature in each sample. The calibration coefficients are calculated by operating the GAT-RWLS method on the 126 formaldehyde samples between the slave and the master. The obtained calibration coefficients in Eq. 19.1 $\mathbf{a} = [1.001, 0.916, 0.971, 1.009]^T$, $\mathbf{b} = [-0.01, 0.092, 0.003, -0.02]^T$ for TGS2620, TGS2602, TGS2201A, and TGS2201B between the slave and the master, respectively.

Based on the obtained calibration coefficients of the four gas sensors, Figs. 19.2, 19.3, 19.4, and 19.5 illustrate the calibration results of benzene, toluene, carbon monoxide, ammonia, and nitrogen dioxide tested on the same slave electronic nose, respectively. We can see that the sensor response in slave can be calibrated to the master very well. Therefore, we can say that the calibration coefficients obtained by using reference gas (formaldehyde) are also effective in correction of other gases samples.

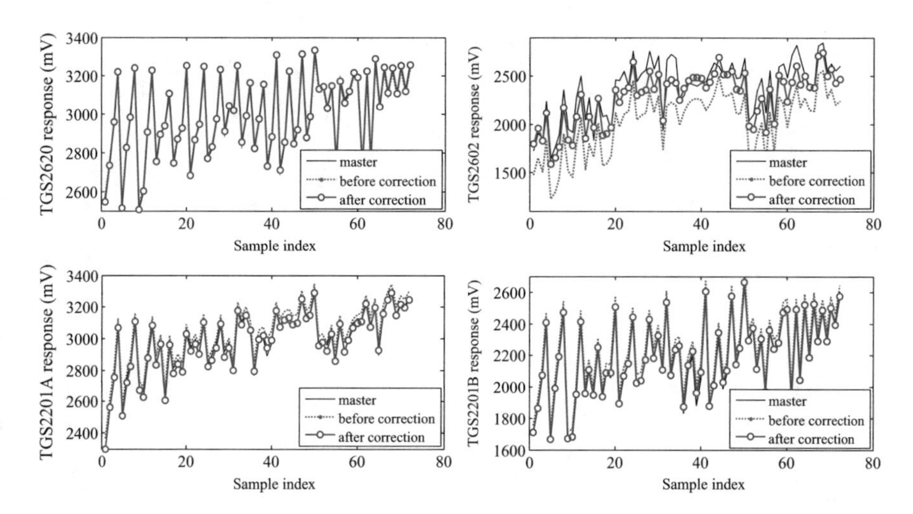

Fig. 19.2 Correction of 72 benzene samples using the calibration coefficients obtained with reference gas (formaldehyde)

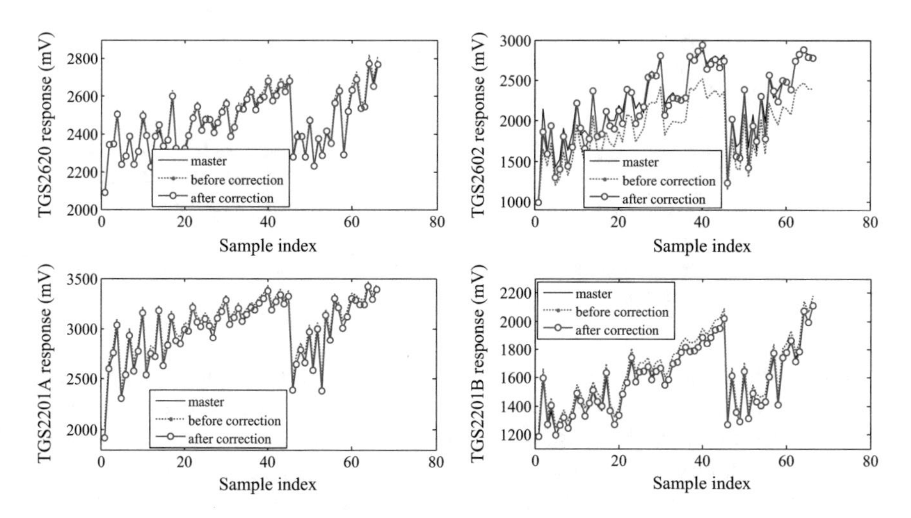

Fig. 19.3 Correction of 66 toluene samples using the calibration coefficients obtained with reference gas (formaldehyde)

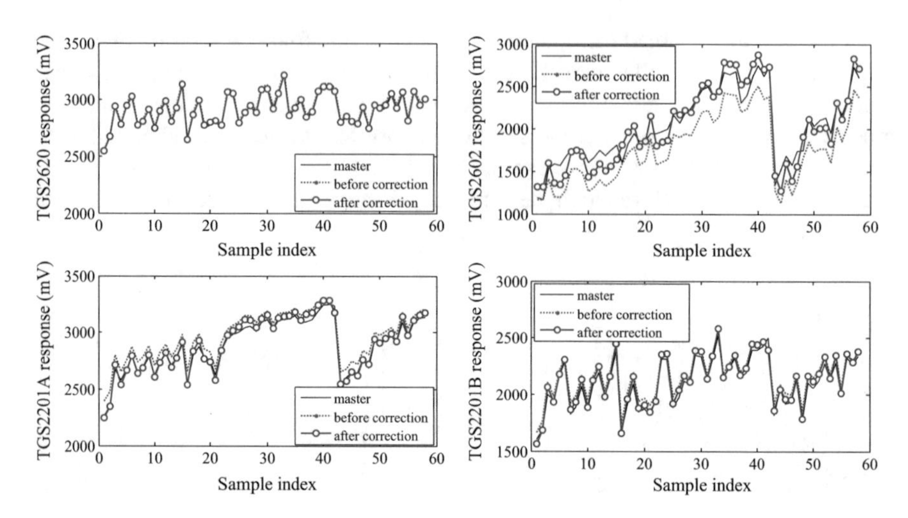

Fig. 19.4 Correction of 58 carbon monoxide samples using the calibration coefficients obtained with reference gas (formaldehyde)

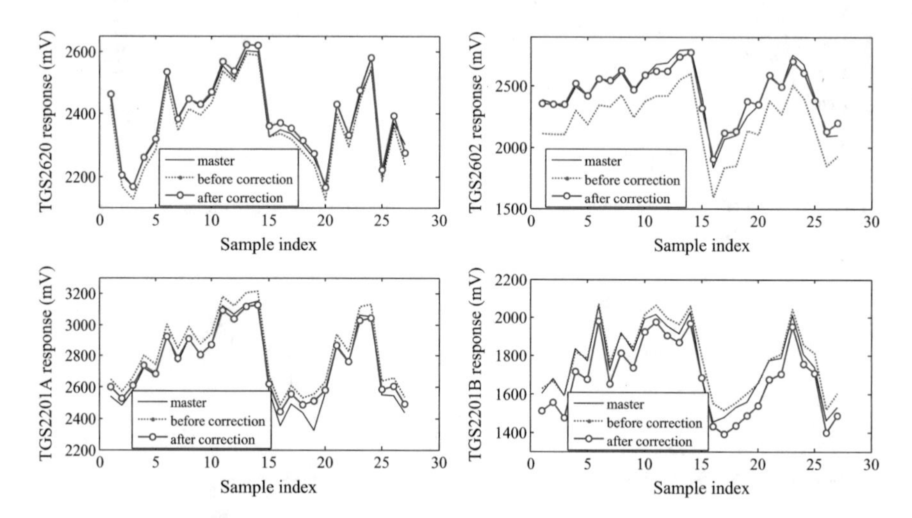

Fig. 19.5 Correction of 27 ammonia samples using the calibration coefficients obtained with reference gas (formaldehyde)

The results from Figs. 19.2, 19.3, 19.4, 19.5, and 19.6 demonstrate that one sensor's discreteness is almost constant in a certain experimental condition and it keeps changeless when the sensor is exposed to different kinds of gas components. It also demonstrates that the discreteness of sensors has little relation with the types of gases and the discreteness correction model can be determined using any kind of reference gas. The finding is optimistic that the discreteness problem of sensors can be solved and the reproducibility can be improved in batch-oriented applications of

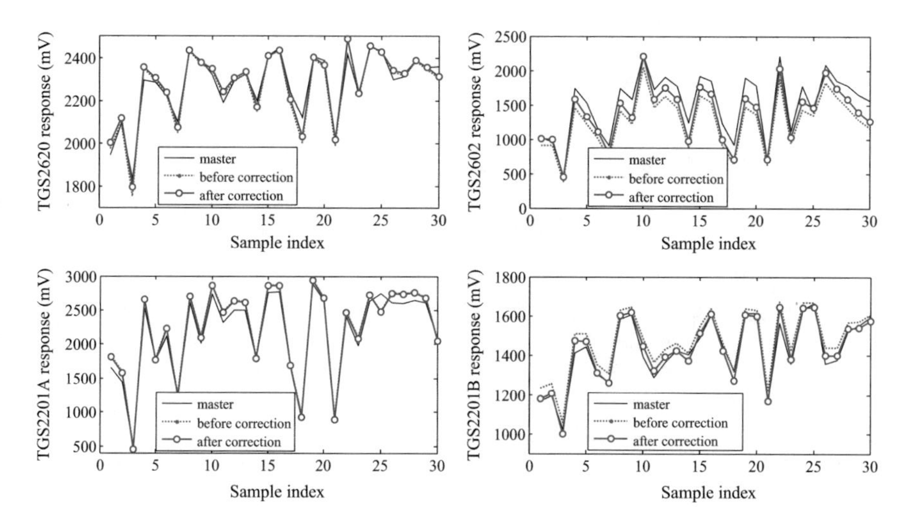

Fig. 19.6 Correction of 30 nitrogen dioxide samples using the calibration coefficients obtained with reference gas (formaldehyde)

sensors by designing an appropriate scheme coupled with some mathematical method.

Then, we have presented the calibration results of the four sensors in one batch of electronic nose instruments. Totally, 11 electronic noses including the *master* were employed in one batch of experiments. Figure 19.7 illustrates the original sensor response curves (approximately 3000 sampling points) of TGS2620, TGS2602, TGS2201A, and TGS2201B, respectively. We can find that the 11 sensor curves in Fig. 19.7 are not in coincidence which demonstrates that the MOS gas sensor discreteness is very serious and significant in the same environment. Note that there are 11 curves which represent 11 sensors with completely the same type in each figure. To validate the proposed discreteness correction scheme in batch-oriented application, the corrected sensor response curves of TGS2620, TGS2602, TGS2201A, and TGS2201B in accordance with Fig. 19.7 have been illustrated in Fig. 19.7. We can see from Fig. 19.7 that the sensor response curves of the 11 sensors with the same type have been coincided together after correction which also demonstrates that the proposed scheme in batch of electronic nose production is very successful and easier to be realized. This shows that the proposed scheme can also be used in industry for large scaled electronic noses manufacture. It can be inferred that the proposed scheme can also be used in the manufacture and production of electronic nose systems based on an array of MOS gas sensors in other applications (i.e., food control, medical diagnosis).

It is worth noting that some time delays of the sensor response exist among electronic nose systems in Fig. 19.7. In detail, the positions in some inflection points such as 1000, 1750, and 2250 in the sensor curves show the time delays which cannot be coincided after correction without appropriate translation. This is

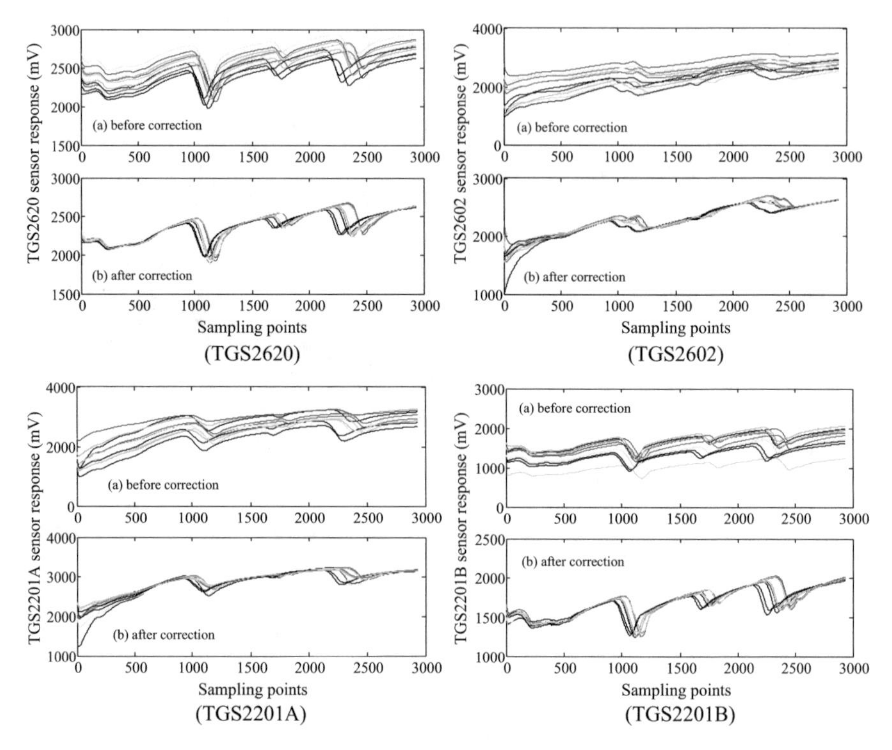

Fig. 19.7 Sensor's discreteness correction of 10 electronic noses using the proposed scheme. **a** Sensor responses before correction; **b** sensor responses after correction

due to the slight differences of sampling frequency which result from the software program running and tiny difference of electronic device in circuits among the electronic systems. Note that the inflection points represent the transient response of sensors. However, it does not influence the calibration of the sensors' discreteness in our work, because the calibration coefficients of each sensor in Eq. 19.1 are obtained using the selected three appropriate features in the steady state response but not the transient response around the inflection points.

19.4 Summary

This chapter proposes an easily realized and rapid scheme for batch-oriented MOS gas sensors' discreteness correction in E-nose instruments production and improves the reproducibility of sensors. The calibration parameters obtained using reference gas (formaldehyde) and the GAT-RWLS method have also been validated on other gases samples (benzene, toluene, carbon monoxide, ammonia, and nitrogen dioxide) and demonstrate that the sensor discreteness is independent and has little relation and with the types of gas components. Experimental results also demonstrate that the proposed correction scheme is very effective and can realize rapid sensor discreteness correction in batch of E-noses production.

References

1. L. Zhang, F. Tian, S. Liu, L. Dang, X. Peng, X. Yin, Chaotic time series prediction of E-nose sensor drift in embedded phase space. Sens. Actuators B **182**, 71–79 (2013)
2. A. Ziyatdinov, S. Marco, A. Chaudry, K. Persaud, P. Caminal, A. Perera, Drift compensation of gas sensor array data by common principal component analysis. Sens. Actuators B **146**, 460–465 (2010)
3. M. Padilla, A. Perera, I. Montoliu, A. Chaudry, K. Persaud, S. Marco, Drift compensation of gas sensor array data by orthogonal signal correction. Chemometr. Intell. Lab. Syst. **100**, 28–35 (2010)
4. S. Di Carlo, M. Falasconi, E. Sanchez, A. Scionti, G. Squillero, A. Tonda, Increasing pattern recognition accuracy for chemical sensing by evolutionary based drift compensation. Pattern Recogn. Lett. **32**, 1594–1603 (2011)
5. A.C. Romain, J. Nicolas, Long term stability of metal oxide-based gas sensors for e-nose environmental applications: an overview. Sens. Actuators B **146**, 502–506 (2010)
6. J. Fonollosa, A. Vergara, R. Huerta, Algorithmic mitigation of sensor failure: Is sensor replacement really necessary? Sens. Actuators B **183**, 211–221 (2013)
7. M.O. Balaban, F. Korel, A.Z. Odabasi, G. Folkes, Transportability of data between electronic noses: mathematical methods. Sens. Actuators B **71**, 203–211 (2000)
8. A. Berna, Metal oxide sensors for electronic noses and their application to food analysis. Sensors **10**, 3882–3910 (2010)
9. S.M. Kanan, O.M. El-Kadri, I.A. Abu-Yousef, M.C. Kanan, Semiconducting metal oxide based sensors for selective gas pollutant detection. Sensors **9**, 8158–8196 (2009)
10. L. Zhang, F. Tian, C. Kadri, B. Xiao, H. Li, L. Pan, H. Zhou, On-line sensor calibration transfer among electronic nose instruments for monitoring volatile organic chemicals in indoor air quality. Sens Actuators B **160**, 899–909 (2011)
11. L. Zhang, F. Tian, X. Peng, L. Dang, G. Li, S. Liu, C. Kadri, Standardization of metal oxide sensor array using artificial neural networks through experimental design. Sens Actuators B **177**, 947–955 (2013)

Chapter 20
Book Review and Future Work

Abstract This chapter first gives a basic introduction and objective of this book on algorithmic challenges in E-nose community. Then, the review and outline of this book are presented by describing the basic research findings and experimental performance of each chapter, respectively. Finally, the future work on the algorithms for further improvement of the odor recognition and estimation in E-nose is presented and summarized.

Keywords Algorithmic challenges · Outline · Odor recognition
Future work

Electronic nose is an electronic olfactory system constructed to mimic the biological olfactory mechanism, which is also an important scientific field of artificial intelligence. Recently, electronic nose has attracted worldwide attention and a number of research findings, including sensor arrays, bionic systems, pattern recognition algorithms, have been developed. These findings proved the feasibility and effectiveness of electronic nose in odor recognition, medical diagnosis, food quality monitoring, etc. However, there are several fundamental problems in electronic nose community being solved, such as sensor drift issue, disturbance issue, and discreteness issue, which seriously prohibit the scientific progress and industrial development of olfactory intelligence. Unfortunately, there is very few research contributing to the addressed challenges. Therefore, this book systematically describes and defines these challenges in electronic nose community and presents specific methods, models, and techniques how to effectively and efficiently address these important issues. To our best knowledge, this is the first book to review and address the newest algorithmic challenges that researchers over the world can launch and achieve in the future.

Specifically, this book focuses on four algorithmic challenges. First, in part II, for general electronic nose, the algorithmic challenge in pattern recognition and machine learning (challenge I) is described, including the odor recognition algorithms and concentration estimation algorithms. Second, in part III, the sensor drift issue (challenge II) that leads to aging of E-nose is introduced and described,

L. Zhang et al., *Electronic Nose: Algorithmic Challenges*,
https://doi.org/10.1007/978-981-13-2167-2_20

with specific solving schemes of transfer learning-guided models and algorithms. Third, in part IV, the disturbance issue (challenge III) that breaks the general use of E-nose is introduced and described, with specific approaches and techniques. Finally, in part V, the discreteness issue (challenge IV) that prohibits the large-scale industrial application of E-nose instruments is introduced and described, with effective models and algorithms.

20.1 Introduction

The first chapter of this book describes the background and motivation of electronic nose and the unsolved issues and challenges. This is followed by a statement of the objective of the research, a brief summary of the work, and a general outline of the overall structure of the present study.

Chapter 2 reviews the existing work in electronic nose and the algorithms addressing the new challenges including odor recognition, drift compensation, interference elimination, and discreteness correction. This is followed by a detailed review of existing approaches in electronic nose.

Chapter 3 describes a heuristic and bio-inspired neural network model for gases concentration estimation and guide readers to optimize the weights of artificial neural network in the training process. The proposed method focuses on the bionic intelligent optimization model, which integrates the advantages of particle optimization algorithm and genetic search, such that the optimization can converge to a global optimum quickly. Quantitative experiments on gases concentration estimation show the effectiveness.

Chapter 4 presents a new chaos-based neural network optimization algorithm for gases prediction, in which the neural network weights are imposed to be chaotic characteristic, and chaos operator is introduced to update the network weights. Experiments demonstrate the chaos-based neural network can achieve a particularly good prediction performance.

In Chap. 5, we systemically present how to predict gas concentration using multilayer perceptrons, such that researchers and engineers can use in real application. Specifically, two network structures including single multi-input multi-output (SMIMO) and multiple multi-input single-output (MMISO) are introduced for gases concentration estimation and prediction. The heuristic optimization algorithms from Chap. 3 are used for network optimization. Experiments show that the MMISO structure achieves the best performance in predicting the concentrations of multiple kinds of gases.

Chapter 6 introduces a discriminative support vector machine model which works together with linear discriminant analysis, such that the odor recognition accuracy can be improved benefiting from the discriminative subspace projection of data. Experiments on the electronic nose datasets of multiple kinds of odors show the effectiveness of the proposed method.

In Chap. 7, a new local kernel discriminant analysis method for discriminative feature subspace learning is proposed. In this chapter, the local affinity matrices of intra-class and inter-class are formulated based on the manifold learning, which strengthen the local similarity of data. This benefits to the odor recognition accuracy. Experiments show that after learning a local kernel discriminative subspace with local similarity constraints the odor recognition performance is much improved.

In Chap. 8, a classifier ensemble method for classification of multiple kinds of odors is introduced. First, in order to improve the accuracy of base classifiers, kernel principal component analysis (KPCA) method is used for nonlinear feature extraction of E-nose data; second, in the process of establishing classifiers ensemble, a new fusion approach which conducts an effective base classifier weighted method is proposed. Experiments show the effectiveness of classifier ensemble in classification.

Chapter 9 presents a novel sensor drift prediction approach based on chaotic time series. This method realizes a long-term prediction of sensor baseline and drift based on phase space reconstruction (PSR) and radial basis function (RBF) neural network. PSR can memory all of the properties of a chaotic attractor and clearly show the motion trace of a time series; thus, PSR makes the long-term drift prediction using RBF neural network possible. Results demonstrate that the proposed model can make long-term and accurate prediction of chemical sensor baseline and drift time series.

Chapter 10 presents a very new perspective of machine learning for effectively addressing the drift compensation issue, by using classifier transfer learning idea. This chapter aims to propose a unified framework of ELMs with domain adaptation and improve their transfer learning capability and generalization performance of cross domains without loss of the computational efficiency and learning ability of traditional ELMs. Therefore, we integrate domain adaptation into ELMs and two algorithms including source domain adaptation transfer ELM and target domain adaptation transfer ELM are proposed. Experiments verify the effectiveness of transfer learning.

Chapter 11 presents a cross-domain subspace projection approach, for addressing drift reduction issue, by using feature transfer learning idea. The main idea behind is that given two data clusters with a different probability distribution, we tend to find a latent projection $\mathbf{P}$ (i.e., a group of basis), such that the newly projected subspace of the two clusters is with a similar distribution. In other words, drift is automatically removed or reduced by projecting the data onto a new common subspace. Experiments on synthetic data and real E-nose datasets demonstrate the effectiveness and efficiency.

Chapter 12 presents a common subspace learning model for drifted data classification, such that the drifted data can share a common feature space with the non-drifted data. Specifically, a unified subspace transfer framework called cross-domain extreme learning machine (CdELM), which aims at learning a common (shared) subspace across domains, is proposed. Experiments demonstrate that the proposed CdELM method significantly outperforms other compared methods.

Chapter 13 introduces a transfer learning-guided domain correction method for drift compensation. The framework consists of two parts: (1) domain correction (DC) which makes the distributions of two domains close; (2) adaptive extreme learning machine (AELM) which realizes the knowledge transfer at the decision level and makes the robustness of prediction model improved. Experiments show the effectiveness of DC-AELM in drift compensation and interference insensitiveness.

In Chap. 14, a multi-feature jointly semi-supervised learning approach is introduced, which aims to improve the robustness of electronic nose to outliers and drift by multi-feature joint learning with local similarity preservation for unlabeled data usage. Specifically, a multi-feature kernel semi-supervised learning framework nominated as MFKS is proposed for the first time in E-nose community. Experiments on drift dataset and modulated E-nose dataset show the effectiveness.

Chapter 15 presents a pattern recognition-based background interference reduction, which describes a simple idea for interference recognition and then achieves interference reduction. First, two artificial intelligence learners including a multi-class least square support vector machine (learner-1) and a binary classification artificial neural network (learner-2) are developed for discrimination of unwanted odor interferences. Second, a real-time dynamically updated signal matrix is constructed for correction. Experiments demonstrate the effectiveness in abnormal odor detection.

Chapter 16 presents a pattern mismatch-based interference elimination method by sensor reconstruction. Considering that target gases detected by an E-nose can be fixed as invariant information, a novel and effective pattern mismatch-based interference elimination (PMIE) method is proposed in this chapter. It contains two parts: discrimination (i.e., pattern mismatch) and correction (i.e., interference elimination). Experiments show the effectiveness.

Chapter 17 introduces a new self-expression idea for interference recognition and reduction. (1) A self-expression model (SEM) with l_1/l_2-norm regularizer is proposed, which is trained on target odor data for coding, and then, a very few abnormal odor data is used as prior knowledge for threshold learning. (2) Inspired by self-expression mechanism, an extreme learning machine (ELM)-based self-expression (SE^{2}LM) is proposed, which inherits the advantages of ELM in solving a single hidden layer feed-forward neural network. Experiments verify the proposed self-expression models.

Chapter 18 introduces an affine transformation-based discreteness calibration model for the pursuit of signal uniqueness, which is simple yet effective, and benefits to large-scale instrument calibration. GAT is achieved in terms of one single sensor by a robust weighted least square (RWLS) algorithm, and KSS is studied for representative transfer sample subset selection from a large sample space. Experiments show that the proposed method can well solve the challenging problem.

Chapter 19 presents a very effective scheme for large-scale instrument standardization by using the affine calibration shown in Chap. 18. The contribution of this chapter is to solve the discreteness and improve the reproducibility of sensors by designing an effective and easily realized scheme for large-scale instrument calibration. Experiments on many instruments show the desired performance.

20.2 Future Work

The current work in this book for dealing with the key challenges of electronic nose in algorithmic viewpoint has achieved great achievement, and efficient solutions have been provided for these four E-nose challenges. However, there also some limitations of the current research that should be stated. First, in challenge I, for odor recognition and prediction, the current research focuses on single gas but not mixture. In real-world application, the odor in air should be mixture of many kinds of gases. Second, in challenge II, the proposed method for drift compensation is transfer learning oriented but between two batches, which, actually, is not lifelong learning strategy and periodical improvement of the classification model parameters is necessary. Third, in challenge III, the proposed interference elimination is actually an abnormal detection and signal correction method, but not real sensor signal processing with interference signal de-noising. Finally, in challenge IV, the proposed sensor signal alignment model parameters are actually fixed after calibration between two instruments. However, with aging of sensors, the calibration parameters of the affine transformation may be drifted. Therefore, these issues should be further studied and solved in the future work.

Several directions for dealing with the mentioned issues above can be worked to improve the recognition and prediction accuracy of electronic nose in different application fields.

(1) Dataset enhancement of mixtures: Generally, the E-nose data is prepared in single gas. For simulating the real-world case, the experiments with mixtures of several gases under different ambient environment and mixture concentrations can be conducted. Then, in algorithmic viewpoint, multi-label learning model can be effectively formulated for prediction of mixtures.
(2) Formulation of lifelong learning model: In transfer learning-oriented drift compensation, two domains (drifted data vs. non-drifted data) are formulated. However, E-nose data can be sequential with time. For long-term prediction, the algorithm is better lifelong learning model with fewer training times. Therefore, the formulation of multi-target domain adaptation models is a feasible direction in the future.
(3) Incremental calibration model fine-tuning: In calibration for the pursuit of signal uniqueness between two instruments, the affine calibration model parameters can be updated incrementally. Considering that the master instrument will not be available for slave instrument, the rule of model parameter shifting can be further studied periodically, so that the linear parameter fine-tuning strategy can be achieved for each slave instrument.

Printed in the United States
By Bookmasters